Advances in Optimization and Numerical Analysis

AF769081

Mathematics and Its Applications

Managing Editor:

M. HAZEWINKEL

Centre for Mathematics and Computer Science, Amsterdam, The Netherlands

Volume 275

Advances in Optimization and Numerical Analysis

Proceedings of the Sixth Workshop on Optimization and Numerical Analysis, Oaxaca, Mexico

edited by

Susana Gomez
and
Jean-Pierre Hennart

Instituto de Investigaciones en Mátematicas Aplicadas y Sistemas,
Universidad Nacional Autónoma de México,
México

KLUWER ACADEMIC PUBLISHERS
DORDRECHT / BOSTON / LONDON

Library of Congress Cataloging-in-Publication Data

Workshop on Optimization and Numerical Analysis (6th : 1992 : Oaxaca, Mexico)
Advances in optimization and numerical analysis : proceedings of the sixth Workshop on Optimization and Numerical Analysis, Oaxaca, Mexico / edited by Susana Gomez and Jean-Pierre Hennart.
p. cm. -- (Mathematics and its applications ; 275)

1. Mathematical optimization--Congresses. 2. Numerical analysis--Congresses. I. Gómez, S. (Susana) II. Hennart, J. P. (Jean Pierre), 1942- . III. Title. IV. Series: Mathematics and its applications (Kluwer Academic Publishers) ; 275.
QA402.5.W68 1992
519.4--dc20 93-45679

ISBN 978-90-481-4358-0

Published by Kluwer Academic Publishers,
P.O. Box 17, 3300 AA Dordrecht, The Netherlands.

Kluwer Academic Publishers incorporates
the publishing programmes of
D. Reidel, Martinus Nijhoff, Dr W. Junk and MTP Press.

Sold and distributed in the U.S.A. and Canada
by Kluwer Academic Publishers,
101 Philip Drive, Norwell, MA 02061, U.S.A.

In all other countries, sold and distributed
by Kluwer Academic Publishers Group,
P.O. Box 322, 3300 AH Dordrecht, The Netherlands.

Printed on acid-free paper

All Rights Reserved
© 1994 Kluwer Academic Publishers
Softcover reprint of the hardcover 1st edition 1994
No part of the material protected by this copyright notice may be reproduced or utilized in any form or by any means, electronic or mechanical, including photocopying, recording or by any information storage and retrieval system, without written permission from the copyright owner.

TABLE OF CONTENTS

PREFACE

In January 1992, the Sixth Workshop on Optimization and Numerical Analysis was held in the heart of the Mixteco-Zapoteca region, in the city of Oaxaca, México, a beautiful and culturally rich site in ancient, colonial and modern Mexican civilization.

The Workshop was organized by the Numerical Analysis Department at the Institute of Research in Applied Mathematics of the National University of Mexico in collaboration with the Mathematical Sciences Department at Rice University, as were the previous ones in 1978, 1979, 1981, 1984 and 1989. As were the third, fourth, and fifth workshops, this one was supported by a grant from the Mexican National Council for Science and Technology, and the US National Science Foundation, as part of the joint Scientific and Technical Cooperation Program existing between these two countries.

The participation of many of the leading figures in the field resulted in a good representation of the state of the art in Continuous Optimization, and in an overview of several topics including Numerical Methods for Diffusion-Advection PDE problems as well as some Numerical Linear Algebraic Methods to solve related problems. This book collects some of the papers given at this Workshop.

Susana Gómez
Instituto de Investigaciones en Matemáticas Aplicadas y Sistemas,
Universidad Nacional Autónoma de México, México
and during the edition of this book at:
Institut National de Recherche en Informatique et en Automatique,
Rocquencourt, France

Jean-Pierre Hennart
Instituto de Investigaciones en Matemáticas Aplicadas y Sistemas,
Universidad Nacional Autónoma de México, México

ANALYSIS OF INTERIOR-POINT METHODS FOR LINEAR PROGRAMMING PROBLEMS WITH VARIABLE UPPER BOUNDS

MICHAEL J. TODD*
School of Operations Research and Industrial Engineering
Engineering and Theory Center, Cornell University, Ithaca, NY 14853

Abstract.
We describe path-following and potential-reduction algorithms for linear programming problems with variable upper bounds. Both methods depend on a barrier function for the cone of solutions to the variable upper bounds, and we establish the key properties of this barrier that allow the complexity of the algorithms to be analyzed. These properties mostly follow from the self-concordance of the function, a notion introduced by Nesterov and Nemirovsky. Our analysis follows that of Freund and Todd for problems with (possibly two-sided) simple bounds.

Key words: linear programming, variable upper bounds, barrier functions, interior-point algorithms, self-concordance.

1. Introduction

The aim of this paper is to describe two polynomial-time interior-point algorithms for linear programming problems with variable upper bounds, one based on path-following and the other on potential reduction. In this endeavor, we are continuing the work of Freund and Todd [5] for the case of simple upper and lower bounds; the idea is to highlight the role of the barrier function and elucidate in a simple setting the notion of self-concordance introduced by Nesterov and Nemirovsky [12]. At the same time we show how variable upper bounds can be handled implicitly, without increasing the size of the linear system to be solved at each iteration.

Exploiting the structure of variable upper bounds has been discussed by Schrage [16] and Todd [17] in the context of the simplex method, by Todd [18] for Karmarkar's projective algorithm, and by Choi and Goldfarb [2] for a short-step primal-dual path-following method. In addition, since variable upper bounds define a cone, the general primal-dual potention-reduction method of Nesterov and Nemirovsky (see [11] or Chapter 3 of [12]) can be adapted to this problem. However, as we argue in Section 2.5, their algorithm is also restricted to relatively short step sizes.

Let $N = \{1, 2, \ldots, n\}$ index the variables of a linear programming problem, and suppose $J \cup K \cup L \cup F = N$ is a partition of N. A variable upper bound is a constraint of the form $x_j \leq x_{k(j)}$ where $j \in J$ and $k(j) \in K$. We call j (or x_j) the *child* of its *parent* $k(j)$ (or $x_{k(j)}$); each child has only one parent, but a parent may have several (but at least one) children. Variables in L (as well as those in $J \cup K$) are required to be nonnegative, while those in F are free. We assume there are no simple upper bounds. Let $J(k) = \{j \in J : k(j) = k\}$. Then our problem can be

* Research supported in part by NSF, AFOSR, and ONR through NSF Grant DMS-8920550.

S. Gomez and J-P. Hennart (eds.), Advances in Optimization and Numerical Analysis, 1–23.
© 1994 *Kluwer Academic Publishers.*

written

$$(P)\qquad \begin{aligned} \min_x \ & c^\top x \\ Ax &= b \\ x &\in C, \end{aligned}$$

where

$$C := \{x \in \Re^n : 0 \le x_j \le x_k,\ j \in J(k), \quad k \in K, \qquad 0 \le x_\ell, \qquad \ell \in L\} \tag{1.1}$$

and A is an $m \times n$ matrix. If we added the constraints $x_j \le x_k$ with slack variables explicitly to the constraint matrix, m and n would increase by $|J|$, a substantial increase if $m \ll n$ and $|J|$ is of the same order as n.

Writing the dual problem directly, we obtain

$$(D)\qquad \begin{array}{llll} \max_{y,s}\ b^\top y & & & \\ a_j^\top y - & t_j & \le c_j, & j \in J, \\ a_k^\top y + & \sum_{j \in J(k)} t_j & = c_k, & k \in K, \\ a_\ell^\top y & & \le c_\ell, & \ell \in L, \\ a_f^\top y & & = c_f, & f \in F, \\ & t_j \ge 0, & & j \in J. \end{array}$$

It is easy to see that y is feasible in (D) with some t iff $s = c - A^\top y$ lies in

$$\begin{aligned} C^* := \{s \in \Re^n : \ & s_k + \sum_{i \in I(k)} s_i \ge 0,\ I(k) \subseteq J(k), \quad k \in K, \\ & s_\ell \ge 0, \qquad \ell \in L, \\ & s_f = 0, \qquad f \in F\}, \end{aligned} \tag{1.2}$$

the dual or polar cone of C: $C^* := \{s \in \Re^n : x^\top s \ge 0 \text{ for all } x \in C\}$. Note that C can also be described in terms of its generators,

$$\begin{aligned} C = \text{cone}\{\ & e^k + \sum_{i \in I(k)} e^i,\ I(k) \subseteq J(k), \quad k \in K, \\ & e^\ell, \qquad \ell \in L, \\ & e^f, -e^f, \qquad f \in F\}, \end{aligned} \tag{1.3}$$

where e^j denotes the jth unit vector in $\Re^n$, and similarly

$$\begin{aligned} C^* = \text{cone}\{\ & e^j, e^{k(j)} - e^j,\ j \in J, \\ & e^\ell, \qquad \ell \in L\}. \end{aligned} \tag{1.4}$$

We can then describe (D) compactly as

$$(D)\qquad \begin{array}{rr} \max_{y,s} & b^\top y \\ & A^\top y + s = c \\ & s \in C^*, \end{array}$$

and this is (easily seen to be equivalent to) the dual problem of Nesterov and Nemirovsky ([11] or Chapter 3 of [12]).

For any x feasible in (P) and (y,s) feasible in (D), the duality gap is seen to be

$$c^\top x - b^\top y = (A^\top y + s)^\top x - (Ax)^\top y = x^\top s \geq 0, \tag{1.5}$$

so $x^\top s$ is the gap as in the standard case of nonnegativities only in C.

We make the following assumptions throughout the paper:
(A1) $F^0(P) := \{x \in \text{int } C : Ax = b\} \neq \emptyset$;
(A2) the set of optimal solutions of (P) is nonempty and bounded; and
(A3) A has rank m.
The assumption (A3) is for convenience only; it can easily be relaxed and only minor modifications are necessary. Of course, given feasibility of (P), linearly dependent rows of A give redundant constraints, which can be deleted.

If there were a linear dependence among the columns of A corresponding to the free variables, say $A_F x_F = 0$, $x_F \neq 0$, then necessarily $c_F^\top x_F = 0$ (else there would be no optimal solutions to (P)) and hence the set of optimal solutions of (P) would be unbounded. Hence (A2) implies that

$$A_F \quad \text{has full column rank.} \tag{1.6}$$

Of course, this assumption in A_F is fairly harmless; as long as (P) has an optimal solution, $A_F x_F = 0$ implies $c_F^\top x_F = 0$, and so linearly dependent columns of A_F could be deleted without changing the problem. (This is dual to removing dependent rows.) (A2) is stronger than this requirement; as in the standard case, it is not hard to show that it implies

$$F^0(D) := \{(y,s) \in \Re^m \times \text{ri } C^* : A^\top y + s = c\} \neq \emptyset, \tag{1.7}$$

where ri C^* is the relative interior of C^*, obtained by making all the inequalities in (1.2) strict. Conversely, (1.6) and (1.7) imply (A2), and similarly (A1) and (A3) imply that the set of optimal solutions of (D) is nonempty and bounded.

In Section 2 we describe and study a barrier function for C and show how it determines primal and dual norms and projections as well as central trajectories. These form the basis for the algorithms given in Section 3.

2. Analysis

In this section we define a barrier function for int C, and develop its key properties for our algorithms. Section 2.1 defines the barrier and computes its gradient and

Hessian matrix. In Section 2.2 we use these to define primal and dual metrics. Then we establish key self-concordance properties and Taylor approximation results in Section 2.3. (The reader may prefer to skip the proofs here on a first reading.) Section 2.4 shows how the metrics are used to define projections. The central path is defined in Section 2.5, and Section 2.6 uses the projections defined in Section 2.4 to analyze near-central points.

2.1. A BARRIER FUNCTION FOR INT C

Here we describe a barrier function for int C, a convex function that tends to $+\infty$ if the argument converges to a point of $C \backslash (\text{int } C)$, and compute its first and second derivatives. Our barrier is a simple logarithmic function, following the standard techniques for constructing barriers, with one term for each inequality defining C. We group these as follows. For each x in

$$\begin{aligned} \text{int } C = \{x \in \Re^n : \ & 0 < x_j < x_k, \quad j \in J(k), k \in K, \\ & 0 < x_\ell, \qquad\qquad \ell \in L\}, \end{aligned} \tag{2.8}$$

define

$$\Psi^j(x) := -\ln x_j - \ln(x_{k(j)} - x_j), \quad j \in J, \tag{2.9}$$

and

$$\Psi^\ell(x) := -\ln x_\ell, \quad \ell \in L, \tag{2.10}$$

and then set

$$\Psi(x) := \sum_{j \in J} \Psi^j(x) + \sum_{\ell \in L} \Psi^\ell(x). \tag{2.11}$$

From (2.9) we have, for $j \in J$,

$$\begin{aligned} \nabla \Psi^j(x) &= ((x_k - x_j)^{-1} - x_j^{-1})e^j - (x_k - x_j)^{-1} e^k \\ &= -x_j^{-1} e^j - (x_k - x_j)^{-1}(e^k - e^j) \end{aligned} \tag{2.12}$$

and

$$\begin{aligned} \nabla^2 \Psi^j(x) &= ((x_k - x_j)^{-2} + x_j^{-2}) e^j (e^j)^\top \\ &\quad + (-(x_k - x_j)^{-2})(e^j (e^k)^\top + e^k (e^j)^\top) \\ &\quad + (x_k - x_j)^{-2} e^k (e^k)^\top, \\ &= x_j^{-2} e^j (e^j)^\top + (x_k - x_j)^{-2}(e^k - e^j)(e^k - e^j)^\top, \end{aligned} \tag{2.13}$$

where $k := k(j)$, and from (2.10) we have

$$\nabla \Psi^\ell(x) = -x_\ell^{-1} e^\ell \tag{2.14}$$

and

$$\nabla^2 \Psi^\ell(x) = x_\ell^{-2} e^\ell (e^\ell)^\top. \tag{2.15}$$

From these elements we can assemble the gradient and Hessian of Ψ. (Note that we use superscripts for both powers and indices; no confusion should result.) For example, if $n = 7$, $J = \{1,2,4\}$, $K = \{3,5\}$, $k(1) = k(2) = 3$, $k(4) = 5$, and $L = \{6\}$, we find

$$\nabla^2\Psi(x) = \begin{bmatrix} \alpha_1^{-2}+\beta_1^{-2} & & -\beta_1^{-2} & & & & \\ & \alpha_2^{-2}+\beta_2^{-2} & -\beta_2^{-2} & & & & \\ -\beta_1^{-2} & -\beta_2^{-2} & \beta_1^{-2}+\beta_2^{-2} & & & & \\ & & & \alpha_4^{-2}+\beta_4^{-2} & -\beta_4^{-2} & & \\ & & & -\beta_4^{-2} & \beta_4^{-2} & & \\ & & & & & \alpha_6^{-2} & \\ & & & & & & 0 \end{bmatrix},$$

where $\alpha_i = x_i$, $i = 1,2,4,6$; $\beta_1 = x_3 - x_1$, $\beta_2 = x_3 - x_2$; and $\beta_4 = x_5 - x_4$. All unmarked entries are zero. We see that $\nabla^2\Psi$ has a block diagonal structure where each block has an arrowhead pattern if "families" are grouped together, with the parent appearing last. (The same structure appears in the matrix S in Choi and Goldfarb [2]; see their (4.2).)

2.2. Primal and Dual Metrics

Let us fix some $\hat{x} \in \operatorname{int} C$ and denote $\nabla^2\Psi(\hat{x})$ by $\hat{\Theta}^2$. We note from (2.13) and (2.15) that $\nabla^2\Psi(\hat{x})$ is symmetric positive semi-definite; thus this notation is appropriate if we use $\hat{\Theta}$ for the symmetric positive semi-definite square root of $\nabla^2\Psi(\hat{x})$. We further note that $\hat{\Theta}^2$ is positive definite on

$$\Re^{J\cup K\cup L} := \{x \in \Re^n : x_f = 0 \quad \text{for all} \quad f \in F\}. \tag{2.16}$$

Let us write $v^{J\cup K\cup L}$ for the projection of a vector $v \in \Re^n$ into $\Re^{J\cup K\cup L}$, obtained by setting its components indexed by F to zero. Let V be any subspace of $\Re^n$ satisfying

$$v \in V \quad \text{and} \quad v^{J\cup K\cup L} = 0 \quad \text{imply} \quad v = 0. \tag{2.17}$$

By the discussion above,

Proposition 2.1 $\|\cdot\|_{\hat{x}}$, *defined by*

$$\|v\|_{\hat{x}} := \|\hat{\Theta}v\|_2, \tag{2.18}$$

is a norm on V. ∎

From (1.6), we see that the null space of A, $\mathcal{N}(A)$, satisfies (2.17), so $\|\cdot\|_{\hat{x}}$ provides a norm on this space, and hence a metric on $F^0(P)$.

Note that the lineality space of C (the largest subspace contained in it) is

$$\Re^F := \{x \in \Re^n : x^{J\cup K\cup L} = 0\},$$

while its linear span is $C - C = \Re^n$. Also, the lineality space of C^* is $\{0\}$ and its span is $C^* - C^* = \Re^{J\cup K\cup L}$. The difference of s's for any two feasible solutions in $F^0(D)$ lies in $C^* - C^* = \Re^{J\cup K\cup L}$, and this subspace also contains the ranges of $\nabla\Psi$ and of $\hat{\Theta}^2$. We now define a dual norm on this space.

Since $\hat{\Theta}^2$ is nonsingular (and positive definite) as an operator on $\Re^{J\cup K\cup L}$, it has an inverse, also nonsingular and positive definite, which we denote $\hat{\Theta}^{-2}$. We also use $\hat{\Theta}^{-2}$ to denote the matrix that is block diagonal, representing the operator $\hat{\Theta}^{-2}$ on $\Re^{J\cup K\cup L}$, and diagonal with diagonal entries equal to $+\infty$ on $\Re^F$. Our convention is that $(+\infty)0 = 0$, so $\hat{\Theta}^{-2}v$ is well-defined for $v \in \Re^{J\cup K\cup L}$, whether $\hat{\Theta}^{-2}$ is viewed as an operator or a matrix. We define $\hat{\Theta}^{-1}$ similarly.

Proposition 2.2 $\|\cdot\|_{\hat{x}}^*$, *defined by*

$$\|v\|_{\hat{x}}^* = \|\hat{\Theta}^{-1}v\|_2, \tag{2.19}$$

is a norm on $C^* - C^* = \Re^{J\cup K\cup L}$. ∎

2.3. Self-concordance and Taylor approximations to Ψ

Here we show that Ψ is self-concordant in the sense of Nesterov and Nemirovsky [12] and also establish some key bounds on the errors in Taylor approximations to Ψ and $\nabla\Psi$.

A convex function Φ on an open subset Q of $\Re^n$ is said to be *self-concordant* (with parameter 1) if Φ is C^3 and for every $x \in Q$ and $d \in \Re^n$,

$$|D^3\Phi(x)[d,d,d]| \le 2(d^T\nabla^2\Phi(x)d)^{3/2}. \tag{2.20}$$

Here $D^3\Phi(x)$ denotes the third derivative of Φ. We easily find

$$D^3\Psi(x)[d,d,d] = \sum_{k\in K}\sum_{j\in J(k)}\left[-2\frac{d_j^3}{x_j^3} - 2\frac{(d_k-d_j)^3}{(x_k-x_j)^3}\right] + \sum_{\ell\in L}\left(-2\frac{d_\ell^3}{x_\ell^3}\right) \tag{2.21}$$

while

$$d^\top \nabla^2 \Psi(x) d = \sum_{k \in K} \sum_{j \in J(k)} \left[\frac{d_j^2}{x_j^2} + \frac{(d_k - d_j)^2}{(x_k - x_j)^2} \right] + \sum_{\ell \in L} \frac{d_\ell^2}{x_\ell^2}, \tag{2.22}$$

so (2.20) holds for Ψ by the inequality $\sum \mu_i^3 \leq (\sum \mu_i^2)^{3/2}$ for nonnegative μ_i's.

Note that Ψ also satisfies

$$\Psi(\lambda x) = \Psi(x) - (2|J| + |L|) \ln \lambda, \tag{2.23}$$

for $\lambda > 0$ and $x \in \text{int } C$, directly from the definition. For ease of notation, we define

$$p := |J| + \frac{1}{2}|L|, \tag{2.24}$$

and note that $p \leq n - |F|$, the number of non-free variables. Equation (2.23) is then the defining relationship for Ψ to be logarithmically homogeneous with homogeneity parameter $2p$ (Nesterov and Nemirovsky [11]). Hence we have:

Theorem 2.1 *Ψ is a $(2p)$-logarithmically homogeneous self-concordant barrier for int C.* ∎

(The terminology $(2p)$-normal barrier is used in ([12], Chapter 3).)

We can now follow the development of [11, 12]: (2.23) implies (differentiating with respect to λ at $\lambda = 1$)

$$\nabla \Psi(x)^\top x = -2p \tag{2.25}$$

and (differentiating with respect to x)

$$\nabla \Psi(\lambda x) = \lambda^{-1} \nabla \Psi(x), \tag{2.26}$$

and hence (differentiating (2.26) with respect to λ at $\lambda = 1$)

$$\nabla^2 \Psi(x) \cdot x = -\nabla \Psi(x), \tag{2.27}$$

and

$$\nabla^2 \Psi(x) \cdot x^{J \cup K \cup L} = -\nabla \Psi(x). \tag{2.28}$$

From this we obtain (similar to [5]):

Proposition 2.3 *For any $x \in$ int C, $\nabla \Psi(x) \in -ri\ C^* \subseteq \Re^{J \cup K \cup L}$ and*

$$\|\nabla \Psi(x)\|_x^* = (2p)^{1/2} \leq (2n)^{1/2}. \tag{2.29}$$

Proof. (2.12) and (2.14) show that $-\nabla \Psi(x)$ is a positive combination of the generators of C^* (see (1.4)), establishing the first part. For the second,

$$\begin{aligned} \nabla \Psi(x)^\top (\nabla^2 \Psi(x))^{-1} \nabla \Psi(x) &= -\nabla \Psi(x)^\top x^{J \cup K \cup L} \\ &= -\nabla \Psi(x)^\top x \\ &= 2p, \end{aligned}$$

where the first equation comes from (2.28) and the last from (2.25). This proves (2.29). ■

The following result is essentially a general consequence of self-concordance (see Theorem 1.1 of Nesterov and Nemirovsky [12]) but we provide a direct proof because it is so simple.

Proposition 2.4 *If $\hat{x} \in int\ C$ and $d \in \Re^n$ with $||d||_{\hat{x}} < 1$, then $\hat{x} + d \in int\ C$.*

(Note that we can use $||\cdot||_{\hat{x}}$, defined in (2.18), even for vectors not lying in a subspace V satisfying (2.17).)

Proof. Using (2.13) and (2.15) we find

$$\begin{aligned} 1 > ||d||_{\hat{x}}^2 &= d^\top \nabla^2 \Psi(\hat{x}) d \\ &= \sum_J \left[\frac{d_j^2}{\hat{x}_j^2} + \frac{(d_{k(j)} - d_j)^2}{(\hat{x}_{k(j)} - \hat{x}_j)^2} \right] + \sum_L \frac{d_\ell^2}{\hat{x}_\ell^2}, \end{aligned}$$

whence $|d_j| < \hat{x}_j$ and $|d_{k(j)} - d_j| < \hat{x}_{k(j)} - \hat{x}_j$ for each $j \in J$ and $|d_\ell| < \hat{x}_\ell$ for each $\ell \in L$. This implies $\hat{x} + d$ lies in int C. ■

We can use this property of self-concordance to eliminate another natural candidate for a self-concordant barrier. Indeed, by analogy with [5], we might suppose that Ψ^j in (2.9) could be replaced by

$$\tilde{\Psi}^j(x) := \frac{\min\{x_j, x_k - x_j\}}{x_k/2} - \ln(\min\{x_j, x_k - x_j\})$$

for $k = k(j)$. This $\tilde{\Psi}^j$ can be shown to be twice continuously differentiable (but not thrice), and its Hessian at a point x with $x_j < x_k/2$ is

$$\nabla^2 \tilde{\Psi}^j(x) = x_j^{-2} e^j (e^j)^\top - 2x_k^{-2}(e^j (e^k)^\top + e^k (e^j)^\top) + 4x_j x_k^{-3} e^k (e^k)^\top.$$

Suppose $k \in K$ and $\hat{x} \in$ int C has $\hat{x}_j < \hat{x}_k/2$ for all $j \in J(k)$. Let $d = d_k e^k$, and note that

$$d^\top \nabla^2 \tilde{\Psi}(\hat{x}) d = 4(\sum_{J(k)} \hat{x}_j / \hat{x}_k)(d_k/\hat{x}_k)^2,$$

where $\tilde{\Psi} := \sum_J \tilde{\Psi}^j + \sum_L \Psi^\ell$ is the new barrier function. It follows that we may have $d^\top \nabla^2 \tilde{\Psi}(\hat{x}) d$ less than 1 while $d_k < -\hat{x}_k$ (so $\hat{x}_k + d_k < 0$) as long as all $\hat{x}_j$'s, $j \in J(k)$, are sufficiently small. Hence Proposition 2.4 fails for $\tilde{\Psi}$, and we conclude that $\tilde{\Psi}$ is not self-concordant. Indeed, for any positive constant ϵ, $d^\top \nabla^2 \tilde{\Psi}(\hat{x}) d < \epsilon$ does not imply $d_k \leq -\hat{x}_k$ for all $\hat{x}$, so no positive multiple of $\tilde{\Psi}$ is self-concordant (or in other words, $\tilde{\Psi}$ is not self-concordant with any positive parameter [12]).

We now establish a key result on the first-order Taylor approximation of $\nabla\Psi$.

Theorem 2.2 *Let $\hat{x}, x \in int\ C$ and let $\bar{d} := x - \hat{x}$ and $\hat{\Theta}^2 := \nabla^2\Psi(\hat{x})$. Then*

$$\|\nabla\Psi(x) - \nabla\Psi(\hat{x}) - \hat{\Theta}^2\bar{d}\|_x^* \leq \|\bar{d}\|_{\hat{x}}^2. \tag{2.30}$$

(Note that the vector appearing on the left-hand side of (2.30) is the error in the first-order Taylor approximation to $\nabla\Psi(x)$ based on the point $\hat{x}$. Since all vectors in this expression lie in $\Re^{J \cup K \cup L}$, its dual norm is defined, and indeed appropriate—gradients of Ψ lie in dual space. On the right-hand side we have the square of the primal norm of the primal displacement $\bar{d}$. Note also that the dual norm is with respect to the "new point" x, while the primal norm corresponds to the base point $\hat{x}$.)

Proof. We obtain a bound on the norm of each constituent of the vector appearing on the left in (2.30). For ease of notation, let

$$\begin{aligned}
&\alpha_j = x_j, \quad \hat{\alpha}_j = \hat{x}_j, \quad \beta_j = x_{k(j)} - x_j, \quad \hat{\beta}_j = \hat{x}_{k(j)} - \hat{x}_j, \\
&\alpha_\ell = x_\ell, \quad \hat{\alpha}_\ell = \hat{x}_\ell, \\
&\delta_j = x_j - \hat{x}_j = \alpha_j - \hat{\alpha}_j, \quad \epsilon_j = x_{k(j)} - x_j - (\hat{x}_{k(j)} - \hat{x}_j) = \beta_j - \hat{\beta}_j, \\
&\delta_\ell = x_\ell - \hat{x}_\ell = \alpha_\ell - \hat{\alpha}_\ell.
\end{aligned}$$

We also omit subscripts when they are clear from the context. Then

$$\begin{aligned}
&\nabla\Psi^j(x) - \nabla\Psi^j(\hat{x}) - \nabla^2\Psi^j(\hat{x})(x - \hat{x}) \\
&= -\alpha_j^{-1}e^j - \beta_j^{-1}(e^k - e^j) + \hat{\alpha}_j^{-1}e^j + \hat{\beta}_j^{-1}(e^k - e^j) \\
&\quad -\hat{\alpha}_j^{-2}\delta_j e^j - \hat{\beta}_j^{-2}\epsilon_j(e^k - e^j) \\
&= \alpha^{-1}\hat{\alpha}^{-2}(-\hat{\alpha}^2 + \alpha\hat{\alpha} - \alpha\delta)e^j + \beta^{-1}\hat{\beta}^{-2}(-\hat{\beta}^2 + \beta\hat{\beta} - \beta\epsilon)(e^k - e^j) \\
&= -\alpha^{-1}\hat{\alpha}^{-2}\delta^2 e^j - \beta^{-1}\hat{\beta}^{-2}\epsilon^2(e^k - e^j),
\end{aligned} \tag{2.31}$$

where $k = k(j)$.

Now if P and Q are symmetric positive semi-definite matrices with $P \geq Q$ ($P - Q$ positive semi-definite), and $v \in Im(Q) \subseteq Im(P)$, then we can show that

$$v^\top P^{-1} v \leq v^\top Q^{-1} v,$$

where $P^{-1}v$ denotes the vector in the range of P with $P(P^{-1}v) = v$, and similarly for $Q^{-1}v$. Indeed, this is a standard result when P and Q are positive definite, and the general result follows by a limiting argument, using $P_k := P + \frac{1}{k}I$ and similarly for Q_k. (I am grateful to C. Van Loan for this line of reasoning, which simplifies my earlier argument using Schur complements.)

Applying this inequality to the vector in (2.31), with $P = \nabla^2\Psi(x)$ and $Q = \nabla^2\Psi^j(x)$, we obtain

$$\begin{aligned}
&(\|\nabla\Psi^j(x) - \nabla\Psi^j(\hat{x}) - \nabla^2\Psi^j(\hat{x})(x-\hat{x})\|_x^*)^2 \\
&\leq \begin{pmatrix} -\alpha^{-1}\hat{\alpha}^{-2}\delta^2 + \beta^{-1}\hat{\beta}^{-2}\epsilon^2 \\ -\beta^{-1}\hat{\beta}^{-2}\epsilon^2 \end{pmatrix}^\top \begin{pmatrix} \alpha^{-2}+\beta^{-2} & -\beta^{-2} \\ -\beta^{-2} & \beta^{-2} \end{pmatrix}^{-1} \\
&\quad \begin{pmatrix} -\alpha^{-1}\hat{\alpha}^{-2}\delta^2 + \beta^{-1}\hat{\beta}^{-2}\epsilon^2 \\ -\beta^{-1}\hat{\beta}^{-2}\epsilon^2 \end{pmatrix} \\
&= \begin{pmatrix} -\alpha^{-1}\hat{\alpha}^{-2}\delta^2 + \beta^{-1}\hat{\beta}^{-2}\epsilon^2 \\ -\beta^{-1}\hat{\beta}^{-2}\epsilon^2 \end{pmatrix}^\top \begin{pmatrix} \alpha^2 & \alpha^2 \\ \alpha^2 & \alpha^2+\beta^2 \end{pmatrix} \\
&\quad \begin{pmatrix} -\alpha^{-1}\hat{\alpha}^{-2}\delta^2 + \beta^{-1}\hat{\beta}^{-2}\epsilon^2 \\ -\beta^{-1}\hat{\beta}^{-2}\epsilon^2 \end{pmatrix} \\
&= \hat{\alpha}^{-4}\delta^4 + \hat{\beta}^{-4}\epsilon^4 \leq (\hat{\alpha}^{-2}\delta^2 + \hat{\beta}^{-2}\epsilon^2)^2 \\
&= (\bar{d}^\top \nabla^2\Psi^j(\hat{x})\bar{d})^2.
\end{aligned}$$

In the same way, using $\nabla^2\Psi(x) \geq \nabla^2\Psi^\ell(x)$, we get

$$(\|\nabla\Psi^\ell(x) - \nabla\Psi^\ell(\hat{x}) - \nabla^2\Psi^\ell(\hat{x})(x-\hat{x})\|_x^*)^2 \leq \hat{\alpha}_\ell^{-4}\delta_\ell^4 = (\bar{d}^\top\nabla^2\Psi^\ell(\hat{x})\bar{d})^2.$$

Adding the square roots of all these inequalities gives the desired inequality (2.30). ■

From this result we can prove, exactly as in [5],

Theorem 2.3 *Let $\hat{x} \in$ int C. If $\bar{d} \in \Re^n$ and $\gamma > 0$ are such that $\gamma\|\bar{d}\|_{\hat{x}} < 1$, then $\hat{x} + \gamma\bar{d} \in$ int C and*

$$\begin{aligned}
&\Psi(\hat{x}) + \gamma\nabla\Psi(\hat{x})^\top\bar{d} \leq \Psi(\hat{x}+\gamma\bar{d}) \\
&\leq \Psi(\hat{x}) + \gamma\nabla\Psi(\hat{x})^\top\bar{d} + \tfrac{\gamma^2\|\bar{d}\|_{\hat{x}}^2}{2(1-\gamma\|\bar{d}\|_{\hat{x}})}.
\end{aligned} \tag{2.32}$$

Proof. The first part follows from Proposition 2.4, while (2.32) is derived from (2.30) and the fundamental theorem of calculus exactly as in [5], using the following consequence of the self-concordance property established in Theorem 2.1: For every $\in$ int $C, d \in \Re^n$, and $h \in \Re^n$, if $\gamma \in \Re$ satisfies $|\gamma|\|d\|_x < 1$, then

$$(1-|\gamma|\|d\|_x)^2 h^\top\nabla^2\Psi(x)h \leq h^\top\nabla^2\Psi(x+\gamma d)h \leq (1-|\gamma|\|d\|_x)^{-2}h^\top\nabla^2\Psi(x)h.$$

(This implication is Theorem 1.1 of Nesterov and Nemirovsky [12].) ■

2.4. Projections

We can use the metrics in Section 2.2 to perform projections. Note that the Euclidean projection of a vector $v \in \Re^n$ onto $\mathcal{N}(A)$ can easily be seen to be the unique solution to

$$\max_d \{v^\top d - \frac{1}{2}\|d\|^2 : Ad = 0\}.$$

Correspondingly, we have (as in [5]):

Theorem 2.4 *Fix $\hat{x} \in int\ C$. For each $v \in \Re^n$, there is a unique solution to the problem*

$$\begin{array}{rl} \max_d & v^\top d - \frac{1}{2}\|d\|_{\hat{x}}^2 \\ & Ad = 0. \end{array} \tag{2.33}$$

Moreover, this solution $\bar{d}$ is part of the solution $(\bar{d}, \bar{y})$ to the system

$$\begin{pmatrix} \hat{\Theta} & A^\top \\ A & 0 \end{pmatrix} \begin{pmatrix} d \\ y \end{pmatrix} = \begin{pmatrix} v \\ 0 \end{pmatrix}, \tag{2.34}$$

where $\hat{\Theta}^2 := \nabla^2 \Psi(\hat{x})$, and satisfies

$$\|\bar{d}\|_{\hat{x}}^2 = v^\top \bar{d}; \tag{2.35}$$

$$\|\bar{d}\|_{\hat{x}} = \|v - A^\top \bar{y}\|_{\hat{x}}^*; \tag{2.36}$$

and, if $v \in \Re^{J \cup K \cup L}$,

$$\|\bar{d}\|_{\hat{x}} \leq \|v\|_{\hat{x}}^*. \tag{2.37}$$

Proof. Exactly as in [5]. The system (2.34) forms the Karush-Kuhn-Tucker conditions for the concave maximization problem (2.33), and these conditions are necessary and sufficient for optimality. The homogeneous system corresponding to (2.34) has only the trivial solution (using Proposition 2.1 and the remark following it, and assumption (A3)), so (2.34) has a unique solution. Then (2.35) – (2.37) follow from the first set of equations in (2.34). ■

We write $P_{\hat{x}}(v)$ for the solution to (2.33). Simple algebra shows that (2.34) also forms the KKT conditions for

$$\begin{array}{rl} \min_{y,s} & \frac{1}{2} s^\top \hat{\Theta}^{-2} s \\ & A^\top y + s = v \\ & s \in \Re^{J \cup K \cup L}, \end{array}$$

which is feasible since A has rank m. Hence we obtain the dual result:

Theorem 2.5 *There is a unique solution to*

$$\begin{array}{ll}\min_y & \frac{1}{2}(\|v - A^\top y\|_{\hat{x}}^*)^2 \\ & v - A^\top y \in \Re^{J \cup K \cup L}.\end{array} \tag{2.38}$$

Moreover, this solution $\bar{y}$ is part of the solution $(\bar{d}, \bar{y})$ to (2.34). ■

Hence (2.33) and (2.38) can be viewed as dual problems with equal optimal values.

Since computing projections is the basic task at each iteration, we conclude this section by describing how (2.34) can be solved efficiently. Of course, (2.34) is a sparse symmetric indefinite system of size $m+n$, and it can be solved by general techniques as in the standard case—see, for instance, Fourer and Mehrotra [3] or Vanderbei and Carpenter [19]. However, here we show how it can be reduced to a positive definite system of order m (assuming for simplicity there are no free variables).

Note that solving the system (2.34) reduces to the solution of a smaller system with coefficient matrix $A\hat{\Theta}^{-2}A^\top$. Since we are assuming $F = \emptyset$, $\hat{\Theta}^2$ is positive definite and hence its inverse is finite and also positive definite. However, $\hat{\Theta}^{-2}$ is not diagonal when variable upper bounds are present, so our matrix is more complicated than $AD^2A^\top = \sum_j d_{jj}^2 a_j a_j^\top$ as in the standard case.

As noted in Section 2.1, $\hat{\Theta}^2$ is block diagonal. If we set

$$\begin{array}{lll}\hat{\alpha}_j = \hat{x}_j, & \hat{\beta}_j = \hat{x}_k - \hat{x}_j, & \hat{\gamma}_j = (\hat{\alpha}_j^{-2} + \hat{\beta}_j^{-2})^{1/2}, \\ \hat{\eta}_j = -\hat{\beta}_j^{-2}\hat{\gamma}_j^{-1}, & \text{and} & \hat{\kappa} = (\sum_{J(k)} \hat{\alpha}_j^{-2}\hat{\beta}_j^{-2}\hat{\gamma}_j^{-2})^{1/2},\end{array} \tag{2.39}$$

for $j \in J(k)$, the kth block of $\hat{\Theta}^2$ is

$$\begin{pmatrix} \ddots & & & \vdots \\ & \hat{\alpha}_j^{-2} + \hat{\beta}_j^{-2} & & -\hat{\beta}_j^{-2} \\ & & \ddots & \vdots \\ \cdots & -\hat{\beta}_j^{-2} & \cdots & \sum_{J(k)} \hat{\beta}_j^{-2} \end{pmatrix}, \tag{2.40}$$

which has Cholesky factorization $WW^\top$ with

$$W = \begin{pmatrix} \ddots & & & \\ & \hat{\gamma}_j & & \\ & & \ddots & \\ \cdots & \hat{\eta}_j & \cdots & \hat{\kappa} \end{pmatrix}, \tag{2.41}$$

as can easily be checked. Thus the corresponding part of $A\hat{\Theta}^{-2}A^\top$ is

$$(\cdots a_j \cdots a_k) W^{-\top} W^{-1} \begin{pmatrix} \vdots \\ a_j^\top \\ \vdots \\ a_k \end{pmatrix}. \tag{2.42}$$

From (2.41) we find that

$$(\cdots a_j \cdots a_k)W^{-\top} = (\cdots \hat{\gamma}_j^{-1} a_j \cdots \hat{\kappa}^{-1}(a_k - \sum_{J(k)} \hat{\eta}_j \hat{\gamma}_j^{-1} a_j)). \tag{2.43}$$

Assembling all these pieces, and substituting back from (2.39), we find

$$\begin{aligned} A\hat{\Theta}^{-2}A^\top &= \textstyle\sum_{k\in K} \left[\sum_{j\in J(k)} \frac{a_j a_j^\top}{\hat{x}_j^{-2} + (\hat{x}_k - \hat{x}_j)^{-2}} + \frac{\tilde{a}_k \tilde{a}_k^\top}{\sum_{J(k)} \frac{1}{\hat{x}_j^2 + (\hat{x}_k - \hat{x}_j)^2}} \right] \\ &+ \textstyle\sum_{\ell\in L} \hat{x}_\ell^2 a_\ell a_\ell^\top, \end{aligned} \tag{2.44}$$

where

$$\tilde{a}_k := a_k + \sum_{j\in J(k)} \frac{(\hat{x}_k - \hat{x}_j)^{-2}}{\hat{x}_j^{-2} + (\hat{x}_k - \hat{x}_j)^{-2}} a_j. \tag{2.45}$$

Notice that $A\hat{\Theta}^{-2}A^\top$ in (2.44) is expressed as $\tilde{A}D^2\tilde{A}^\top$ with D diagonal, where $\tilde{A}$ differs from A only in that its parent columns a_k are augmented by a linear combination of their children columns. (If each $\hat{x}_j$ converges to 0 or to $\hat{x}_k$, then $\tilde{a}_k$ converges to $a_k + \sum\{a_j : j \in J(k),\ \ \hat{x}_j \text{ converges to } \hat{x}_k\}$, and this is exactly the modification of parent columns in the working basis scheme of [17].)

A similar analysis of the simplifications resulting from the special structure of $\hat{\Theta}^2$ appears in Choi and Goldfarb [2], but without noting the relationship to perturbing the parent columns as above.

2.5. Central paths

Consider the barrier problem

$$(BP) \qquad \begin{aligned} \min_x \quad c^\top x &+ \mu\Psi(x) \\ Ax &= b, \\ x &\in \operatorname{int} C, \end{aligned}$$

for $\mu > 0$. By our assumption that $F^0(P)$ is nonempty, (BP) has a feasible solution. Since we also assume that (P) has a nonempty bounded set of optimal solutions, along any direction in $\mathcal{N}(A) \cap C$ we see that $c^\top x$ increases linearly while Ψ decreases at most logarithmically. Hence standard arguments from convex analysis (e.g., see Rockafellar [14]) imply that (BP) has an optimal solution. Moreover, Ψ is strictly convex in $\Re^{J\cup K\cup L}$, while the columns of A_F are linearly independent, so the optimal solution is unique. We denote it by $x(\mu)$, and define the primal central trajectory to be $\{x(\mu) : \mu > 0\}$.

The KKT conditions are necessary and sufficient for (BP), so $x(\mu)$ together with some $y(\mu), s(\mu)$ satisfies uniquely

$$Ax = b, \tag{2.46}$$

$$A^{\mathsf{T}}y + s = c, \tag{2.47}$$

$$\mu\nabla\Psi(x) + s = 0. \tag{2.48}$$

Condition (2.48) implies that $s = -\mu\nabla\Psi(x) \in \text{ri } C^*$ by Proposition 2.3 so that $(y,s) \in F^0(D)$, and that $-s/\mu = \nabla\Psi(x) \in \partial\Psi(x)$ where $\partial\Psi$ denotes the subdifferential of the convex function Ψ. We define the convex conjugate of Ψ by

$$\Psi^*(s) := \sup_x\{-s^{\mathsf{T}}x - \Psi(x)\}.$$

(Note that we use $-s^{\mathsf{T}}x$ instead of the more usual $s^{\mathsf{T}}x$ here, in order to get the usual formula when there are only nonnegativity constraints.) Then (2.48) is equivalent [14] to

$$-x \in \partial\Psi^*(s/\mu).$$

This can be made even more symmetric with (2.48). Indeed, since Ψ is a $2p$-logarithmically homogeneous barrier function (recall $2p := 2|J| + |L|$), so is Ψ^* (Nesterov and Nemirovsky [11, 12]), and hence (or directly) $\partial\Psi^*(s/\mu) = \mu\partial\Psi^*(s)$. So (2.48) is equivalent to

$$\mu\partial\Psi^*(s) + x \ni 0. \tag{2.49}$$

Thus conditions (2.46) – (2.48) also form the optimality conditions for

$$(BD) \qquad \begin{aligned} \max_{y,s} \quad & b^{\mathsf{T}}y - \mu\Psi^*(s), \\ & A^{\mathsf{T}}y + s = c, \\ & \qquad s \in \text{ri } C^*. \end{aligned}$$

(As in [11, 12], Ψ^* is finite exactly on ri C^*.) Hence $(y(\mu), s(\mu))$ lies on the central trajectory for (D), defined as the set of solutions to (BD) for $\mu > 0$.

We could use Ψ and Ψ^* to construct a primal-dual potential-reduction algorithm for (P) and (D) following the general scheme of Nesterov and Nemirovsky [11, 12]. However, in our case it turns out to be impossible to obtain Ψ^* in closed form, and this precludes the possibility of line searches in the dual space. (This is in interesting contrast to [5], where Ψ^* could be obtained explicitly, but because Ψ and Ψ^* were not logarithmically homogeneous (the problems treated were not "conical"), we could not use the simplification (2.49) and (BD) involved $\mu\Psi^*(s/\mu)$.)

Hence in Section 3, we will confine ourselves to primal algorithms.

2.6. Duality gaps and near-central points

For every $\mu > 0$, we have $x(\mu) \in F^0(P)$ and $(y(\mu), s(\mu)) \in F^0(D)$. Our first result concerns the corresponding duality gap.

Proposition 2.5 *We have* $x(\mu)^\top s(\mu) = 2p\mu$. *(Recall that* $p = |J| + \frac{1}{2}|L|$.)

Proof. Since $s \in \Re^{J\cup K\cup L}$ (we omit the argument μ for ease of notation),

$$x^\top s = (x^{J\cup K\cup L})^\top s = [-(\nabla^2\Psi(x))^{-1}\nabla\Psi(x)]^\top[-\mu\nabla\Psi(x)]$$

(using (2.28) and (2.48)), so

$$x^\top s = \mu\nabla\Psi(x)(\nabla^2\Psi(x))^{-1}\nabla\Psi(x) = \mu(\|\nabla\Psi(x)\|_x^*)^2.$$

The result now follows from Proposition 2.3. ∎

Hence, if we could follow the path $\{x(\mu)\}$, we could get arbitrarily close to optimal. Unfortunately, we cannot follow the path exactly. Thus we will be interested in pairs $x \in F^0(P)$ and $(y,s) \in F^0(D)$ satisfying (2.46) and (2.47) exactly but (2.48) only approximately. The following result allows us to bound the duality gap of such a pair. (Think of t as s/μ.)

Theorem 2.6 *Suppose* $t = -\nabla\Psi(x) + h$, *where* $x \in \mathit{int}\ C$ *and* $h \in \Re^{J\cup K\cup L}$ *with* $\|h\|_x^* \le \beta < 1$. *Then* $t \in \mathit{ri}\ C^*$ *and*

$$x^\top t \le 2p + \beta\sqrt{2p}. \tag{2.50}$$

Proof. From Proposition 2.3 it is immediate that $t \in \Re^{J\cup K\cup L}$. To show $t \in \mathrm{ri}\ C^*$ we show that $v^\top t > 0$ for each generator v (other than e^j or $-e^j$) of C (see (1.3)), or in other words

$$-v^\top h < -v^\top\nabla\Psi(x)$$

for such v. Since $h \in \Re^{J\cup K\cup L}$, this holds if

$$\|v\|_x\|h\|_x^* < -v^\top\nabla\Psi(x),$$

hence if

$$\beta\|v\|_x < -v^\top\nabla\Psi(x). \tag{2.51}$$

For $v = e^\ell$, the left-hand side is βx_ℓ^{-1}, which is less than x_ℓ^{-1}, the right-hand side. Now suppose $v = e^k + \sum_{i\in I(k)} e^i$, where $I(k) \subseteq J(k)$, $k \in K$. Then

$$\begin{aligned}\|v\|_x^2 &= v^\top\nabla^2\Psi(x)v \\ &= \sum_{j\in J}\left[\frac{v_j^2}{x_j^2} + \frac{(v_{k(j)} - v_j)^2}{(x_{k(j)} - x_j)^2}\right] + \sum_{\ell\in L}\frac{v_\ell^2}{x_\ell^2} \\ &= \sum_{k\in K}\left[\sum_{i\in I(k)} x_i^{-2} + \sum_{j\in J(k)\setminus I(k)} (x_k - x_j)^{-2}\right].\end{aligned}$$

On the other hand, (2.12) and (2.14) show that

$$-v^{\mathsf{T}}\nabla\Psi(x) = \sum_{k\in K}\left[\sum_{i\in I(k)} x_i^{-1} + \sum_{j\in J(k)\setminus I(k)} (x_k - x_j)^{-1}\right].$$

Thus (2.51) follows from $\beta < 1$ and the inequality between the 2-norm and the 1-norm of a vector.

To prove (2.50), we have as in Proposition 2.5

$$\begin{aligned} x^{\mathsf{T}}t &= \nabla\Psi(x)(\nabla^2\Psi(x))^{-1}(\nabla\Psi(x) - h) \\ &= (\|\nabla\Psi(x)\|_x^*)^2 - \nabla\Psi(x)(\nabla^2\Psi(x))^{-1}h \\ &\leq (\|\nabla\Psi(x)\|_x^*)^2 + \|\nabla\Psi(x)\|_x^*\|h\|_x^* \\ &\leq 2p + \sqrt{2p}\beta, \end{aligned}$$

as required. ■

Note that the first conclusion of the theorem can be viewed as a dual version of Proposition 2.4.

As mentioned above, t should be thought of as s/μ, where $s = c - A^{\mathsf{T}}y$ for some y. Then $\mu h = \mu t + \mu\nabla\Psi(x) = c + \mu\nabla\Psi(x) - A^{\mathsf{T}}y$, and choosing y to make h small is an instance of problem (2.38). Combining Theorems 2.4, 2.5 and 2.6 gives us the following important result, which we call the *approximately-centered theorem*. It allows us to obtain a feasible dual solution from a sufficiently central primal solution.

Theorem 2.7 *Suppose $\hat{x} \in F^0(P)$ is given. Choose $\hat{\mu} > 0$, and let*

$$v := c + \hat{\mu}\nabla\Psi(\hat{x}). \tag{2.52}$$

Let $(\hat{d}, \hat{y})$ be the solution to (2.34) for this v, and hence define

$$\hat{s} := c - A^{\mathsf{T}}\hat{y}. \tag{2.53}$$

Then $\|\hat{d}\|_{\hat{x}} = \|\hat{s} + \hat{\mu}\nabla\Psi(\hat{x})\|_{\hat{x}}^*$. If

$$\|\hat{d}/\hat{\mu}\|_{\hat{x}} = \|\hat{s}/\hat{\mu} + \nabla\Psi(\hat{x})\|_{\hat{x}}^* \leq \beta, \tag{2.54}$$

where $\beta < 1$, then

(i) $(\hat{y}, \hat{s}) \in F^0(D)$;

(ii) the duality gap is $\hat{x}^{\mathsf{T}}\hat{s} \leq \hat{\mu}(2p + \beta\sqrt{2p})$.

If (2.54) holds, we say $\hat{x}$ is *β-close* to $x(\hat{\mu})$.

Proof. The equality of the norms follows from (2.36). Now define $\hat{t} := \hat{s}/\hat{\mu}$ and $\hat{h} := \nabla\Psi(\hat{x}) + \hat{t}$. Then we find $\hat{h} \in \Re^{J\cup K\cup L}$ and $\|\hat{h}\|_{\hat{x}}^* \leq \beta < 1$. From Theorem 2.6, $\hat{t} \in \text{ri } C^*$, so $\hat{s} \in \text{ri } C^*$ and $(\hat{y}, \hat{s}) \in F^0(D)$; and $\hat{x}^\top \hat{t} \leq 2p + \beta\sqrt{2p}$, whence (ii) follows. ∎

To conclude this section, we give as in [5] a sufficient condition for $\hat{x} \in F^0(D)$ to be β-close to $x(\hat{\mu})$. This follows from Theorem 2.5.

Proposition 2.6 *Suppose $\hat{x} \in F^0(P)$ and $\hat{\mu} > 0$ are given. If there exist (y, s) satisfying*

(i) $A^\top y + s = c, \quad s \in \Re^{J\cup K\cup L}$,

(ii) $\|s/\hat{\mu} + \nabla\Psi(\hat{x})\|_{\hat{x}}^* \leq \beta$,

then $\hat{x}$ is β-close to $x(\hat{\mu})$. ∎

3. Algorithms

Here we describe two algorithms for problem (P) based on the barrier function Ψ. The first is a path-following method, using the measure of closeness given in the approximately-centered theorem. Progress sufficient for polynomiality is assured by Proposition 2.3 and Theorem 2.2. The second algorithm is a potential-reduction method, using the barrier function Ψ as part of the potential function. Constant decrease of the latter is guaranteed by Theorems 2.3 and 2.7. We do not deal with initialization of the methods; techniques similar to those in [5] can be employed.

3.1. A PATH-FOLLOWING METHOD

Here we generate a sequence of points approximating $x(\mu)$ for a geometrically decreasing sequence of values of μ. The idea is similar to that in the algorithms of Renegar [13], Gonzaga [6], and Roos and Vial [15], for instance; see also Gonzaga [8]. Our argument follows [5].

Suppose we have some $\hat{\mu} > 0$ and $\hat{x} \in F^0(P)$ that is β-close to $x(\hat{\mu})$ for some $\beta < 1$. We generate a new value of x by applying Newton's method to (BP), and then shrink $\hat{\mu}$ to $\mu := \alpha\hat{\mu}$ for some $\alpha < 1$. We then want to show that x is again β-close to $x(\mu)$.

With $\hat{x}$, $\hat{\mu}$ and β as above, define

$$v := c + \hat{\mu}\nabla\Psi(\hat{x}) \tag{3.1}$$

as in the approximately-centered Theorem 2.7. Let $(\hat{d}, \hat{y})$ be the solution to (2.34) for this v, so that $\hat{d} = P_{\hat{x}}(v)$, and let

$$\hat{s} := c - A^\top \hat{y}. \tag{3.2}$$

From our assumption and the theorem, $\|\hat{d}/\hat{\mu}\|_{\hat{x}} = \|\hat{s}/\hat{\mu} + \nabla\Psi(\hat{x})\|_{\hat{x}}^* \leq \beta$. Now note that v is the gradient of the objective function of (BP) at $\hat{x}$, while its Hessian is $\hat{\mu}\nabla^2\Psi(\hat{x}) = \hat{\mu}\hat{\Theta}^2$. It follows that $-\hat{d}/\hat{\mu}$ is the Newton step for (BP) at $\hat{x}$. We write

$$\bar{d} := -\hat{d}/\hat{\mu}. \tag{3.3}$$

Thus, being β-close to $x(\hat{\mu})$ means precisely that the length of the Newton step for (BP) at $\hat{x}$, measured in the primal norm associated with $\hat{x}$, is at most β. Now let x be the Newton iterate

$$x := \hat{x} + \bar{d}. \tag{3.4}$$

Proposition 3.1 *With the notation above, $x \in F^0(P)$ is β^2-close to $x(\hat{\mu})$.*

Proof. From (2.34), $\bar{d}$ lies in the null space of A, so that $Ax = A\hat{x} = b$. Also, since $\|\bar{d}\|_{\hat{x}} \le \beta < 1$, Proposition 2.4 guarantees that $x \in \operatorname{int} C$; hence $x \in F^0(P)$ as desired.

Now to show that x is β^2-close to $x(\hat{\mu})$, it is enough by Proposition 2.6 to find (y, s) with

$$A^\top y + s = c, \quad s \in \Re^{J\cup K\cup L}, \quad \|s/\hat{\mu} + \nabla\Psi(x)\|_x^* \le \beta^2. \tag{3.5}$$

We prove that (3.5) holds for $(y, s) = (\hat{y}, \hat{s})$. This vector certainly satisfies the first two conditions, and we only need the norm inequality.

From (2.34),

$$\hat{\Theta}^2\hat{d} + A^\top\hat{y} = c + \hat{\mu}\nabla\Psi(\hat{x}),$$

where $\hat{\Theta}^2 := \nabla^2\Psi(\hat{x})$, so we find

$$\begin{aligned}\hat{s}/\hat{\mu} = (c - A^\top\hat{y})/\hat{\mu} &= -\nabla\Psi(\hat{x}) - \hat{\Theta}^2(-\hat{d}/\hat{\mu}) \\ &= -\nabla\Psi(\hat{x}) - \hat{\Theta}^2\bar{d}.\end{aligned}$$

Hence the norm in (3.5) is exactly that on the left-hand side of (2.30) in Theorem 2.2, and thus at most $\|\bar{d}\|_{\hat{x}}^2$. But by hypothesis, this is at most β^2 and the proof is complete. ∎

Now we show how $\hat{\mu}$ can be decreased:

Proposition 3.2 *Let $\hat{x}, \hat{\mu}, \hat{\beta}$, and x be as above. Let*

$$\alpha := 1 - \frac{\beta - \beta^2}{\beta + \sqrt{2p}} = \frac{\beta^2 + \sqrt{2p}}{\beta + \sqrt{2p}}, \quad \mu = \alpha\hat{\mu}. \tag{3.6}$$

Then x is β-close to $x(\mu)$. (Recall, $p \le n$ is defined in (2.24).)

Proof. Again, we use Proposition 2.6 with $(y, s) = (\hat{y}, \hat{s})$. We have

$$\begin{aligned}\|\hat{s}/\mu + \nabla\Psi(x)\|_x^* &= \|\hat{s}/(\alpha\hat{\mu}) + \nabla\Psi(x)\|_x^* \\ &= \|\tfrac{1}{\alpha}\left(\tfrac{\hat{s}}{\hat{\mu}} + \nabla\Psi(x)\right) - \left(\tfrac{1}{\alpha} - 1\right)\nabla\Psi(x)\|_x^* \\ &\leq \tfrac{1}{\alpha}\|\tfrac{\hat{s}}{\hat{\mu}} + \nabla\Psi(x)\|_x^* + \left(\tfrac{1}{\alpha} - 1\right)\|\nabla\Psi(x)\|_x^* \\ &\leq \tfrac{1}{\alpha}\beta^2 + \left(\tfrac{1}{\alpha} - 1\right)\sqrt{2p} \\ &= \beta,\end{aligned} \tag{3.7}$$

where the second inequality follows from (3.5) with $s = \hat{s}$ and Proposition 2.3. ■

Thus μ can be reduced by a constant factor at each iteration, and Theorem 2.7 translates this into a geometrically decreasing bound on the duality gap at each iteration. Let us use $\beta = \frac{1}{2}$, so $\alpha = 1 - \frac{1}{2+4\sqrt{2p}} < 1 - \frac{1}{8\sqrt{p}}$. Thus repeating the Newton procedure k times, we reduce the bound by the factor at most $(1-1/8\sqrt{p})^k$, and hence $O(\sqrt{p})$ iterations reduce it by a constant factor. From this discussion and the propositions above, we have

Theorem 3.1 *Suppose x^0 is β-close to some $x(\mu^0)$, where $\mu^0 > 0$ and $\beta = \frac{1}{2}$. Let $\alpha := 1 - \frac{1}{2+4\sqrt{2p}}$, and define the iterates (x^k, y^k, s^k) as follows. For each $k = 0, 1, \ldots$, let $\hat{\mu} := \mu^k$ and $\hat{x} := x^k$ and define $\hat{d}, \hat{y}$, and $\hat{s}$ as in the approximately-centered Theorem 2.7. Let $(y^k, s^k) := (\hat{y}, \hat{s})$, define $x^t odd.new.tex k + 1 := x$ from (3.3) and (3.4), and set $\mu^{k+1} := \alpha\mu^k$. Then, for each k,*
(i) $x^k \in F^0(P), (y^k, s^k) \in F^0(D)$;
(ii) $\|s^k/\mu^k + \nabla\Psi(x^k)\|_{x^k}^* \leq \beta$; and
(iii) the duality gap is bounded by $(x^k)^\top s^k \leq (\alpha)^k\mu^0(2p + \beta\sqrt{2p})$.
Moreover, $(x^k)^\top s^k \leq \epsilon$ within $O(p\ln(p\mu^0/\epsilon))$ iterations. ■

To conclude the section, we remark that from the proofs of Theorems 2.6 and 2.7 we can deduce that $(x^k)^\top s^k \geq (\alpha)^k\mu^0(2p - \beta\sqrt{2p})$. Thus the only way to accelerate the algorithm is to decrease μ faster. However, note that, as in (3.7),

$$\begin{aligned}\|\hat{s}/\mu + \nabla\Psi(x)\|_x^* &\geq \left(\frac{1}{\alpha} - 1\right)\|\nabla\Psi(x)\|_x^* - \frac{1}{\alpha}\|\frac{\hat{s}}{\hat{\mu}} + \nabla\Psi(x)\|_x^* \\ &\geq \left(\frac{1}{\alpha} - 1\right)\sqrt{2p} - \frac{1}{\alpha}\beta^2,\end{aligned}$$

so that we can only prove that x is β-close to $x(\mu)$ using $\hat{s}$ if $\alpha \geq 1 - \frac{\beta+\beta^2}{\beta+\sqrt{2p}}$. Therefore, without solving another linear system, the rate of decrease of μ to guarantee path-following is severely restricted.

3.2. A POTENTIAL-REDUCTION METHOD

Let us suppose that $c^\top x$ is not constant on $F(P)$. (If it were, c would be in the row space of A; then solving (2.34) for any $\hat{x} \in F^0(P)$ with $v = c$ would give $\bar{d} = 0$, confirming that $\hat{x}$ is optimal.) In this case, $c^{\mathsf{T}}x$ is greater than the optimal value z^* of (P) at any $x \in F^0(P)$, and we can thus define

$$\phi(x, z) := q \ln(c^{\mathsf{T}}x - z) + \Psi(x), \tag{3.8}$$

where $z \leq z^*$ and q is a positive parameter. Our algorithm is based on reducing this potential function (closely related to that of Karmarkar [10]) as in Gonzaga [7], Ye [20], or Freund [4].

Suppose at the start of an iteration we have $\hat{x} \in F^0(P)$ and $\hat{z} \leq z^*$. Then the gradient of ϕ with respect to x at $(\hat{x}, \hat{z})$ is

$$\nabla_x \phi(\hat{x}, \hat{z}) =: \tilde{v} = \frac{q}{c^{\mathsf{T}}\hat{x} - \hat{z}} c + \nabla\Psi(\hat{x}). \tag{3.9}$$

(Note the similarity to v in (3.1).) Let $(\tilde{d}, \tilde{y})$ solve (2.34) for $v = \tilde{v}$, so that $\tilde{d} = P_{\hat{x}}(\tilde{v})$. Then we show, as in [5], that the potential function can be reduced by a constant by taking a step in the direction $-\tilde{d}$, as long as $\|\tilde{d}\|_{\hat{x}}$ is sufficiently large.

Proposition 3.3 *Suppose $\hat{x}, \hat{z}, \tilde{v}, \tilde{d}$ and $\tilde{y}$ are as above. Then, if $\|\tilde{d}\|_{\hat{x}} \geq \frac{4}{5}$ and $\gamma \in (0, 1)$,*

$$x(\gamma) := \hat{x} - \gamma\tilde{d}/\|\tilde{d}\|_{\hat{x}} \in F^0(P)$$

and

$$\phi(x(\gamma), \hat{z}) \leq \phi(\hat{x}, \hat{z}) - \frac{4}{5}\gamma + \frac{\gamma^2}{2(1-\gamma)}. \tag{3.10}$$

In particular, $x(\frac{2}{5}) \in F^0(P)$ and

$$\phi(x(\frac{2}{5}, \hat{z}) \leq \phi(\hat{x}, \hat{z}) - \frac{1}{6}. \tag{3.11}$$

Proof. As in the proof of Proposition 3.1, $\tilde{d}$ lies in the null space of A, so $Ax(\gamma) = A\hat{x} = b$, and $\|x(\gamma) - \hat{x}\|_{\hat{x}} = \gamma < 1$, so $x(\gamma) \in F^0(P)$ by Proposition 2.4. Also, we have

$$\begin{aligned}
\phi(x(\gamma), \hat{z}) - \phi(\hat{x}, \hat{z}) &= q \ln\left(1 - \frac{\gamma c^{\mathsf{T}}\tilde{d}}{\|\tilde{d}\|_{\hat{x}}(c^{\mathsf{T}}\hat{x} - \hat{z})}\right) + \Psi(x(\gamma)) - \Psi(\hat{x}) \\
&\leq -\frac{\gamma q c^{\mathsf{T}}\tilde{d}}{\|\tilde{d}\|_{\hat{x}}(c^{\mathsf{T}}\hat{x} - z)} + \Psi(\hat{x} - \gamma\tilde{d}/\|\tilde{d}\|_{\hat{x}}) - \Psi(\hat{x}) \\
&\quad \text{(from the concavity of the logarithm function)} \\
&\leq -\frac{\gamma q c^{\mathsf{T}}\tilde{d}}{\|\tilde{d}\|_{\hat{x}}(c^{\mathsf{T}}\hat{x} - z)} - \frac{\gamma\nabla\Psi(\hat{x})^{\mathsf{T}}\tilde{d}}{\|\tilde{d}\|_{\hat{x}}} + \frac{\gamma^2}{2(1-\gamma)} \\
&\quad \text{(from Theorem 2.3)} \\
&= -\frac{\gamma}{\|\tilde{d}\|_{\hat{x}}}\left(\frac{q}{c^{\mathsf{T}}\hat{x} - \hat{z}}c + \nabla\Psi(\hat{x})\right)^{\mathsf{T}} \tilde{d} + \frac{\gamma^2}{2(1-\gamma)}
\end{aligned}$$

$$= -\frac{\gamma}{\|\tilde{d}\|_{\hat{x}}}\tilde{v}^{\mathsf{T}}\tilde{d} + \frac{\gamma^2}{2(1-\gamma)} = -\gamma\|\tilde{d}\|_{\hat{x}} + \frac{\gamma^2}{2(1-\gamma)}$$

(from (2.35) in Theorem 2.4)

$$\leq -\frac{4}{5}\gamma + \frac{\gamma^2}{2(1-\gamma)}.$$

This proves (3.10), and (3.11) follows by substituting $\gamma = \frac{2}{5}$. ∎

Suppose now $\|\tilde{d}\|_{\hat{x}} < \frac{4}{5}$. Then let

$$\begin{aligned} \hat{\mu} &:= (c^{\mathsf{T}}\hat{x} - \hat{z})/q, \\ \hat{v} &:= \hat{\mu}\tilde{v} = c + \hat{\mu}\nabla\Psi(\hat{x}), \\ \hat{y} &:= \hat{\mu}\tilde{y}, \quad \text{and} \\ \hat{s} &:= c - A^{\mathsf{T}}\hat{y}. \end{aligned} \tag{3.12}$$

Let $\hat{d} := P_{\hat{x}}(\hat{v}) = \hat{\mu}P_{\hat{x}}(\tilde{v}) = \hat{\mu}\tilde{d}$, and note that $\|\hat{d}/\hat{\mu}\|_{\hat{x}} = \|\tilde{d}\|_{\hat{x}} < \frac{4}{5}$, so $\hat{x}$ is β-close to $x(\hat{\mu})$ for $\beta = \frac{4}{5}$. We can therefore apply the approximately-centered Theorem 2.7 to obtain

$$(\hat{y}, \hat{s}) \in F^0(D) \quad \text{and} \tag{3.13}$$

$$\hat{x}^{\mathsf{T}}\hat{s} \leq \hat{\mu}(2p + \frac{4}{5}\sqrt{2p}). \tag{3.14}$$

Hence

$$z := b^{\mathsf{T}}\hat{y} = c^{\mathsf{T}}\hat{x} - \hat{x}^{\mathsf{T}}\hat{s} \tag{3.15}$$

is a valid lower bound on the optimal value z^* of (P) and (D). We show as in [5] that this update provides a sufficient decrease in ϕ as long as q is sufficiently large.

Proposition 3.4 *Suppose $q \geq 2p + \sqrt{2p}$. If $\hat{x} \in F^0(P)$, and, with the notation above, $\|\tilde{d}\|_{\hat{x}} < \frac{4}{5}$, then $(\hat{y}, \hat{s}) \in F^0(D)$, $z \leq z^*$ and*

$$\phi(\hat{x}, z) \leq \phi(\hat{x}, \hat{z}) - \frac{1}{6}. \tag{3.16}$$

Proof. From the discussion above, we only need to establish (3.16). From (3.12) – (3.15),

$$c^{\mathsf{T}}\hat{x} - z = \hat{x}^{\mathsf{T}}\hat{s} \leq \hat{\mu}(2p + \frac{4}{5}\sqrt{2p}) = \frac{2p + \frac{4}{5}\sqrt{2p}}{q}(c^{\mathsf{T}}\hat{x} - \hat{z}).$$

Hence,

$$\begin{aligned}\phi(\hat{x}, z) - \phi(\hat{x}, \hat{z}) &= q \ln\left(\frac{c^{\mathsf{T}}\hat{x} - z}{c^{\mathsf{T}}\hat{x} - \hat{z}}\right) \\ &\leq q \ln\left(\frac{2p + \frac{4}{5}\sqrt{2p}}{q}\right) \\ &\leq q \ln\left(1 - \frac{\frac{1}{5}\sqrt{2p}}{q}\right) \\ &\quad (\text{since} \quad q \geq 2p + \sqrt{2p}) \\ &\leq -\frac{1}{5}\sqrt{2p} \leq -\frac{1}{6}.\end{aligned}$$

■

These two results naturally suggest an algorithm for which we obtain the following convergence result.

Theorem 3.2 *Let $x^0 \in F^0(P)$ and $z^0 \leq z^*$ be given, and choose $q \geq 2p + \sqrt{2p}$. Suppose $\{x^k\} \subseteq \Re^n$ and $\{z^k\} \subseteq \Re$ are obtained as follows. For each k, let $\hat{x} := x^k$ and $\hat{z} := z^k$ and define $\tilde{v}$ by (3.9). Compute $\tilde{d} := P_{\hat{x}}(\tilde{v})$. If $\|\tilde{d}\|_{\hat{x}} \geq \frac{4}{5}$, set $x^{k+1} := x(\frac{2}{5})$ as in Proposition 3.3, and let $z^{k+1} := z^k$; else, set $x^{k+1} := x^k$ and update $z^{k+1} := z$ as in Proposition 3.4. Then, for each k,*

(i) $x^k \in F^0(P)$ and $z^k \leq z^*$;

(ii) $\phi(x^k, z^k) \leq \phi(x^0, z^0) - k/6$; and

(iii) for some constant C, $c^{\mathsf{T}}x^k - z^k \leq C\exp(-k/6q)$. ■

The last part follows from the same analysis used in the proofs of Theorem 4.1 and Lemma 4.2 of [1].

If we choose $q = O(p)$, this yields a bound of $O(p)$ iterations to reduce the bound on the optimality gap by a constant factor. This is worse than the bound for the path-following method; but the present algorithm allows considerably more flexibility which might improve its practical performance. For instance, as in Freund [4] and Gonzaga [7], we can try to improve the lower bound as in Proposition 3.4 at every iteration, performing a line search on $\mu = \hat{\mu}$ (of which $\hat{y}$ and $\hat{s}$ are linear functions) to obtain the best bound, and we can similarly perform a line search on γ to approximately minimize $\phi(x(\gamma), \hat{z})$, even allowing $\gamma > 1$ as long as $x(\gamma)$ remains feasible.

References

1. K. M. Anstreicher. A combined phase I – phase II scaled potential algorithm for linear programming. *Mathematical Programming*, 52:429–439, 1991.
2. I. C. Choi and D. Goldfarb. Exploiting special structure in a primal–dual path–following algorithm. *Mathematical Programming*, 58:33–52, 1993.
3. R. Fourer and S. Mehrotra. Performance of an augmented system approach for solving least–squares problems in an interior–point method for linear programming. *COAL Newsletter*, 19:26–31, August 1991.
4. R. M. Freund. Polynomial–time algorithms for linear programming based only on primal scaling and projected gradients of a potential function. *Mathematical Programming*, 51:203–222, 1991.
5. R. M. Freund and M. J. Todd. Barrier functions and interior–point algorithms for linear programming with zero–, one–, or two–sided bounds on the variables. Technical Report 1016, School of Operations Research and Industrial Engineering, Cornell University, Ithaca, NY 14853–3801, USA, 1992.
6. C. C. Gonzaga. An algorithm for solving linear programming problems in $O(n^3L)$ operations. In N. Megiddo, editor, *Progress in Mathematical Programming : Interior Point and Related Methods*, pp. 1–28. Springer Verlag, New York, 1989.
7. C. C. Gonzaga. Polynomial affine algorithms for linear programming. *Mathematical Programming*, 49:7–21, 1990.
8. C. C. Gonzaga. Large steps path–following methods for linear programming, Part I : Barrier function method. *SIAM Journal on Optimization*, 1:268–279, 1991.
9. C. C. Gonzaga. Large steps path–following methods for linear programming, Part II : Potential reduction method. *SIAM Journal on Optimization*, 1:280–292, 1991.
10. N. K. Karmarkar. A new polynomial–time algorithm for linear programming. *Combinatorica*, 4:373–395, 1984.
11. Y. E. Nesterov and A. S. Nemirovsky. Conic formulation of a convex programming problem and duality. *Optimization Methods and Software*, 1(2):95–115, 1992.
12. Y. E. Nesterov and A. S. Nemirovsky. *Interior Point Polynomial Methods in Convex Programming*. SIAM, Philadelphia, USA, 1993.
13. J. Renegar. A polynomial–time algorithm, based on Newton's method, for linear programming. *Mathematical Programming*, 40:59–93, 1988.
14. R. T. Rockafellar. *Convex Analysis*. Princeton University Press, Princeton, 1970.
15. C. Roos and J. P. Vial. A polynomial method of approximate centers for linear programming. *Mathematical Programming*, 54:295–305, 1992.
16. L. Schrage. Implicit representation of variable upper bounds in linear programming. *Mathematical Programming Study*, 4:118–132, 1975.
17. M. J. Todd. An implementation of the simplex method for linear programming problems with variable upper bounds. *Mathematical Programming*, 23:34–49, 1982.
18. M. J. Todd. Exploiting special structure in Karmarkar's algorithm for linear programming. *Mathematical Programming*, 41:97–113, 1988.
19. R. J. Vanderbei and T. J. Carpenter. Symmetric indefinite systems for interior point methods. *Mathematical Programming*, 58:1–32, 1993.
20. Y. Ye. An $O(n^3L)$ potential reduction algorithm for linear programming. *Mathematical Programming*, 50:239–258, 1991.

ON THE COMPLEXITY OF THE SIMPLEX METHOD*

DONALD GOLDFARB
Department of Industrial Engineering and Operations Research
Columbia University
New York, NY 10027

Abstract.
Although the efficiency of the simplex method in practice is well documented, the theoretical complexity of the method is still not fully understood. In this paper we briefly review what is known about the worst-case complexity of variants of the simplex method for both general linear programs and network flow problems and the expected behavior of some of these variants based upon probabilistic analysis. We also give a new proof of the fact that the parametric-objective simplex algorithm, which is known to have an expected complexity that is polynomial in the dimensions of the problems being solved, performs an exponential number of pivots in the worst case.

1. Introduction

The simplex method, developed by George Dantzig in 1947 [19] for solving linear programming problems, is one of the best known, most successful, and most extensively used methods in numerical computing. Using advanced pivot selection techniques, current implementations of the simplex method are able to solve linear programs with tens of thousands of rows and hundreds of thousands of columns on a workstation in a reasonable amount of time [13, 14, 25].

In spite of the practical efficiency of the simplex method attested to by the enormous amount of empirical evidence that has been accumulated in solving real-world linear programming problems, the theoretical efficiency of the method is still not fully understood. In particular, the following fundamental question remains unanswered: "Is there a variant of the simplex method that solves linear programs in polynomial time?"

In the next section of this paper we briefly review what is known about the theoretical complexity of the simplex method. Our discussion focuses on the worst-case complexity of different variants of the simplex method, although we also summarize what is known about the performance of the method in practice and mention what is known from a probabilistic analysis of some of these variants. We also consider the special case of network flow problems. In section 3 we examine the worst-case behavior of a particular variant of the simplex method, the *parametric-objective* simplex algorithm of Gass and Saaty [26]. This algorithm is of special interest because its *average-case* complexity has been shown to be polynomial under two different probabilistic models by Borgwardt [16, 17] and Haimovich [41]. The worst-case

* This research was supported in part by the National Science Foundation under Grants MCS-8006064 and DMS-91-06195 by the Army Research Office under Contract DAAG-29-82-K-0163 and by the Department of Energy under Grant DE-FG02-92ER2516.

S. Gomez and J-P. Hennart (eds.), Advances in Optimization and Numerical Analysis, 25–38.
© 1994 *Kluwer Academic Publishers.*

complexity of the algorithm was first shown to be exponential by Murty [46]. Here we present a different family of linear programs on which the parametric-objective simplex algorithm requires a number of pivots that grows exponentially with the size of the problem. Also, our proof of a lower bound for this variant is based upon the geometric interpretation of it as a "shadow-vertex" simplex algorithm given by Borgwardt [17]. Finally, we offer some concluding remarks in section 4.

2. Complexity of Simplex Variants

The complexity of an algorithm is usually defined as the running time, or number of operations, required by that algorithm to solve a problem as a function of the input size of that problem in the worst case, where the input size of a problem is defined as the number of binary bits needed to represent all of the problem' data. When discussing the simplex method, however, the number of pivots required by it in the worst case is often of most interest. As long as each execution of the pivot rule can be implemented in polynomial time, then these two performance measures can readily be related to one another. Also, most upper bounds for the number of pivots required by the simplex method are given in terms of the dimensions (m and n) of the problem.

2.1. General Linear Programs

Starting with the well-known paper of the Klee and Minty [44], several variants of the simplex method have been shown to have a worst-case complexity which is exponential in the size of the linear programming problem being solved. Klee and Minty constructed, for each $n > 2$, a linear programming problem whose feasible region is a polytope that is combinatorially equivalent to the n-cube, and for which the number of pivots required by the simplex variant that uses the most negative reduced cost (i.e. Dantzig) pivot rule, is $2^n - 1$.

Using a more complicated construction, Jeroslow [43] proved that the *greatest change* (i.e., *largest improvement*) pivot rule can require an exponential number of pivots. Bland's pivot rule [15] was analyzed by Avis and Chvatal [8], who showed that it can require more than F_m pivots on standard form problems with m rows and $2m$ columns, where F_m is the m-th Fibonacci number.

The polytopes constructed by Klee and Minty [44] do not, at least in their shape, appear to be *pathological* although they result in pathological behavior. Not only are they combinatorially equivalent to the n-cubes, they are metrically arbitrarily close to the n-cubes. By tilting the faces of the n-cube by large, rather than arbitrarily small amounts, and by choosing the linear objective functions appropriately, Goldfarb and Sit [40] were able to prove that the *steepest-edge* simplex pivot rule [39, 25] requires $2^n - 1$ pivots on such modified Klee-Minty linear programs. Finally Murty [45] proved that the parametric-objective simplex pivot rule can require $2^n - 1$ pivots to solve an $m \times 2m$ standard form linear program. The worst-case complexity of this rule was also analyzed by Goldfarb [30], and it is this analysis that is presented in the next section.

There is still hope that some simplex variant requires at most a polynomial (in n and m) number of pivots. One possible candidate is the *least recently basic* pivot rule, which has been conjectured by Zadeh to be polynomial [59].

In contrast to the exponential worst-case complexity of variants of the simplex method described above, a vast amount of experience gained in using variants of the simplex method in practice indicates that in general the number of pivots required grows linearly with m and sublinearly with n, typically being between m and $4m$ to complete both phase I and phase II, where m and n are the numbers of rows and columns, respectively, in a standard form problem. For a review of some of these empirical results and the results obtained in several Monte Carlo experiments, see Shamir [53]. Almost all of this empirical evidence is for primal or dual simplex variants which use the Dantzig pivot rule or some *partial pricing* pivot rule. However, recent computational tests on very difficult large linear programs by Forrest and Goldfarb [25], using IBM's OSL software package, have shown that the primal or dual Dantzig pivot rule can take as much as $25m$ to $150m$ pivots to solve such problems.

Further evidence that the typical behavior attributed to the simplex method does not apply to very difficult linear programs is provided by the following table which gives results obtained on a family of hard fleet assignment models by Bixby [14] using a primal partial pricing simplex variant in CPLEX.

TABLE I
Performance of a Primal Partial Pricing Simplex Algorithm on Hard Fleet Assignment Models

Problem	Rows, m (rel. to FL4)	Columns, n	$n-m$ (rel. to FL4)	Pivots (rel. to FL4)
FL4	7,772 (1)	16,958	9,186 (1)	325,179 (1)
FL7	15,299 (1.97)	42,593	27,294 (2.97)	3,063,722 (9.42)
FL12	21,971 (2.83)	68,136	46,165 (5.03)	14,707,771 (45.23)

In these experiments the number of pivots grows more than cubically with m and more than quadratically with $n-m$, the dimension of the null space of the constraint matrix in the problem, and hence a measure of the degrees of freedom in the problem. These problems are indeed very difficult for the simplex method; it took CPLEX more than 572 hours to solve Problem $FL12$ on an HP9000/730 workstation using its primal partial pricing option [14]. It should be noted, however, that primal and dual steepest-edge algorithms solve even these very difficult linear programs in less than $2(n+m)$ pivots [25].

About ten years ago there was a flurry of activity to explain the practical performace of the simplex method using probabilistic analysis and, hence, to close the large gap between practice and theory. This sort of analysis determines the expected number of pivots required by a given simplex algorithm under a probability model on the set of linear programs of a given size that are solved.

Borgwardt [16, 17] was the first to show that the expected number of steps required by a simplex algorithm was polynomial in the size (m and n) of the problems. He analyzed the parametric-objective variant for phase II, combined with a special variable-dimension method for phase I, applied to problems of the form max

$[c^T x : Ax \leq e]$, where e is a vector of ones, and c and the rows of A are generated independently from a spherically symmetric distribution. Later Haimovich [40] and Adler [2] proved that the expected number of pivots required by the parametric objective variant in phase II is linear in the problem size, under a simple sign invariant probabilistic model — specifically one in which the distribution is invariant to sign changes in both the rows and columns of the data and satisfies a certain nondegeneracy assumption.

Another parametric variant, Dantzig's self-dual simplex algorithm [20], was analyzed by Smale [54] assuming that the problem data are generated from a spherically symmetric distribution. Subsequently, Adler, Karp and Shamir [3], Adler and Megiddo [4] and Todd [56] proved that a lexicographic variant of the self-dual algorithm requires at most $0(min[m^2, n^2])$ pivots under a sign invariant probabalistic model. Adler and Megiddo [4] also showed that there cannot be a subquadratic upper bound for this variant and model.

2.2. Network Flow Problems

According to Cunningham [18], Edmonds [23] showed that the simplex method applied to the shortest path problem requires an exponential number of pivots. Notice that Edmonds unpublished report predates the publication of Klee and Minty's celebrated paper [44]. Zadeh [57], shortly thereafter, gave a family of minimum cost network flow problems on which the simplex method with the Dantzig pivot rule performs an exponential number of pivots, and Cunningham [18] gave a family of transshipment problems on which both the Dantzig rule and Bland's pivot rule [15] require an exponential number of pivots. It is easy to modify Zadeh's "bad" family of problems to show that Dantzig's pivot rule and rules which choose pivots based upon shortest augmenting or pseudoaugmenting paths can take an exponential number of pivots to solve maximum flow problems [33].

In contrast to the above negative results and what can currently be proved about the simplex method for general linear programs, many simplex variants for solving network flow problems have been shown to be polynomial and, in fact, strongly polynomial (i.e. polynomial in the dimensions of the problems). To describe these results we shall use n, m, and C, to denote the number of nodes, number of arcs, and maximum absolute value of an arc cost, respectively, in the network.

For the shortest path problem, Orlin [48] proved that the Dantzig rule performs at most $0[min(n^2 \log(nC), n^2 m \log m)]$ pivots. Dial, Glover, Karney and Klingman [21] and Zadeh [58] showed that when applied to problems with all nonnegative arc costs (distances), this variant, started from an all artificial basis, can be interpreted as an implementation of Dijkstra's [22] algorithm and Akgul [5] showed that it requires at most $n - 1$ pivots. The most efficient simplex variants for the general shortest path problem have been proposed by Akgul [5] and Goldfarb, Hao and Kai [36]. These variants require at most $\frac{1}{2}n^2$ pivots and Akgul has shown how to implement one of them to run in $0(nm + n^2 \log n)$ time using Fibonacci heaps. Goldfarb, Hao and Kai [37, 38] give other rules which require at most $0(nm)$ or $0(n^3)$ pivots.

The assignment problem can also be solved by the simplex method in at most $\frac{1}{2}n^2$ pivots. The methods for doing this are dual simplex algorithms proposed by Balinski [9, 10] and Goldfarb [31], all of which are based upon what Balinski calls the *signature* of the basis tree. Balinski's first algorithm [9], and the modified versions of it proposed in [31] which can be implemented to run in $0[nm + n^2 \log n]$ time, allow pivots where the objective function does not change. Kleinschmidt, Lee and Schannath [45] have extended Balinski's algorithm to solve certain classes of transportation problems with m sources and n destinations and constant demands in $0(mn)$ pivots and $0(nm^2 + n^2m)$ time, and Balinski and Rispoli [11] have completely characterized the class of transportation problems to which signature algorithms may be applied.

Polynomial-time primal simplex algorithms have also been proposed for the assignment problem. Orlin [48] showed that the primal algorithm using the Dantzig pivot rule requires at most $0[n^2 \log(nC)]$ pivots. Roohy-Laleh [52], Hung [41], Akgul [6], and Ahuja and Orlin [1], give other primal variants that require at most $0(n^3)$, $0[n^3 \log(nC)]$, $0(n^2)$ and $0(n^2 \log C)$ pivots, respectively.

The first polynomial time simplex algorithms for the maximum flow problem were proposed by Goldfarb and Hao [32, 33]. These algorithms perform at most nm pivots and run in $0(n^2m)$ time if implemented straightforwardly and $0(nm \log n)$ time (see Goldberg, Grigoriadis, and Tarjan [27]) if sophisticated data structures are used.

Dual analogs of these algorithms that require at most 2nm pivots and $0(n^2m)$ time have recently been developed by Goldfarb and Chen [32] and equally efficient variants that are based upon "valid" distance labels introduced by Goldberg and Tarjan [28] have been given by Armstrong and Jin [7] and Goldfarb and Chen [32]. $0(n^3)$ implementations of modified versions of these algorithms have also been propossed by these authors. A strongly polynomial primal simplex algorithm that is based upon maximum capacity augmenting paths has also been proposed by Goldfarb and Hao [36].

The only genuine simplex algorithms for the general minimum cost network flow problem that are known to be polynomial are dual simplex algorithms. The first such algorithm was developed by Orlin [47] and was based upon Edmonds-Karp [24] (i.e., right-hand side) scaling. It performs at most $0(n^3 \log n)$ pivots for the transshipment problem and $0(m^3 \log n)$ pivots for the general minimum cost network flow problem. Recently, Plotkin and Tardos [45] gave a dual simplex algorithm that solves the transshipment problem in at most $O(mn \log n)$ pivots and the general problem in at most $0(m^2 \log n)$ pivots and $0(m^3 \log n)$ time. These results are essentially combined in a recent paper [49] by Orlin, Plotkin and Tardos.

Tarjan [55] and Goldfarb and Hao [35] have shown that it is possible to solve minimum cost network flow problems in a polynomial number of primal simplex pivots. Their algorithms, however, are not genuine simplex algorithms, since they allow pivots that increase the value of the objective function. Tarjan's algorithm, which requires at most $0(nm^2 \log n)$ pivots, and one of Goldfarb and Hao's are essentially simplex implementations of the *minimum mean augmenting cycle canceling* method of Goldberg and Tarjan [29]. Goldfarb and Hao's other method is a combination of the cost scaling technique of Rock [51] with the primal network simplex method of Goldfarb and Hao [33] for the maximum flow problem. The development of a genuine (i.e., monotonically decreasing objective value) polynomial-time primal simplex algorithm for the minimum cost network flow problem remains an open problem.

3. Complexity of the Parametric-Objective Simplex Algorithm

In this section we prove that the worst-case behavior of the parametric-objective simplex algorithm is exponential in the size of the problem being solved.

To geometrically describe the parametric-objective simplex algorithm, we define the two-dimensional subspace $U = Span(c, c^o)$, where c is the gradient of the objective function of the linear program

$$(P) \qquad maximize \quad z = c^T x$$
$$subject \quad to \ \ Ax \leq b,$$

where the $A \in R^{m \times n}$, $m \geq n$, and c^0 is chosen so that a given initial starting vertex x_0 is optimal for problem (P) with c replaced by c^o. If we consider the orthogonal projection $\Gamma(X)$ of the feasible polyhedron $X = \{x \in R^n \mid Ax \leq b\}$ into U, then those vertices of X which are mapped into vertices of $\Gamma(X)$ are called *shadow vertices.* In the nondegenerate case, the parametric-objective variant of the simplex method moves from shadow vertex to adjacent shadow vertex (in both X and $\Gamma(X)$), strictly decreasing the value of z until an optimal solution is obtained or unboundedness is detected. Hence it is clear why Borgwardt [16, 17] calls this algorithm the *shadow vertex* simplex algorithm.

3.1. Construction of the Polytopes

We shall now construct polytopes P_n in R^n which are combinatorially equivalent to the n-cube and whose orthogonal projection on the (x_{n-1},x_n)-plane has 2^n vertices. Our construction can be described geometrically as follows. We start with $P_2 = \{x \in R^2 | 0 \leq x_1 \leq 1, \beta x_1 \leq x_2 \leq \delta - \beta x_1\}$ where $\beta \geq 2$ and $\delta > 2\beta$. P_2 is a trapezoid which can be thought of as having been obtained from the rectangle $C_2 = \{x \in R^2 | 0 \leq x_1 \leq 1, 0 \leq x_2 \leq \delta\}$, by rotating its "lower face" (actually a facet) $x_2 = 0$ through an angle $\theta = \tan^{-1} \beta$ in the x_2-direction and its "upper face" $x_2 = \delta$ by the same amount in the opposite direction. For $n > 2$, we obtain P_n from P_{n-l} recursively by first forming the prism $C_n = \{x \in R^n | (x_1, \ldots, x_{n-1}) \in P_{n-1}, 0 \leq x_n \leq \delta^{n-1}\}$, which has lower and upper faces, $x_n = 0$ and $x_n = \delta^{n-1}$ equal to P_{n-1}. These faces are then rotated in the (x_{n-1}, x_n)-plane, through angles θ and $-\theta$, respectively, in the x_n-direction and rotated in the (x_{n-2}, x_n)-plane through angles $\phi = \tan^{-1} 1$ and $-\phi$, respectively, in the x_n-direction.

Klee and Minty [44] and Goldfarb and Sit [40] constructed polytopes using only rotations in the (x_{n-1}, x_n)-plane. The orthogonal projection of the lower and upper faces of such polytopes in the (x_{n-1}, x_n)-plane are line segments and hence the orthogonal projection of the polytopes themselves in this plane are trapezoids. The additional rotation in the (x_{n-2}, x_n)-plane ensures that all vertices of P_n are shadow vertices. We now give an explicit representation for P_n and its vertex set and prove that it is combinatorially equivalent to the n-cube.

We define the polyhedron, P_n, as the set of points $x = (x_1, \ldots, x_n) \in R^n$ which satisfy the inequalities:

$$0 \leq \ x_1 \ \leq 1, \tag{1}$$

$$\beta x_1 \leq \ x_2 \ \leq \delta - \beta x_1, \tag{2}$$

$$\beta x_j - x_{j-1} \leq x_{j+1} \leq \delta^j - \beta x_j + x_{j-1}, j = 2, \ldots, n-l. \tag{3}$$

where $\beta > 2$ and $\delta > 2\beta$. Since P_n is clearly a bounded polyhedron, it is the convex hull of its vertex set. To generate this set, let $B = \{0, 1\}$ and for $i = 1, 2, \ldots, n$ let $\pi_i^i : R^i \to R$ and $\pi_{i-1}^i : R^i \to R$ be, respectively, the i-th and $(i-1)$-st projections. Define $v_i : B^i \to R^i$ recursively by

$$\begin{aligned} v_1(0) &= 0, & v_1(1) &= 1, \\ v_2(0,0) &= (0,0), & v_2(1,0) &= (1,\beta), \\ v_2(0,1) &= (0,\delta), & v_2(1,1) &= (1,\delta-\beta), \end{aligned}$$

and for $i = 2, \ldots, n$ and $a \in B^i$,

$$v_{i+1}(a,0) = (v_i(a), (\beta\pi_i^i - \pi_{i-1}^i)v_i(a)),$$

and

$$v_{i+1}(a,1) = (v_i(a), \delta^i - (\beta\pi_i^i - \pi_{i-1}^i)v_i(a)).$$

Note that for notational convenience, we have written an element of B^{i+1} by concatenating its coordinates. Henceforth, we shall denote v_n simply by v.

Theorem 1 *(a)* $v(B^n) \equiv \{v(a) \mid a \in B^n\}$ *is the vertex set of* P_n
(b) P_n *is combinatorially equivalent to the* n*-dimensional cube* I^n *, where* $I = [0, 1]$.

To prove this theorem we first must establish some preliminary results. For this purpose let

$$\phi_0 = 0, \quad \phi_1 = 1, \tag{4}$$

$$\phi_{j+1} = \beta\phi_j - \phi_{j-1}, \quad for \quad j \geq 1, \tag{5}$$

$$x_0 = 0, \quad x_1 \geq 0, \tag{6}$$

and

$$x_{j+1} \geq \beta x_j - x_{j-1}, for \quad j \geq 1. \tag{7}$$

Note that (6)-(7) is just a restatement of the inequalities of the left in (1)-(3).

Lemma 1 *If* ϕ_j *,* $j \geq 0$ *is defined by (4)-(5) and* x_j *,* $j \geq 0$ *, satisfies (6)-(7), then if* $\beta \geq 2$ *,*

$$\begin{aligned} &a\ ) \quad \phi_{j+1} \geq \phi_j \geq 0, \quad for \quad all j \geq 0, \\ &b\ ) \quad \phi_i x_j - \phi_{i-1}x_{j-1} > \phi_{i+j-1}x_1, \quad for \quad all \quad i \geq 1 \quad and \quad j \geq 1, \\ &c\ ) \quad \beta x_j - x_{j-1} \geq \phi_{j+1}x_1, \quad for \quad all \quad j \geq 1. \end{aligned}$$

Proof. The proofs of (a) and (b) are by induction on j. For $j = 0$, (a) follows from (4). Assume now that (a) is true for $j = k - 1$ then from (5)

$$\phi_{k+1} = (\beta - 1)\phi_k + \phi_k - \phi_{k-1} \geq (\beta - 1)\phi_k \geq \phi_k \geq 0.$$

To prove (b), observe that it reduces to $\phi_i x_1 \geq \phi_i x_1$ for all $i \geq 1$, when $j = 1$, since $x_0 = 0$. Now assume that (b) is true for $j = k$ and all $i \geq 1$. Then, since $\phi_i \geq 0$ by (a), it follows from (5) and (7) that

$$\phi_i x_{k+1} - \phi_{i-1} x_k \geq \phi_i(\beta x_k - x_{k-1}) - \phi_{i-1} x_k = \phi_{i+1} x_k - \phi_i x_{k-1} \geq \phi_{i+k} x_1$$

for all $i \geq 1$, proving (b) and (c). Part (c) is an immediate consequence of (b) and the facts that $\phi_2 = \beta$ and $\phi_1 = 1$.

Proof of Theorem 1. A necessary and sufficient condition for $x \in P_n$ to be a vertex is that it satisfy a linearly independent set of n of the constraints (1)-(3) as equalities. First let us show that both inequalities in (1), (2) or (3) for a fixed j, cannot be simultaneously satisfied as equalities by a point $x \in P_n$. The case (1) is trivial. If equality holds in (2) then $x_1 = \frac{\delta}{2\beta} > 1$, which contradicts (1). If equality holds in (3) for some fixed $j \geq 2$, then

$$2\beta x_j - 2x_{j-1} = \delta^j . \tag{8}$$

From the right inequality in (2) and the left inequality in (1) we have that

$$2\beta x_2 - 2x_1 \leq 2\beta(\delta - \beta x_1) \leq 2\beta\delta < \delta^2,$$

which contradicts (8) for $j = 2$. For $j \geq 3$, it follows from the inequalities in (1)-(3) and Lemma 1 that

$$\begin{aligned} 2\beta x_j - 2x_{j-1} &\leq 2\beta(\delta^{j-1} - \beta x_{j-1} + x_{j-2}) - 2(\beta x_{j-2} - x_{j-3}) \\ &\leq 2\beta\delta^{j-1} - 2\beta\phi_j x_1 - 2\phi_{j-1} x_1 \\ &= 2\beta\delta^{j-1} - 2\phi_{j+1} x_1 \\ &< \delta^j , \end{aligned}$$

which contradicts (8).

Thus, each vertex of P_n corresponds to a choosing n of the constraints in (1)-(3) to be satisfied as equalities, where precisely one constraint is chosen from each pair. Clearly such a choice gives an $x \in P_n$ and a linearly independent set of n active constraints. It is also obvious that the set of 2^n vertices so defined is just $v(B^n)$.

To prove (b), observe that $v : B^n \to R^n$ is a bijection between the vertex set B^n of I^n and the the vertex set $v(B^n)$ of P_n. Moreover, the facets of $P_n, x_j = \delta^{j-1} - \beta x_{j-1} + x_{j-2}$ and $x_j = \beta x_{j-1} - x_{j-2}, j = 1, 2, \ldots, n$, where $x_{-1} = x_0 = 0$, corresponds to the facets $x_j = 1$ and $x_j = 0, j = 1, 2, \ldots, n$, of I^n. Since these facets are, respectively, the convex hulls $CH\{a \in B^n \mid a_j = 1\}, CH\{a \in B^n \mid a_j = 0\}, CH\{v(a) \mid a \in B^n, a_j = 1\}$ and $CH\{v(a) \mid a \in B^n, a_j = 0\}$, it follows from a result of Klee and Minty [44] that P^n and I^n are combinatorially equivalent.

3.2. Proof that all Vertices of P_n are Shadow Vertices

We now shall show that all vertices of P_n are "shadow vertices" of P_n in the two-dimensional subspace $U \equiv Span\{x_{n-1}, x_n\}$.

First, observe that for each vertex $v(a) \in v(B^n)$, the vertex set of P_n, the matrix whose columns are the outward normals corresponding to the active constraints at $v(a)$, has the form

$$N(a) = \begin{bmatrix} \sigma_1 & \beta & -1 & & & & \\ & \sigma_2 & \beta & -1 & & & \\ & & \ddots & \ddots & \ddots & & \\ & & & \ddots & \ddots & \ddots & \\ & & & & \ddots & \ddots & \ddots \\ & & & & \sigma_{n-2} & \beta & -1 \\ & & & & & \sigma_{n-1} & \beta \\ & & & & & & \sigma_n \end{bmatrix}$$

where $a = a_1, a_2, \ldots, a_n$ and

$$\sigma_j = \begin{cases} 1 & \text{if } a_j = 1 \\ -1 & \text{if } a_j = 0 \end{cases}$$

For a given $a \in B^n$, let us define $\alpha_1, \alpha_2, \ldots, \alpha_{n+1}$ for $n \geq 3$, recursively as follows:

$$\begin{aligned} \alpha_1 &= 1, \alpha_2 = \beta \\ \alpha_{j+2} &= n_{j,j+1}\alpha_{j+1} + n_{j,j}\alpha_j \\ &= \beta\alpha_{j+1} + \sigma_j\alpha_j, \qquad j = 1, 2, \ldots, n-1, \end{aligned} \tag{9}$$

where $n_{i,j}$ is the (i, j)-th element of $N(a)$. From this it immediately follows that $\omega(a) = N(a)\alpha$, where $\alpha = (\alpha_1, \alpha_2, \ldots, \alpha_n)$, has the form

$$\omega(a) = (0, 0, \ldots, 0, \alpha_{n+1}, \sigma_n\alpha_n)^T. \tag{10}$$

Lemma 2

$$0 < \alpha_1 < \alpha_2 < \cdots < \alpha_n$$

Proof. We shall prove the above by induction. Clearly it is true for $n = 2$. Now assume that it is true for $n = k$. From this and the fact that $\beta \geq 2$, it follows from the recurrence for α that

$$\alpha_{k+1} \geq 2\alpha_k + \sigma_{k-1}\alpha_{k-1} > \alpha_k$$

since $\sigma_{k-1} = \pm 1$ and $\alpha_k \pm \alpha_{k-1} > 0$.

This lemma and (10) show that for each vertex $v(a)$ of P_n, there exists a nonzero vector $\omega \in Span\{x_{n-1}, x_n\}$ which is in the convex cone generated by the outward normals to the constraints active at $v(a)$. From this it follows that the vertices $v(B^n)$ of P_n are vertices of the orthogonal projection $\Gamma(P_n)$ of P_n into $U \equiv Span\{x_{n-1}, x_n\}$; that is, they are all shadow vertices. Moreover, the mapping $\Gamma(v(B^n))$ of the vertices $v(B^n)$ is one-to-one.

To prove the latter statement, let us assume that two vertices $v(a)$ and $v(a')$ correspond to the same vertex of $\Gamma(P^n)$. If follows that any $\omega \in Span\{x_{n-1}, x_n\}$ that is in the convex cone generated by the set of active normal at $v(a)$ is also in the convex cone generated by the active normals at $v(a')$. In particular, let $\omega(a)$ be constructed as in (9)-(10).

$$\text{Let } a_k \neq a'_k \text{ and } a_j = a'_j\ , \quad j = k+1, \cdots, n.$$

Now since $\omega(a) = N(a)\alpha = N(a')\alpha'$, it follows that $\alpha_j = \alpha'_j$, $j = k+1, \cdots, n$, and hence that

$$\sigma_j \alpha_j = \sigma'_j \alpha'_j$$

$$\text{with } \sigma_j = -\sigma'_j = \pm 1; \text{ that is, } \alpha_j = -\alpha'_j.$$

But by construction (i.e. Lemma 2) $\alpha_j > 0$. Hence $\alpha'_j > 0$ which proves that $v(a)$ and $v(a')$ cannot map onto the same vertex of $\Gamma(P_n)$, if $a \neq a'$. Thus we have proved:

Theorem 2 *All vertices $v(B^n)$ of P_n are shadow vertices with respect to the orthogonal projection $\Gamma(P_n)$ of P_n into $U \equiv Span\{x_{n-1}, x_n\}$. Moreover, the mapping $\Gamma(v(B^n))$ is one-to-one.*

3.3. The Shadow Vertex Simplex Path

Let us number the vertices $v(a)$ of P_n according to the following Gray code: given $a \in B^n$ we define $P_n(a)$ to be the n-bit binary number $d_n \cdots d_1$, where $d_j = 0$ if there are an even number of 1's amongst $a_j, \cdots, a_n$ and $d_j = 1$ otherwise. In decimal notation $P_n(a) = \sum_{j=1}^{n} d_j 2^{j-1}$. Specifically, we shall denote by $v(a^k)$ that vertex of P_n corresponding to $a \in B^n$ with $P_n(a) = k$. It is easily verified that if $k \equiv p_n(a^k) = d_n \cdots d_1$, where $d_m = 0$ and $d_j = 1$, for all $j < m$, then $k+1 \equiv p_n(a^{k+1}) = d_n \cdots d_{m+1} d_m \cdots d_1$, where a^{k+1} differs from a^k only in that $a_m^{k+1} = \bar{a}_m^k$. (The overbar denotes binary complementation.)

We shall now show that by choosing $c^0 = \omega(a^0)$ and $c = \omega(a^{2n-1})$, where $\omega(a)$ is defined by (9) and (10) with $\sigma_j = a_j - \bar{a}_j$, the path followed by the parametric-objective simplex algorithm passes sequentially through the vertices $a^0, a^1, \cdots, a^{2n-1}$. First observe that the slope $m_n(a) = \sigma_n \alpha_n / \alpha_{n+1}$ of the orthogonal projection of $\omega(a)$ into $U \equiv Span\{x_{n-1}, x_n\}$ can be expressed as a continued fraction

$$\begin{aligned} m_n(a) &= \sigma_n/(\beta + \sigma_{n-1}/(\beta + \sigma_{n-2}/(\ldots(\beta + \sigma_1/\beta)\ldots))) \\ &= 1/(\rho_n\beta + 1/(\rho_{n-1}\beta + 1/(\ldots(\rho_2\beta + 1/\rho_1\beta)\ldots))) \\ &\equiv /\rho_n\beta, \rho_{n-1}\beta, \ldots, \rho_1\beta/ \end{aligned}$$

where $\rho_j = \prod_{i=j}^{n} \sigma_j . (\rho_j = -1$ if there are an odd number zeros amongst $a_j, \cdots, a_n$, and $\rho_j = 1$ otherwise.)

Lemma 3 *$m_n(a^k), k = 0, \cdots, 2^n - 1, a^k \in B^n$, is an increasing function of k, and $\mid m_n(a^k) \mid < 1$ for $k = 0, \cdots, 2^n - 1$.*

Proof. Our proof is by induction on n. Since $m_2(a^0) = -1/(\beta - 1/\beta), m_2(a^1) = -1/(\beta + 1/\beta), m_2(a^2) = 1/(\beta + 1/\beta)$, and $m_2(a^3) = 1/(\beta - 1/\beta)$ and $\beta \geq 2$, the statement of the lemma is true for $n = 2$. Now assume that the lemma is true for $n = q$. If $p_q(a) = d_q \cdots d_1$ then $p_{q+1}(a \enspace 0) = 0 \enspace d_q \cdots d_1 = p_q(a)$ and $p_{q+1}(a \enspace 1) = 1 \enspace \bar{d}_q \cdots \bar{d}_1 = 2^q + 2^q - 1 - p_q(a) = 2^{q+1} - 1 - p_q(a)$. Hence, the Gray code ordering of $(a \enspace a_{q+1}) \in B^{q+1}$, where $a \in B^q$, is $(a^0 \enspace 0, a^1 \enspace 0, \cdots, a^{2^q-1} \enspace 0, a^{2^q-1} \enspace 1, \cdots, a^1 \enspace 1, a^0 \enspace 1)$. Since $\beta \geq 2$ and $\mid m_q(a^k) \mid < 1, \beta + m_q(a^k) > 1$. Furthermore, since $m_q(a^k)$ is an increasing function of k for $k = 0, 1, \cdots, 2^q - 1$,

$$m_{q+1}(a^k \enspace 0) = -1/(\beta + m_q(a^k))$$

is negative and increases as k increases and

$$m_{q+1}(a^k \enspace 1) = 1/(\beta + m_q(a^k))$$

is positive and increases and k decreases and $\mid m_{q+1}(a^k \enspace a_{q+1}) \mid < 1$. Thus the lemma is proved.

¿From this lemma our main result follows.

Theorem 3 *Given the polyhedron P_n defined by (2) and $c^0 = w(a^0)$ and $c = w(a^{2^n-1})$, where $w(a)$ is defined by (6) and (7), the shadow vertex simplex path proceeds through all 2^n vertices of P_n in the order $a^0, a^1, \cdots, a^{2^n-1}$.*

Thus, the worst-case complexity of the parametric-objective simplex algorithm is exponential.

4. Concluding Remarks

Since the parametric-objective simplex algorithm is a special case of the self-dual simplex algorithm of Dantzig [19], the results in the previous section also show that the self-dual algorithm will perform an exponential number of pivots in the worst case. As discussed in the introduction these algorithms are the only simplex variants that have been analyzed probabilistically. Under the sign invariant probability model these variants have been shown to perform at most $\min\{m, n\} + 1$ Phase II pivots in the worst case on an problem with constraints of the form $Ax \leq b$, $x \geq 0$ [2, 40]. When applied to a particular problem, the sign invariant model generates 2^{n+m} problem instances. When these observations are combined with the results of the previous section, we see that almost all other problem instances (there are a total of 2^{2n} instances) generated by the sign invariant model from the pathological polytope Pn are either infeasible or are easily solved by the parametric-objective simplex algorithm. Thus we see that the problems constructed in section 3 are indeed pathological.

References

1. Ahuja, R.K. and J.B. Orlin. 1992a. "The Scaling Network Simplex Algorithm," *Operations Research,* 40, Supplement 1, S5-13.
2. Adler, I. "The Expected Number of Pivots Needed to Solve Parametric Linear Programs and the Efficiency of the Self-Dual Simplex Method." Working Paper, Department of Industrial Engineering and Operations Research, University of California, Berkeley, CA, 1983.
3. Adler, I., Karp, R.M., and Shamir, R. "A Simplex Variant Solving an mxd Linear Program in $O(min(m^2, d^2))$ Expected Number of Pivot Steps." Report UCB/CSD 83/158, Computer Science Division, University of California, Berkeley, CA, 1983.
4. Adler, I. and Megiddo, N. "A Simplex Algorithm whose Average Number of Steps is Bounded Between Two Quadratic Functions of the Smaller Dimension." *Journal of the Association for Computing Machinery,* 32, (871-895), 1984.
5. Akgul, M. "Shortest Path and Simplex Method." Research Report, Department. of Computer Science and Operations Research, North Carolina State University, Raleigh, NC, 1985.
6. Akgul, M. "A Genuinely Polynomial Primal Simplex Algorithm for the Assignment Problem." Research Report, Department of Computer Science and Operations Research, North Carolina State University, Raleigh, NC, 1985.
7. Armstrong, R.D. and Z. Jin. "A Strongly Polynomial Dual Network Simplex Algorithm," Technical Report, Graduate School of Management, Rutgers University, New Brunswick, NJ, 1993.
8. Avis, D. and V. Chvátal. "Notes on Bland's Pivoting Rule," *Math. Programming Study,* 8 (24-34) 1978.
9. Balinski, M.L. "Signature Methods for the Assignment Problem," *Operations Research,* 33 (527-536), 1985.
10. Balinski, M.L. "A Competitive (Dual) Simplex Method for the Assignment Problem," *Mathematical Programming,* 34 (125-141) 1986.
11. Balinski, M.L. and F.J. Rispoli. "Signature Classes of the Transportation Polytopes," *Math. Programming,* 60, (127-144), 1993.
12. Bixby, R.E. "Implementing the Simplex Method: The Initial Basis." Technical Report TR90-32, Department of Mathematical Sciences, Rice University, Houston, TX, 1990.
13. Bixby, R.E. "The Simplex Method: It Keeps Getting Better." Presented at the 14th International Symposium on Mathematical Programming, Amsterdam, The Netherlands, 1991.
14. Bixby, R.E. Private communication, 1993.
15. Bland, R. G. "New Finite Pivoting Rules For the Simplex Method," *Math. Oper. Res.,* 29, 6 (1039-1091), November 1981.
16. Borgwardt, K.H. "Some Distribution-Independent Results About the Asymptotic Order of the Average Number of Pivot Steps of the Simplex Method." *Math. Oper. Res.,* Vol.7, (441-462), 1982.
17. Borgwardt, K.H. "The Average Number of Pivot Steps Required by the Simplex Method is Polynomial." *Zeitschrift for Operations Research,* Vol. 26, (157-177), 1982.
18. Cunningham, W.H. "Theoretical Properties of the Network Simplex Method," *Math. Oper. Res.,* 4, 2 (196-203), May 1979.
19. Dantzig, G.B. "Application of the Simplex Method to a Transportation Problem." *Activity Analysis and Production and Allocation,* Edited by T.C. Koopmans, Wiley, New York, (359-373) 1951.
20. Dantzig, G.B. Linear Programming and Extensions, Princeton University Press, Princeton, NJ, 1963.
21. Dial, R., F. Glover, D. Karney, and D. Klingman. "A Computational Analysis of Alternative Algorithms and Labeling Techniques for Finding Shortest Path Trees." *Networks,* 9, (215-248), 1979.
22. Dijkstra, E. "A Note on Two Problems in Connexion with Graphs." *Numeriche Mathematic,* 1, (269-271), 1959.
23. Edmonds, J. "Exponential Growth of the Simplex Method for the Shortest Path Problem." Unpublished paper, University of Waterloo, 1970.
24. Edmonds, J., and R.M. Karp. "Theoretical Improvements in Algorithmic Efficiency for Network Flow Problems." *Journal of ACM,* 19, (248-264), 1972.
25. Forrest, J.J. and D. Goldfarb. "Steepest-edge Simplex Algorithms for Linear Programming,"

Mathematical Programming, 57, (341-374), 1992.

26. Gass, S.I. and E.L. Saaty. "The Computational Algorithm for the Parametric Objective Function," *Naval Res. Logist. Quart.,* 2, 1 (39-45), June 1955.
27. Goldberg, A.V., M.D. Grigoriadis and R.E. Tarjan, "Use of Dynamic Trees in a Network Simplex Algorithm for the Maximum Flow Problem," *Mathematical Programming,* 50, (277-290), 1991.
28. Goldberg, A.V., and R.E. Tarjan. "A New Approach to the Maximum Flow Problem." Proceedings of the 18th ACM Symposium on the Theory of Computing, (136-146), 1986.
29. Goldberg, A.V. and R.E. Tarjan. "Finding Minimum-Cost Circulations by canceling negative cycles." *Journal of ACM,* 36 (873-886), 1989.
30. Goldfarb, D. "Worst-Case Complexity of the Shadow Vertex Simplex Algorithm." Report, Department of Industrial Engineering and Operations Research, Columbia University, May 1983.
31. Goldfarb, D. "Efficient Dual Simplex Algorithms for the Assignment Problem." *Mathematical Programming,* 33, (187-203), 1985.
32. Goldfarb, D. and W. Chen. "Strongly Polynomial Dual Simplex Algorithms for the Maximum Flow Problem," Technical Report, Department of Industrial Engineering and Operations Research, Columbia University, New York, NY, 1993.
33. Goldfarb, D., and J. Hao. "A Primal Simplex Algorithm that Solves the Maximum Flow Problem in at Most nm Pivots and $O(n^2m)$ Time." *Mathematical Programming,* 47, (353-365), 1990.
34. Goldfarb, D., and J. Hao. "On Strongly Polynomial Variants of the Network Simplex Algorithm for the Maximum Flow Problem," *Operations Research Letter,* 10, (383-387), 1991.
35. Goldfarb, D., and J. Hao. "Polynomial-time Primal Simplex Algorithms for the Minimum Cost Network Flow Problem." *Algorithmica,* 8, (145-160), 1992.
36. Goldfarb, D., and J. Hao. "On the Maximum Capacity Augmenting Path Algorithm for the Maximum Flow Problem." Technical Report, Department of Industrial Engineering and Operations Research, Columbia University, New York, NY, 1993.
37. Goldfarb, D., J. Hao, and S. Kai. "Efficient Shortest Path Simplex Algorithms." *Operations Research,* 38, (624-628), 1990.
38. Goldfarb, D., J. Hao, and S. Kai. "Anti-stalling Pivot Rules for the Network Simplex Algorithm." *Networks,* 20, (79-91), 1990.
39. Goldfarb, D., and J.K. Reid. "A Practical Steepest-Edge Simplex Algorithm." *Mathematical Programming,* 12, (361-371), 1977.
40. Goldfarb, D. and W.Y. Sit. "Worst Case Behavior of the Steepest Edge Simplex Method." *Discrete Applied Mathematics,* Vol. 1, (277-285), 1979.
41. Haimovich, M. "The Simplex Algorithm is Very Good!On the Expected Number of Pivot Steps and Related Properties of Random Linear Programs." Draft, 415 Uris Hall, Columbia University, New York, NY, April 1983.
42. Hung, M.S. "A Polynomial Simplex Method for the Assignment Problem." *Operations Research,* 31, 3, (595-600), May 1983.
43. Jeroslow, R.G. "The Simplex Algorithm with the Pivot Rule of Maximizing Criterion Improvement." *Discrete Mathematics,* Vol. 4, (367-377), 1973.
44. Klee, V. and G.J. Minty. "How Good is the Simplex Algorithm?" Inequalities III (O. Shisha Ed.), Academic Press, New York, (159-175), 1972.
45. Kleinschmidt, P., C.W. Lee, and H. Schannath. "Transportation Problems Which Can Be Solved By the Use of Hirsch-Paths for the Dual Problems." *Mathematical Programming,* 37 (153-168), 1987.
46. Murty, G.K. "Computational Complexity of Parametric Linear Programming," *Mathematical Programming,* 19 (213-219), 1980.
47. Orlin, J.B. "Genuinely Polynomial Simplex and Non-Simplex Algorithms for the Minimum Cost Flow Problem." Technical Report 1615-84, Sloan School of Management, MIT, Cambridge, MA, 1984.
48. Orlin, J.B. "On the Simplex Algorithm for Networks and Generalized Networks." *Mathematical Programming Study,* 24, (166-178), 1985.
49. Orlin, J.B., Plotkin, S., and E. Tardos. "Polynomial Dual Network Simplex Algorithms." *Math. Programming,* 60, (255-276), 1993.
50. Plotkin, S., and E. Tardos. "Improved Dual Network Simplex." Proceedings of the First

ACM-SIAM Symposium on Discrete Algorithms, (367-376), 1990.

51. Rock, H. "Scaling Techniques for Minimal Cost Network Flows." In Discrete Structures and Algorithms, Edited by V. Page, Carl Hanser, Munich, (181-191), 1980.
52. Roohy-Laleh, E. "Improvements to the Theoretical Efficiency of the Network Simplex Mehod." Unpublished Ph.D. dissertation, Carleton University, Ottawa, Canada, 1980.
53. Shamir, Ron. "The Efficiency of the Simplex Method: A Survey." *Management Science*, Vol. 33, No. 3, (301-334), March 1987.
54. Smale, S. "On the Average Number of Steps of the Simplex Method of Linear Programming." *Mathematical Programming*, 27, (241-262), 1983.
55. Tarjan, R.E. "Efficiency of the Primal Network Simplex Algorithm for the Minimum-Cost Circulation Problem." *Mathematics of Operations Research*, 16, (272-291), 1991.
56. Todd, M.J. "Polynomial Expected Behavior of a Pivoting Algorithm for Linear Complementarity and Linear Programming Problems." *Mathematical Programming*, 35, (173-192), 1986.
57. Zadeh, N. "A Bad Network Example for the Simplex Method and Other Minimum Cost Flow Algorithms," *Mathematical Programming*, 5, (255-266), 1973.
58. Zadeh, N. "Near-Equivalence of Network Flow Algorithms." Technical Report 26, Stanford University, Stanford, CA, 1979.
59. Zadeh, N. "What is the Worst-Case Behavior of the Simplex Algorithm?," Technical Report, Department of Operations Research, Stanford University, Stanford, CA, 1980.

THE LINEAR COMPLEMENTARITY PROBLEM

PANOS M. PARDALOS
303 Weil Hall, Department of Industrial and Systems Engineering
University of Florida, Gainesville, FL 32611

Abstract. This paper discusses a number of observations and conclusions drawn from ongoing research into more efficient algorithms for solving nonconvex linear complementarity problems (LCP). We apply interior point approaches and partitioning techniques to classes of problems that can be solved efficiently. Using the potential reduction algorithm, we characterize some classes of problems that can be solved in polynomial time. The same algorithm is used for the solution of problems with a row-sufficient matrix. The algorithm also generates a stationary point for the LCP in fully polynomial approximation time. When the problem data has no structure, we show equivalence of mixed integer programming problems and the linear complementarity problems.

Key words: Linear Complementarity Problem, NP-hard, Integer Programming, Interior Point Methods

1. Introduction

We consider the general linear complementarity problem (LCP) of finding a vector $x \in R^n$ such that

$$Mx + q \geq 0, x \geq 0, x^T Mx + q^T x = 0 \qquad (1)$$

(or proving that such an x does not exist) where M is an $n \times n$ rational matrix and $q \in R^n$ is a rational vector. For given data M and q, the problem is generally denoted by LCP(M, q).

The LCP unifies a number of important problems in operations research. In particular, it generalizes the primal-dual linear programming problem, convex quadratic programming, and bimatrix games. It has many important applications in science and technology including fluid flow problems, economic equilibrium analysis, and numerical solution of differential equations [1], [7]. For this reason, Cottle and Dantzig [5] refer to the linear complementarity problem as the "Fundamental Problem". Most of the work on the LCP has focused on the convex case or problems with a special structure. The books by K. Murty [13] and by Cottle et al [4] are a compendium of LCP developments over the past years. In addition, the recent monograph by Kojima et al [10] provides a unified approach to interior point algorithms for linear complementarity problems.

In this paper, we use interior point approaches and partitioning techniques to solve certain nonconvex LCPs in polynomial time. We prove that a row-sufficient matrix LCP, under certain conditions, can be solved in polynomial time. The general linear complementarity problem (no structure on M) is a very difficult problem, since (as shown below) it contains the general integer feasibility problem.

S. Gomez and J-P. Hennart (eds.), Advances in Optimization and Numerical Analysis, 39–49.
© 1994 *Kluwer Academic Publishers.*

2. Problems solvable in polynomial time

If the $LCP(M, q)$ has a finite number of solutions, then all solutions are extreme points of the polyhedron $S = \{x : Mx + q \geq 0, x \geq 0\}$. If there are an infinite number of solutions, then there is one that occurs at an extreme point of S.

For the $LCP(M, q)$, it is natural to consider the case where solving LCP is reducible to solving a linear programming problem. Various necessary and sufficient conditions for this were derived in [11] and [2].

Mangasarian [11] proved that the $LCP(M, q)$ has a solution iff there is a $p \in R^n$ such that the linear program

$$\min_{x \in S} p^T x \tag{2}$$

and its dual have optimal solutions $\bar{x}$, $\bar{y}$, respectively, where

$$(I - M^T)\bar{y} + \tau > 0, p = M^T \sigma + \tau, \sigma, \tau > 0.$$

Under these conditions, the optimal solution $\bar{x}$ of (2) also solves the $LCP(M, q)$.

The vector p can be easily determined for a number of special cases. Mangasarian [11] proved that certain classes of linear complementarity problems can be solved by a single linear program. For many important cases, such as the case where M is positive definite or $M > 0$, it is not easy to find such a vector p.

In those cases where the vector p can be found in polynomial time, the corresponding LCP can be solved in polynomial time by solving the related linear problem. For example, when M is a Z-matrix the vector p can be found easily, and therefore the LCP is solved in polynomial time.

When the matrix M is positive semidefinite, then the LCP is a convex quadratic program and can be solved in polynomial time using an interior point approach. A potential reduction algorithm that solves the LCP in $O(L\sqrt{n})$ iterations and $O(Ln^3)$ total arithmetic operations has been proposed in [9] (L is the size of the input data M and q).

An $n \times n$ real matrix M is called a P-matrix if all its principal subdeterminants are positive. Among many equivalent definitions of such a matrix, one characterization states that M is a P-matrix iff for every nonzero $\pi \in R^n$

$$\max_{1 \leq i \leq n} \pi_i (M\pi)_i > 0.$$

Moreover, the $LCP(q, M)$ has a unique solution for every $q \in R^n$ iff M is a P-matrix [13]. It is unknown whether the LCP with a P-matrix, can be solved in polynomial time since it is not known whether a P-matrix can be recognized in polynomial time. Regarding complexity for this case, Megiddo [12] proved that if it is NP-hard to solve the LCP with P-matrix, then $NP = coNP$. However, the potential reduction algorithm solves the LCP (with P-matrix) in $O(n^2 \max(\frac{|\lambda|}{\beta} n, 1)L)$ iterations and each iteration solves a linear system of equations in, at most, $O(n^3)$ operations. Here λ is the smallest eigenvalue of $(M + M^T)/2$ and $\beta > 0$ is the P-matrix number for M^T; that is, for every $\pi \in R^n$, there exists an index j such that

$$\pi_j (M^T \pi)_j \geq \beta \| \pi \|^2.$$

Hence, when $|\lambda|/\beta$ is bounded above by a polynomial of L and n, the LCP with P-matrix is solved in polynomial time [22].

2.1. A class of LCPs solvable in polynomial time

Next, we characterize a new class of LCPs solvable in polynomial time using the potential reduction algorithm (see [8] and [20]). We assume that the interior of S is nonempty. The potential function used is defined by

$$\phi(x,y) = \rho \ln(x^T y) - \sum_{i=1}^{n} \ln(x_i y_i), \rho > n + \sqrt{n}. \tag{3}$$

The potential reduction algorithm generates an interior solution path that terminates at the point (x^k, y^k) such that $\phi(x^k, y^k) \le -(\rho - n)L$. Then, $(x^k)^T y^k \le 2^{-L}$ and the exact solution can be obtained in $O(n^3)$ additional operations.

To achieve this reduction, we use the scaled gradient projection method. The gradient vector of the potential function (3) is

$$\nabla\phi_x = \frac{\rho}{\Delta} y - X^{-1}e, \nabla\phi_y = \frac{\rho}{\Delta} x - Y^{-1}e, (\Delta = x^T y)$$

where X, Y denote the diagonal matrices of x and y, respectively, and e is a vector of ones. Next, solve the linear program with an ellipsoid constraint:

$$\min \nabla\phi_{x^k}\delta x + \nabla\phi_{y^k}\delta y \tag{4}$$

$$s.t.\ \delta y = M\delta x, \| (X^k)^{-1}\delta x \|^2 + \| (Y^k)^{-1}\delta y \|^2 \le \beta^2 < 1,$$

for some constant $0 < \beta < 1$. Then, we can show that the solution $\delta\bar{x}$ and $\delta\bar{y}$ of the linear program (4) satisfies

$$\begin{pmatrix} (X^k)^{-1}\delta\bar{x} \\ (Y^k)^{-1}\delta\bar{y} \end{pmatrix} = -\beta\frac{p^k}{\| p^k \|},$$

where

$$p^k = \begin{pmatrix} \frac{\rho}{\Delta^k} X^k(y^k + M^T\pi^k) - e \\ \frac{\rho}{\Delta^k} Y^k(x^k - \pi^k) - e \end{pmatrix}, \tag{5}$$

and

$$\pi^k = ((Y^k)^2 + M(X^k)^2 M^T)^{-1}(Y^k - MX^k)(X^k y^k - \frac{\Delta^k}{\rho}e).$$

Using $\beta = \min\{\| p^k \| /(\rho+2), 1/(\rho+2)\} \le 1/2$ it can be shown that

$$\phi(x^k + \delta\bar{x}, y^k + \delta\bar{y}) - \phi(x^k, y^k) \le -\min\{\| p^k \|^2/2(\rho+2), 1/2(\rho+2)\}.$$

It has been proved [8], that if M is positive semidefinite and $\rho \ge 2n + \sqrt{2n}$, then $\| p^k \|^2 \ge 1$. More generally, the following result [24] holds:

Lemma 1: If $\| p^k \| < 1$, then $y^k + M^T\pi^k > 0, x^k - \pi^k > 0$ and $\frac{2n-\sqrt{2n}}{\rho}\Delta^k < \bar{\Delta} < \frac{2n+\sqrt{2n}}{\rho}\Delta^k$, where $\bar{\Delta} = (x^k)^T(y^k + M^T\pi^k) + (y^k)^T(x^k - \pi^k)$.

Proof: Suppose that $\bar{y} = y^k + M^T\pi^k$ and $\bar{x} = x^k - \pi^k$ are not positive. Then $\| p^k \|^2 \ge 1$; hence, $\| p^k \| < 1$ implies $\bar{y}, \bar{x} > 0$.

On the other hand, it is easy to see that

$$2n(\frac{\rho\bar{\Delta}}{2n\Delta^k}-1)\leq \| p^k \|^2 < 1,$$

which implies the last inequality.

Using this lemma we can prove the following results:
Theorem 1 [24]: Given the LCP(M, q), suppose that the set

$$\Omega^+ = \{(x,y) : y = Mx + q > 0, x > 0\}$$

is nonempty, and the set

$$\Sigma^+ = \{\pi : x^T y - q^T \pi < 0, x - \pi > 0, y + M^T \pi > 0, \text{ for some } (x,y) \in \Omega^+\}$$

is empty. Then the potential reduction algorithm (with $\rho = 2n + \sqrt{2n}$) solves the LCP in polynomial time.

Corollary 1:
[a] If Ω^+ is nonempty and the set

$$\{\pi : x^T y - q^T \pi > 0, x - \pi > 0, y + M^T \pi > 0, \text{ for some } (x,y) \in \Omega^+\}$$

is empty, then the potential reduction algorithm (with $\rho = 2n - \sqrt{2n}$) solves the LCP in polynomial time.
[b] If Ω^+ is nonempty and the set

$$\{\pi : x^T y - q^T \pi < (1/p(n) - 1)x^T y, x - \pi > 0, y + M^T \pi > 0, \text{ for some } (x,y) \in \Omega^+\}$$

is empty, then the potential reduction algorithm (with $\rho = (2n + \sqrt{2n})p(n)$, where $p(n)$ is a polynomial of n) solves the LCP in polynomial time.

These results suggest that positive semidefiniteness of the matrix M may not be the basic line to separate the classes of polynomially solvable LCPs from the ones that are not.

2.2. LCPs WITH ROW-SUFFICIENT MATRICES

A new interesting class of LCPs that was recently defined is based on the notion of a row-sufficient matrix [3]. A matrix $M_{n \times n}$ is row-sufficient if the following is true for any $x \in R^n$:

$$[\max_{1 \leq i \leq n} x_i(M^T x)_i \leq 0] \Rightarrow [x_i(M^T x)_i = 0 \ for \ all \ i = 1, \cdots, n]. \tag{6}$$

Associated with the LCP is the quadratic problem

$$\min_{x \in S} f(x) = q^T x + x^T M x.$$

In [3], it is shown that the matrix M is row-sufficient iff for each vector $q \in R^n$, if (x^*, λ^*) is a Kuhn-Tucker pair of the associated quadratic problem, then x^* solves the LCP(M, q).

In the case of row-sufficient matrix, we can also prove that $\| p^k \| > 0$. Suppose that $\| p^k \| = 0$, i.e., the iterative process jams. Then this implies that

$$X^k M^T \pi = -Y^k \pi = -X^k y^k + \frac{\Delta^k}{\rho},$$

or

$$X^k M^T \pi + Y^k \pi = 0.$$

Then, $\pi_i (M^T \pi)_i \leq 0$ for $i = 1, \cdots, n$ and since M is row-sufficient (6) $\pi^T M^T \pi = 0$, which implies $\| p^k \| \geq 1$, a contradiction to the assumption that $\| p^k \| = 0$. Hence, if the input matrix M is row-sufficient, then the gradient projection vector of the potential function satisfies $\| p^k \| > 0$.

Next we define a "condition number" for a row-sufficient matrix similar to the one in [8]. Let

$$a = \frac{\rho}{\Delta} X y - e$$

and

$$Q = 2I - (XM^T - Y)(Y^2 + MX^2M^T)^{-1}(MX - Y)$$

where $\Delta = x^T y$ and $X = diag(x)$.

Claim 1: The matrix Q is positive semidefinite.

Define the condition number of a row-sufficient matrix M to be

$$\gamma(M) = inf\{(a^T Q a)^{1/2} : (x, y) \in \Omega^+,\ x^T y \geq 2^{-L}\}.$$

Claim 2: If $\rho \geq 3n + \sqrt{2n}$, then

[a] $\gamma(M) \geq 1$ if M is positive semidefinite matrix [8],

[b] $\gamma(M) \geq 1$ if M is in the class defined in [24],

[c] $\gamma(M) \geq \min\{n\beta/|\lambda|, 1\}$ if M is a P-matrix, where β is the P-matrix number of M^T, and λ is the least eigenvalue of $(M + M^T)/2$ [22].

Claim 3: If $\rho \geq 3n + \sqrt{2n}$, M is a row-sufficient matrix and Ω^+ is bounded, then $\gamma(M) > 0$.

The class of matrices M defined and studied in [24] includes matrices that are indefinite, and therefore the corresponding LCP is in general nonconvex.

The proof of the last claim is based in our proof that $p > 0$. As a consequence we have the following result.

Theorem 2: The potential reduction algorithm solves a row-sufficient matrix LCP with bounded Ω^+ in $O(n^2 \gamma L)$ iterations and each iteration solves a system of linear equations in $O(n^3)$ operations.

Hence if γ is bounded by a polynomial of n and L, then the potential reduction algorithm solves the row-sufficient matrix LCP in polynomial time.

2.3. Computing a stationary point of the LCP

The stationary point of the LCP is defined as a point satisfying the first order optimality conditions. More precisely, the stationary point $(\bar{x}, \bar{y}) \in \Omega$ the LCP can be represented by

$$\bar{y}^T \bar{x} + \bar{x}^T \bar{y} \leq \bar{y}^T x + \bar{x}^T y \quad \text{for all} \quad (x, y) \in \Omega,$$

or there exist $\bar{\pi} \in R^m$ satisfying

$$\bar{y} + M^T \bar{\pi} \geq 0, \quad \bar{x} - \bar{\pi} \geq 0,$$

and

$$\bar{x}^T(\bar{y} + M^T \bar{\pi}) = 0 \quad \text{and} \quad \bar{y}^T(\bar{x} - \bar{\pi}) = 0.$$

Finding such a stationary point itself is a linear complementarity problem. We also note that the solution to the LCP can be viewed as a special stationary point with $\bar{\pi} = 0$.

Using Lemma 1 in the potential reduction algorithm, we can prove the following theorem, which is a special case of a theorem developed by Ye ([23]) for the general LCP.

Theorem 3: For any given $0 < \epsilon \leq 1$, let $\rho = (2n + \sqrt{2n})/\epsilon$. Then, the potential reduction algorithm terminates in at most $O(\rho^2 L)$ iterations, generating an ϵ-approximate stationary point $(x^k, y^k) \in \Omega$ and $\pi^k \in R^m$ of the LCP, either

$$(x^k)^T y^k \leq 2^{-L}$$

or

$$y^k + M^T \pi^k > 0, \quad x^k - \pi^k > 0, \quad \text{and}$$
$$\frac{(x^k)^T(y^k + M^T \pi^k) + (y^k)^T(x^k - \pi^k)}{(x^k)^T y^k} < \epsilon.$$

Proof. The proof directly follows Lemma 1. If $\|p^k\| \geq 1$ for all k, then

$$\phi(x^{k+1}, y^{k+1}) - \phi(x^k, y^k) \leq -\Omega(1/\rho).$$

Therefore, in at most $O(\rho^2 L)$ iterations

$$\phi(x^k, y^k) \leq -(\rho - n)L,$$

and from (2.2)

$$(x^k)^T y^k \leq 2^{-L}.$$

As we mentioned before, in this case (x^k, y^k) is a special stationary point with $\bar{\pi} = 0$. Otherwise, we have $\|p^k\| < 1$ for some $k \leq O(\rho^2 L)$. This implies the second case from Lemma 1.

Theorem 3 indicates that the potential reduction algorithm is a fully polynomial-time approximation scheme for computing an ϵ-approximate stationary point of the

LCP. It has been shown that every stationary point of the LCP with a row-sufficient matrix is a solution. Moreover, if the LCP has a nonempty interior feasible region, an initial feasible point can be found in polynomial time via the linear programming Phase-I procedure. Therefore, the potential reduction algorithm is a fully polynomial-time approximation scheme for solving the LCP with a row-sufficient matrix and a nonempty interior feasible region.

The concept of the fully polynomial-time approximation scheme (FPTAS) was introduced in combinatorial optimization. For some combinatorial optimization problems the theory of NP-completeness can be applied to prove not only that they cannot be solved exactly by polynomial-time algorithms (unless $P = NP$), but also that they do not have ϵ-approximate algorithms, for various ranges of ϵ, again unless $P = NP$. Furthermore, approximation algorithms are widely used and accepted in practice.

2.4. A partitioning technique for the LCP

Next, we consider some partitioning techniques which also can be used to characterize some classes of LCPs solvable in polynomial time.

Lemma 2: Suppose that the matrix M satisfies $m_{ij} \leq 0, i \neq j$, $m_{ii} > 0$ and it is strictly diagonally dominant. Then $M^{-1} \geq 0$. (For a proof, see [14].)

Theorem 4: Let M be a strictly diagonally dominant matrix with $m_{ii} < 0$ and $m_{ij} \geq 0, i \neq j$. Let $q \in R^n$ and assume that the corresponding feasible domain S is nonempty, then $x_0 = -M^{-1}q$ is a solution. (For a proof, see [13].)

The next theorem is a generalization of the above result. Some notation is given first: Define $M_{JK} = (m_{jk}), j \in J, k \in K$ where $J, K \subseteq N = \{1, \cdots, n\}$, $\bar{J} = N - J$ and $x_J = (x_{j_1}, \cdots, x_{j_p})$ for $j_i \in J$.

Theorem 5: Suppose that M is an $n \times n$ matrix with $m_{ij} \geq 0, i \neq j$ and $|m_{ii}| > 0$. Let J be the largest index set such that $m_{ii} < 0, i \in J \subseteq N$ and suppose that the submatrix M_{JJ} is a strictly diagonally dominant matrix with $S(M_{JJ}, q_J) \neq \emptyset$. Then the LCP$(M, q)$ has a solution.

Proof: Without loss of generality we may assume that,

$$M = \begin{pmatrix} M_{JJ} & M_{J\bar{J}} \\ M_{\bar{J}J} & M_{\bar{J}\bar{J}} \end{pmatrix}, q = \begin{pmatrix} q_J \\ q_{\bar{J}} \end{pmatrix}.$$

Then the feasibility condition $Mx + q \geq 0$ is equivalent to

$$v_J = M_{JJ}x_J + M_{J\bar{J}}x_{\bar{J}} + q_J \geq 0, \tag{7}$$

$$v_{\bar{J}} = M_{\bar{J}J}x_J + M_{\bar{J}\bar{J}}x_{\bar{J}} + q_{\bar{J}} \geq 0. \tag{8}$$

By the above lemma, we have that $M^{-1}{}_{JJ} \leq 0$. Multiplying (7) by $-M^{-1}{}_{JJ}$ we obtain:

$$-x_J - M^{-1}{}_{JJ}M_{J\bar{J}}x_{\bar{J}} - M^{-1}{}_{JJ}q_J \geq 0$$

which implies

$$x_J \leq -M^{-1}{}_{JJ}M_{J\bar{J}}x_{\bar{J}} - M^{-1}{}_{JJ}q_J. \tag{9}$$

Since $S(M_{JJ}, q_J) \neq \emptyset$, it follows that $-M^{-1}{}_{JJ}q_J \geq 0$. Also, $-M^{-1}{}_{JJ} \geq 0$ and $M_{J\bar{J}} \geq 0$. Then

$$x_J = -M^{-1}{}_{JJ}M_{J\bar{J}}x_{\bar{J}} - M^{-1}{}_{JJ}q_J \text{ (i.e. } v_J = 0). \tag{10}$$

Substituting this value of x_J in (7), we obtain

$$M_{\bar{J}\bar{J}}x_{\bar{J}} + M_{\bar{J}J}(-M^{-1}{}_{JJ}M_{J\bar{J}}x_{\bar{J}} - M^{-1}{}_{JJ}q_J) + q_{\bar{J}} \geq 0$$

which implies that

$$\bar{M}_{\bar{J}\bar{J}}x_{\bar{J}} + \bar{q}_{\bar{J}} \geq 0, \tag{11}$$

where

$$\bar{M}_{\bar{J}\bar{J}} = M_{\bar{J}\bar{J}} - M_{\bar{J}J}M^{-1}{}_{JJ}M_{J\bar{J}}, \tag{12}$$

$$\bar{q}_{\bar{J}} = -M_{\bar{J}J}M^{-1}{}_{JJ}q_J + q_{\bar{J}}. \tag{13}$$

Since $\bar{M}_{\bar{J}\bar{J}} \geq 0$ and it has positive diagonal entries, the $LCP(\bar{M}_{\bar{J}\bar{J}}, \bar{q}_{\bar{J}})$ always has a solution, say $x_{\bar{J}}$. Then $x = (x_J, x_{\bar{J}})$ (where x_J is given by (10)) gives the solution to the original LCP. Note that $v_J = 0$.

Theorem 6: Suppose that M is a strictly diagonally dominant matrix with $m_{ij} \geq 0$, $i \neq j$. Let J be the largest index set such that $m_{ii} < 0$, $i \in J \subset N$ and $S(M_{JJ}, q_J) \neq \emptyset$, then the $LCP(M, q)$ has a solution, and it can be computed in polynomial time.
Proof: As in the case of the previous theorem, we take

$$x_J = -M^{-1}{}_{JJ}M_{J\bar{J}}x_{\bar{J}} - M^{-1}{}_{JJ}q_J, \tag{14}$$

and compute $x_{\bar{J}}$ by solving the $LCP(\bar{M}_{\bar{J}\bar{J}}, \bar{q}_{\bar{J}})$, where $\bar{M}_{\bar{J}\bar{J}}$ and $\bar{q}_{\bar{J}}$ are given by equations (12) and (13). Note that $\bar{M}_{\bar{J}\bar{J}}$ is the Schur complement of M_{JJ} in matrix M. Since M is strictly diagonally dominant, $\bar{M}_{\bar{J}\bar{J}}$ is also strictly diagonally dominant. Also, $\bar{M}_{\bar{J}\bar{J}} \geq 0$. Hence, $\bar{M}_{\bar{J}\bar{J}}$ is a positive definite matrix and therefore $LCP(\bar{M}_{\bar{J}\bar{J}}, \bar{q}_{\bar{J}})$ can be solved in polynomial time. This problem always has a solution, since $\bar{M}_{\bar{J}\bar{J}} \geq 0$ with positive diagonal entries. Once we have computed $x_{\bar{J}}$, we can compute x_J by using equation (14). Q.E.D.

It is clear that partitioning techniques can be used to identify embedded subproblems that can be solved in polynomial time, when the initial problem is nonconvex. This again suggests that convexity may not be the key property to classify LCPs from the complexity point of view.

3. Integer programming and LCP

For the general matrix M, where S can be bounded or unbounded, the LCP can always be solved by solving a specific zero-one, linear, mixed-integer problem with n zero-one variables. Consider the following mixed zero-one integer problem (MIP):

$$\max_{\alpha, y, z} \alpha$$

$$s.t.\ 0 \leq My + \alpha q \leq e - z,$$

$$\alpha \geq 0, 0 \leq y \leq z, \tag{15}$$

$$z \in \{0,1\}^n.$$

Theorem 7: Let (α^*, y^*, z^*) be any optimal solution of (15). If $\alpha^* > 0$, then $x^* = y^*/\alpha^*$ solves the LCP. If in the optimal solution $\alpha^* = 0$, then the LCP has no solution.

The equivalent mixed integer programming formulation (MIP) was first given in [18]. Every feasible point (α, y, z) of MIP, with $\alpha > 0$, corresponds to a solution of LCP. Therefore, solving MIP, we may generate several solutions of the corresponding LCP. Rosen [21] proved that the solution obtained by solving MIP is the minimum norm solution to the linear complementarity problem.

Remark: Computational experience using the integer formulation, shows that the average number of (linear) subproblems of the branch and bound tree required to solve an LCP (with solution) is approximately $n/2 + n^2/60$. However, many of the subproblems require an exponential number of pivot steps. In fact, as the dimension increases, this situation appears more often. The hypercubes embedded in the domain of the relaxed linear programs seems to be the cause of exponential number of pivots (as in the Klee-Minty type examples). As an example, consider the following LCP with data

$$q^T = (-120.535, -103.7422, -333.418), M = \begin{pmatrix} 94.2280 & 43.5854 & -27.4712 \\ 25.5269 & 20.4233 & -49.0630 \\ 87.1616 & 71.0659 & 21.9937 \end{pmatrix}.$$

The solution to this problem is $x^* = (0., 5.07959, 0.)$. Using a branch and bound algorithm to solve the equivalent mixed zero-one program, we solve 7 subproblems. Many subproblems require an exponential number ($2^8, 2^7$ etc) of pivots (depending on the height of the branch and bound tree).

It is difficult to obtain easily verifiable conditions that prove that a given LCP is not solvable. The next theorem [16] gives sufficient conditions for LCPs without solution.

Theorem 8: The $LCP(M, q)$ has no solution if the system of linear inequalities

$$0 \leq Mx + q \leq \omega e - z, \tag{16}$$

$$0 \leq x \leq z, 0 \leq z \leq \omega e, \omega \geq 1,$$

is infeasible.

In general, feasibility of (16) does not imply that the LCP is solvable (see also [15]). However, if the LCP is solvable, then the system in (3.2) is feasible.

Next, we show that the mixed integer feasibility problem can be formulated as an LCP. Given matrices $A_{n \times n}$, $B_{n \times l}$ and a vector $b \in R^n$ with rational entries, the mixed integer feasibility problem is to find (x, z), such that $x \in R^n$, $x \geq 0$, $z \subset \{0,1\}^l$ that satisfy $Ax + Bz = b$.

Theorem 8: The mixed integer feasibility problem can be reduced to the solution of the linear complementarity problem.
Proof: The condition $z_i \in \{0,1\}$ is equivalent to:

$$z_i + w_i = 1, z_i \geq 0, w_i \geq 0, z_i w_i = 0.$$

With this transformation z_i is a continuous variable and for each z_i a new continuous variable w_i is introduced. In addition, let $s, t \in R^n$ be such that

$$s = Ax + Bz - b \geq 0, t = -Ax - Bz + b \geq 0.$$

The only way for these two inequalities to be satisfied is to have $s = t = 0$, which implies that $Ax + Bz = b$. Then, the mixed integer feasibility problem can be reduced to solution of the LCP: Find v, y such that

$$v \geq 0, y \geq 0, v^T y = 0, v = My + q,$$

where

$$y = \begin{pmatrix} z \\ x \\ \theta \end{pmatrix}, v = \begin{pmatrix} w \\ s \\ t \end{pmatrix}, M = \begin{pmatrix} -I & 0 & 0 \\ B & A & 0 \\ -B & -A & 0 \end{pmatrix}, q = \begin{pmatrix} e \\ b \\ -b \end{pmatrix},$$

where $\theta \in R^n$ and $e \in R^l$ is the vector of all 1's.

In addition, similar integer programming techniques apply for the more difficult "integer linear complementarity problem". For applications and algorithms for such problems, see [17].

4. Concluding remarks

In this paper, we considered the general LCP. This problem is computationally very difficult, since it contains among other problems, the mixed integer feasibility problem. Using interior point algorithms and partitioning techniques, we have characterized some classes of nonconvex LCPs that can be solved in polynomial time.

Since the mixed integer programming problem is NP-complete, so is the LCP. Also, since the LCP is reducible to quadratic programming (see also [6], [19]), it follows that the general quadratic programming problem is harder than the mixed integer programming problem.

References

1. Bershchanskii Y.M. and Meerov M.V., The complementarity problem: Theory and Methods of Solution. *Automation and Remote Control* Vol. 44, No 6, Part I (1983), pp. 687-710.
2. Cottle R.W. and Pang J.S., On solving linear complementarity problems as linear programs. *Math. Progr. Study* 7 (1978), pp. 88-107.
3. Cottle R.W., Pang J.S. and Venkateswaran V., Sufficient matrices and the linear complementarity problem. *Linear Algebra and its Applications* 114/115 (1989), pp. 231-249.
4. Cottle R.W., Pang J.S. and Stone, R.E., The Linear complementarity problem. Academic Press (1992).

5. Cottle R.W. and Dantzig G.B., Complementarity pivot theory of mathematical programming. In: Dantzig G.B. and Veinott A.F., Jr., eds., Mathematics of the Decision Sciences, Part 1 (American Mathematical Society , 1968), pp. 115-136.
6. Gupta S. and Pardalos P.M., On a quadratic formulation of linear complementarity problems. *Journal of Optimization Theory and Applications*, Vol. 57, No. 1 (1988), pp. 197-202.
7. Lemke C.E., A survey of complementarity theory. In: Variational Inequalities and Complementarity Problems: Theory and Applications (edited by Cottle R.W., Giannessi F. and Lions J.L., John Wiley & Sons, New York) (1980) pp. 213-239.
8. Kojima M., Megiddo N. and Ye Y., An interior point potential reduction algorithm for the linear complementarity problem. *Math. Progr.* 54, No. 3 (1992), pp. 267-279.
9. Kojima M., Mizuno S. and Yoshise A., An $O(\sqrt{n}L)$ iteration potential reduction algorithm for linear complementarity problems. Research Report on Information Sciences b-127, Tokyo Institute of Technology (Tokyo, 1988).
10. Kojima M., Megiddo N., Noma T. and Yoshise A., A Unified Approach to Interior Point Algorithms for Linear Complementarity Problems. Springer-Verlag, Lecture Notes in Computer Sciences 538 (1991).
11. Mangasarian O.L., Characterization of linear complementarity problems as linear programs. *Math. Progr. Study* 7 (1978), pp. 213-239.
12. Megiddo N., A note on the complexity of P-matrix LCP and computing an equilibrium. Research Report 6439, IBM Almaden Research Center (1988).
13. Murty K.G., Linear complementarity, linear and nonlinear programming. Heldermann Verlag, Berlin (1988).
14. Ortega J.M., Numerical analysis, A second course. Academic Press (1972).
15. Pardalos P.M., Parallel search algorithms in global optimization. *Applied Mathematics and Computation* 29 (1989), pp. 219-229.
16. Pardalos P.M., Linear complementarity problems solvable by integer programming. *Optimization* 19 (1988), pp. 467-474.
17. Pardalos P.M. and Nagurney A., The integer linear complementarity problem. *Intern. J. of Computer Mathematics* 31 (1990), pp. 205-214.
18. Pardalos P.M. and Rosen J.B., Global optimization approach to the linear complementarity problem. *SIAM J. Scient. Stat. Computing* Vol. 9, No. 2 (1988), pp. 341-353.
19. Pardalos P.M. and Rosen J.B., Global optimization: algorithms and applications. Springer-Verlag, Lecture Notes in Computer Sciences 268 (1987).
20. Pardalos P.M., Ye Y., Han C.G., and Kaliski J.A., Solution of P-matrix linear complementarity problems using a potential reduction algorithm, *SIAM J. on Matrix Analysis and Applications Vol. 14, No. 4* (1993).
21. Rosen J.B., Minimum norm solution to the linear complementarity problem, In: Functional Analysis, Optimization and Mathematical Economics (L.J. Leifman, Ed.) Oxford University Press (1990), pp. 208-216.
22. Ye Y., A further result on the potential reduction algorithm for the P-matrix problem. In: Advances in Optimization and Parallel Computing (P. M. Pardalos Ed.) Elsevier Science Publishers (1992), pp. 310-316.
23. Ye Y., A fully polynomial-time approximation algorithm for computing a stationary point of the general linear complementarity problem. Working Paper, Department of Management Sciences, The University of Iowa (1990).
24. Ye Y. and Pardalos P.M., A class of linear complementarity problems solvable in polynomial time. *Linear Algebra and its Applications* 152 (1991), pp. 3-17.

A DIRECT SEARCH OPTIMIZATION METHOD THAT MODELS THE OBJECTIVE AND CONSTRAINT FUNCTIONS BY LINEAR INTERPOLATION

M.J.D. POWELL
Department of Applied Mathematics and Theoretical Physics, University of Cambridge, Silver Street, Cambridge CB3 9EW, England.

Abstract. An iterative algorithm is proposed for nonlinearly constrained optimization calculations when there are no derivatives. Each iteration forms linear approximations to the objective and constraint functions by interpolation at the vertices of a simplex and a trust region bound restricts each change to the variables. Thus a new vector of variables is calculated, which may replace one of the current vertices, either to improve the shape of the simplex or because it is the best vector that has been found so far, according to a merit function that gives attention to the greatest constraint violation. The trust region radius ρ is never increased, and it is reduced when the approximations of a well-conditioned simplex fail to yield an improvement to the variables, until ρ reaches a prescribed value that controls the final accuracy. Some convergence properties and several numerical results are given, but there are no more than 9 variables in these calculations because linear approximations can be highly inefficient. Nevertheless, the algorithm is easy to use for small numbers of variables.

Key words: Direct search, Linear interpolation, Nonlinear constraints, Optimization without derivatives

1. Introduction

John Dennis has provided the best description of a direct search optimization calculation that I have encountered. It is to find the deepest point of a muddy lake, given a boat and a plumb line, when there is a price to be paid for each sounding. A specification of an algorithm that is suitable for solving this problem would probably appeal to geometric intuition, and probably the procedure would require widely spaced measurements to be taken, in order to smooth out any high frequency variations in the depth of the lake. Experience has shown that many computer users find such algorithms attractive for a wide range of optimization calculations.

In particular, the method of Nelder and Mead (1965) is used in very many fields to calculate the least value of a function $F(\underline{x})$, $\underline{x}\in\mathcal{R}^n$, when there are no constraints on the variables. Confirmation of this assertion can be found in the CMCI CompuMath Citation Index, more than 200 different applications being listed during the last 10 years. An iteration of this method is given the value of F at $n+1$ points, $\{\underline{x}^{(j)} : j=0,1,\ldots,n\}$ say, where the points have to satisfy the nondegeneracy condition that the volume of their convex hull in $\mathcal{R}^n$ is positive. Let $\underline{x}^{(\ell)}$ be a vertex of the convex hull at which F is greatest, so ℓ is determined by the equation

$$F(\underline{x}^{(\ell)}) = \max\{F(\underline{x}^{(j)}) : j=0,1,\ldots,n\}. \tag{1}$$

S. Gomez and J-P. Hennart (eds.), Advances in Optimization and Numerical Analysis, 51–67.
© 1994 *Kluwer Academic Publishers.*

Then, because $\underline{x}^{(\ell)}$ is a vertex at which the objective function is worst, the iteration replaces $\underline{x}^{(\ell)}$ by the point

$$\underline{x}^{(\ell)}_{\text{new}} = -\theta\, \underline{x}^{(\ell)} + (1+\theta)\, n^{-1} \sum_{j=0,\, j\neq \ell}^{n} \underline{x}^{(j)}, \tag{2}$$

where the "reflection coefficient" θ is a constant from the open interval $(0,1)$. Further, if $F(\underline{x}^{(\ell)}_{\text{new}})$ is the least calculated function value so far, then a larger θ may be used instead. We see that formula (2) defines $\underline{x}^{(\ell)}_{\text{new}}$ by extrapolation along the straight line that joins $\underline{x}^{(\ell)}$ to the mean value of the other n points. Therefore it is elementary that, if F is a nonconstant linear function, then $F(\underline{x}^{(\ell)}_{\text{new}})$ is less than the average of the numbers $\{F(\underline{x}^{(j)}) : j = 0,1,\ldots,n,\, j\neq \ell\}$, so we expect an iteration to be successful at reducing F in the general case. If the iterations fail to make progress, however, and if an acceptably small value of the objective function has not been found, then the algorithm shrinks the current simplex with the vertices $\{\underline{x}^{(j)} : j = 0,1,\ldots,n\}$ before continuing the sequence of iterations. This kind of technique when $n=2$ might be suitable for seeking the deepest point of the muddy lake. In any case, the method is so straightforward to understand and to program for computer calculations that it is applied frequently.

The Nelder and Mead algorithm was developed from the method of Spendley, Hext and Himsworth (1962), in which every simplex is regular, this property being sustained by the value $\theta=1$ in formula (2). It was found, however, that the other choices of θ that have been mentioned provide much better efficiency by adapting the shape of the simplex to the curvature of the objective function automatically. Further, the Nelder and Mead algorithm is sometimes used to solve constrained problems by the simple expedient of replacing $F(\underline{x})$ by $+\infty$ if and only if $\underline{x}$ is infeasible. Then the simplex flattens itself so that it tends to be close to the active constraint boundaries. A disadvantage of this approach, however, is that an excellent change to the variables may be discarded because it gives an infinite value of the objective function. Therefore it is recommended by Subrahmanyam (1989) that, if a trial $\underline{x}$ is infeasible, then the setting of $F(\underline{x})$ to $+\infty$ should be postponed until the beginning of the next iteration, the usefulness of this technique being shown by numerical examples.

We take the view, however, that it may be possible to develop direct search methods that provide much better efficiency by taking advantage of the available details of the constraints. Therefore we address constrained optimization calculations that are expressed in the form

$$\left.\begin{array}{ll} \text{minimize} & F(\underline{x}), \quad \underline{x}\in\mathcal{R}^n \\ \text{subject to} & c_i(\underline{x}) \geq 0, \quad i=1,2,\ldots,m \end{array}\right\}, \tag{3}$$

and we assume that the objective and constraint functions can be calculated for every $\underline{x}$, but there are no smoothness assumptions. We take from the Nelder and Mead method the idea of generating the next vector of variables from function values at the vertices $\{\underline{x}^{(j)} : j=0,1,\ldots,n\}$ of a nondegenerate simplex in $\mathcal{R}^n$. In this case there are unique linear functions, $\hat{F}$ and $\{\hat{c}_i : i=1,2,\ldots,m\}$ say, that interpolate F

and $\{c_i : i=1,2,\ldots,m\}$ at the vertices, and we approximate the calculation (3) by the linear programming problem

$$\left.\begin{array}{ll}\text{minimize} & \hat{F}(\underline{x}), \quad \underline{x}\in\mathcal{R}^n \\ \text{subject to} & \hat{c}_i(\underline{x}) \geq 0, \quad i=1,2,\ldots,m\end{array}\right\}. \tag{4}$$

On most iterations the problem (4) guides the changes to the variables, but some iterations give higher priority to modifying the shape of the simplex, in order that interpolation at its vertices is likely to yield good linear models of the objective and constraint functions. We are encouraged by the fact that the use of linear approximations to constraints is highly successful in variable metric algorithms (see Powell, 1978, for instance), because such approximations often take up much of the freedom in the variables in a suitable way. Thus it can happen that constrained calculations are easier than unconstrained ones, even when the given constraints are nonlinear.

The iterative use of expression (4) puts our method in the class of "sequential linear programming algorithms" that originated from the work of Griffith and Stewart (1961). A good discussion of the merits and disadvantages of this class is given in Section 6.1 of Himmelblau (1972), in the case when the gradients of the linear approximations are calculated analytically or are difference approximations to derivatives. It is possible, however, that our construction of these gradients is new, because we derive them by interpolation at the vertices of simplices that are analogous to the ones that occur in the Nelder and Mead algorithm.

Our procedure is specified in Section 2. We will find that it has the following properties. Changes to the variables are restricted by a trust region bound, which gives the user some control over the steps that are taken automatically and which responds satisfactorily to the fact that there may be no finite solution to the linear programming problem (4). The trust region radius remains constant until predicted improvements to the objective function and feasibility conditions fail to occur, although the simplex has a good shape. Then the trust region radius is reduced until it reaches a final value that has to be set by the user. The lengths of the trial steps are usually equal to the current trust region bound, in order that little damage is done to the early iterations by any high frequency fluctuations in the objective and constraint functions that are of small amplitude. The shapes of successive simplices can vary greatly, because most changes to the variables would satisfy any linear constraints, so there is often a tendency for the simplices to be squashed onto the constraint boundaries, which we remove explicitly. Indeed, as mentioned already, some iterations pick changes to the variables whose primary purpose is to improve the shape of the simplex. We employ a merit function of the form

$$\Phi(\underline{x}) = F(\underline{x}) + \mu\,[\max\{-c_i(\underline{x}) : i=1,2,\ldots,m\}]_+\,, \quad \underline{x}\in\mathcal{R}^n, \tag{5}$$

in order to compare the goodness of two different vectors of variables. Here μ is a parameter that is adjusted automatically, and the subscript "+" means that the expression in square brackets is replaced by zero if and only if its value is negative, so we have $\Phi(\underline{x}) = F(\underline{x})$ whenever $\underline{x}$ is feasible. We take the view that $\underline{x}\in\mathcal{R}^n$ is better than $\underline{y}\in\mathcal{R}^n$ if and only if the inequality $\Phi(\underline{x}) < \Phi(\underline{y})$ holds. Moreover, it is not difficult to implement the given rules for adjusting the variables.

Our knowledge of the convergence properties of the algorithm is the subject of Section 3. Then Section 4 discusses some of the details of Section 2 and presents a few numerical results. We conclude that the proposed method is suitable for a range of optimization calculations, but that the final accuracy is sometimes severely limited by the use of linear approximations to nonlinear functions. A Fortran implementation of the algorithm is available from the author at the e-mail address mjdp@amtp.cam.ac.uk.

2. The Algorithm

The algorithm includes several strategies, and is summarised in Figure 1. First we consider the vector of variables that is calculated from the linear programming problem (4) by the "Generate $\underline{x}^{(*)}$" box of the figure. This task requires the vertices $\{\underline{x}^{(j)} : j=0,1,\ldots,n\}$ of a nondegenerate simplex, a positive trust region radius ρ, and the current value of the parameter μ of the merit function (5). The vertices have already been ordered so that $\underline{x}^{(0)}$ is optimal, which means that the inequalities

$$\Phi(\underline{x}^{(0)}) \leq \Phi(\underline{x}^{(j)}), \quad j=1,2,\ldots,n, \tag{6}$$

are satisfied. Then the trust region condition on the new vector of variables, $\underline{x}^{(*)}$ say, is the bound

$$\|\underline{x}^{(*)} - \underline{x}^{(0)}\|_2 \leq \rho. \tag{7}$$

If possible, we let $\underline{x}^{(*)}$ minimize the linear approximation $\hat{F}(\underline{x}^{(*)})$ to the objective function subject to the inequality (7) and to the linear constraints

$$\hat{c}_i(\underline{x}^{(*)}) \geq 0, \quad i=1,2,\ldots,m, \tag{8}$$

of the problem (4), picking the $\underline{x}^{(*)}$ that gives the least value of $\|\underline{x}^{(*)} - \underline{x}^{(0)}\|_2$ if these conditions admit more than one $\underline{x}^{(*)}$. Alternatively, it can happen that the inequalities (7) and (8) are contradictory. Then we define $\underline{x}^{(*)}$ by minimizing the greatest of the constraint violations $\{-\hat{c}_i(\underline{x}^{(*)}) : i=1,2,\ldots,m\}$ subject to the trust region bound. Further, any remaining freedom in $\underline{x}^{(*)}$ is used to minimize $\hat{F}(\underline{x}^{(*)})$ and, if some freedom still remains, then we remove the ambiguity by again making $\|\underline{x}^{(*)} - \underline{x}^{(0)}\|_2$ as small as possible. The calculation of $\underline{x}^{(*)}$ has been implemented by imagining that ρ is increased continuously from zero to the current value. The sequence of values of $\underline{x}^{(*)}$ that would occur for this range of ρ is a continuous trajectory that is composed of straight line pieces. It is convenient to follow the trajectory from $\underline{x}^{(0)}$ to the required $\underline{x}^{(*)}$ by identifying and updating the active sets of linear constraints that define the linear pieces.

Next we describe the adjustment of μ, because it depends on the $\underline{x}^{(*)}$ that has just been specified. We set $\mu = 0$ initially, but in this case, when choosing the optimal vertex, it is assumed that μ is a tiny positive number whose value need not be specified. Later we take the view that it is unreasonable to expect the reduction $\Phi(\underline{x}^{(*)}) < \Phi(\underline{x}^{(0)})$ in the merit function (5) if the value of μ does not provide the condition $\hat{\Phi}(\underline{x}^{(*)}) < \hat{\Phi}(\underline{x}^{(0)})$, where $\hat{\Phi}$ is the approximation

$$\hat{\Phi}(\underline{x}) = \hat{F}(\underline{x}) + \mu\,[\max\{-\hat{c}_i(\underline{x}) : i=1,2,\ldots,m\}]_+, \quad \underline{x}\in\mathcal{R}^n, \tag{9}$$

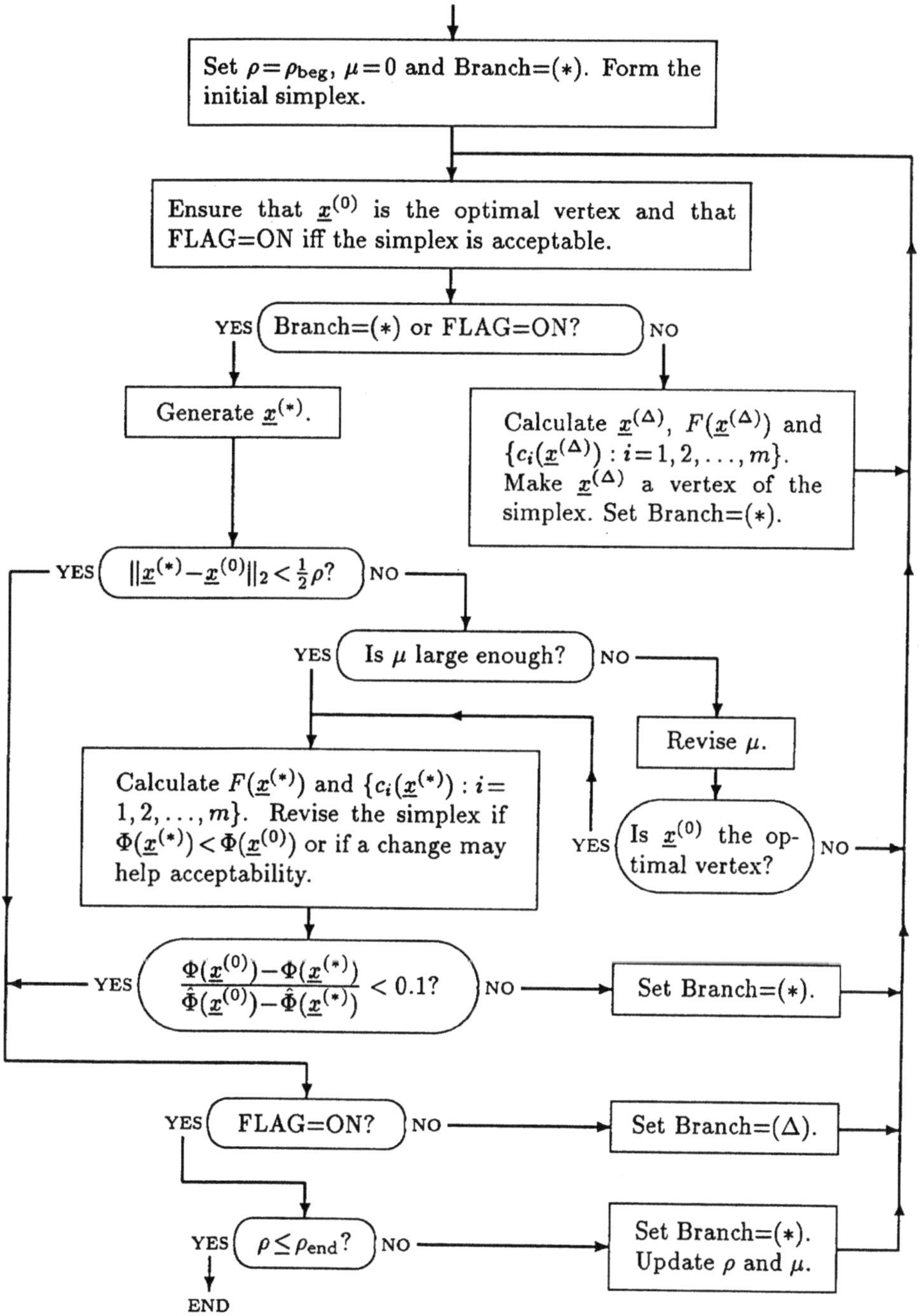

Figure 1: A summary of the algorithm

to Φ that is obtained by replacing F and $\{c_i : i=1,2,\ldots,m\}$ by their current linear approximations. Therefore the following strategy is employed by the "Is μ large enough" and "Revise μ" boxes of the figure. Let $\bar{\mu}$ be the least nonnegative value of μ that would yield $\hat{\Phi}(\underline{x}^{(*)}) \le \hat{\Phi}(\underline{x}^{(0)})$, the existence of $\bar{\mu}$ being an elementary consequence of the definition of $\underline{x}^{(*)}$. We leave μ unchanged if it satisfies $\mu \ge \frac{3}{2}\bar{\mu}$, but otherwise we increase μ to $2\bar{\mu}$. The choice of the factors $\frac{3}{2}$ and 2 in this technique was guided by guesswork and numerical calculations. Many similar factors will occur later, some of their values being discussed in Section 4.

A possible consequence of increasing μ is that condition (6) no longer holds. In this case the optimality of $\underline{x}^{(0)}$ is restored by exchanging two vertices of the simplex. Then the calculation of $\underline{x}^{(*)}$ and any necessary further adjustments to μ are repeated until $\underline{x}^{(0)}$ is optimal and the value of μ is acceptable. This procedure cannot cycle, because each change to $\underline{x}^{(0)}$ must reduce the value of the term $[\max\{-c_i(\underline{x}^{(0)}) : i = 1,2,\ldots,m\}]_+$.

The strategy for adjusting the trust region radius borrows from the Nelder and Mead algorithm the principle that one should continue to use the current ρ until the iterations fail to provide satisfactory reductions in the merit function, and then ρ should be decreased. We have the complication, however, that many different shapes of simplex can occur, so we also require the current simplex to be "acceptable" before decreasing ρ, in case unsuitable shapes have caused the linear programming problem (4) to be a very poor approximation to the main calculation, where "acceptable" is defined later. Therefore the bottom two boxes of Figure 1 are reached if we have an acceptable simplex, and if either the left hand side of inequality (7) is less than $\frac{1}{2}\rho$ or we find the condition

$$\Phi(\underline{x}^{(0)}) - \Phi(\underline{x}^{(*)}) < 0.1 \left[\hat{\Phi}(\underline{x}^{(0)}) - \hat{\Phi}(\underline{x}^{(*)})\right], \tag{10}$$

which means that changing the variables from $\underline{x}^{(0)}$ to $\underline{x}^{(*)}$ fails to provide a tenth of the improvement in the merit function that is predicted by the approximation (9). It can be argued that in both of these cases there is a need for a reduction in the trust region radius.

The initial and final values of ρ, namely ρ_{beg} and ρ_{end}, are given by the user. We recommend that ρ_{beg} be a reasonable change to make to the variables for a coarse exploration of the calculation, while ρ_{end} should be approximately the required distance from the final vector of variables to the solution of the optimization problem. The following action is taken when the conditions for reducing ρ are satisfied. If $\rho \le \rho_{\text{end}}$, then the iterative procedure is terminated, the final vector of variables being the current $\underline{x}^{(0)}$, except that $\underline{x}^{(*)}$ is preferred instead if $\Phi(\underline{x}^{(*)})$ is available and satisfies $\Phi(\underline{x}^{(*)}) < \Phi(\underline{x}^{(0)})$. Alternatively, when $\rho > \rho_{\text{end}}$, the trust region radius is set to the value

$$\rho_{\text{new}} = \begin{cases} \frac{1}{2}\rho, & \rho > 3\rho_{\text{end}} \\ \rho_{\text{end}}, & \rho \le 3\rho_{\text{end}}. \end{cases} \tag{11}$$

Further, because it is shown in Section 3 that the merit function parameter μ can become very large, we estimate whether it would be advantageous to decrease μ when ρ is reduced. Specifically, we take the view that the i-th constraint is important to the merit function if i is in the set

$$\mathcal{I} = \{i : c_i^{(\min)} < \tfrac{1}{2} c_i^{(\max)}\} \cap \{1, 2, \ldots, m\}, \tag{12}$$

where $c_i^{(\min)}$ and $c_i^{(\max)}$ are the least and greatest values of $c_i(\underline{x})$ at the vertices of the current simplex. Then μ is set to zero if $\mathcal{I}$ is empty, but otherwise we replace μ by the number

$$\left[\max_{j=0,1,\ldots,n} F(\underline{x}^{(j)}) - \min_{j=0,1,\ldots,n} F(\underline{x}^{(j)}) \right] \Big/ \min\left\{ [c_i^{(\max)}]_+ - c_i^{(\min)} : i \in \mathcal{I} \right\}, \tag{13}$$

provided that this change reduces μ. An explanation of the ratio (13) is given in Section 4.

A major difference between our algorithm and the method of Nelder and Mead is that we retain the current simplex when ρ is decreased. Therefore any immediate change to $\underline{x}^{(*)}$ is due to the new right hand side of inequality (7). Further, the current simplex will certainly be revised if it is not "acceptable" for the new value of ρ, the definition of "acceptability" being as follows.

For $j = 1, 2, \ldots, n$, let $\sigma^{(j)}$ be the Euclidean distance from the vertex $\underline{x}^{(j)}$ to the opposite face of the current simplex, and let $\eta^{(j)}$ be the length of the edge between $\underline{x}^{(j)}$ and $\underline{x}^{(0)}$. We say that the simplex is "acceptable" if and only if the inequalities

$$\left.\begin{array}{l} \sigma^{(j)} \geq \alpha\rho \\ \eta^{(j)} \leq \beta\rho \end{array}\right\}, \quad j = 1, 2, \ldots, n, \tag{14}$$

hold, where α and β are constants that satisfy the conditions $0 < \alpha < 1 < \beta$. The software picks the values $\alpha = \frac{1}{4}$ and $\beta = 2.1$. Thus the lengths of the edges and the volume of an acceptable simplex are of magnitudes ρ and ρ^n respectively, so they are appropriate to the current trust region radius.

The initial simplex is constructed in the following way from ρ_{beg} and an initial vector of variables, which have to be provided by the user. We let $\underline{x}^{(0)}$ be the given vector and then we cycle through the indices $j = 1, 2, \ldots, n$. For each j we set $\underline{x}^{(j)} = \underline{x}^{(0)} + \rho_{\text{beg}} \underline{e}_j$, where $\underline{e}_j$ is the j-th coordinate vector. Further, $\underline{x}^{(j)}$ is exchanged with $\underline{x}^{(0)}$ before proceeding to the next value of j if and only if the condition $F(\underline{x}^{(j)}) < F(\underline{x}^{(0)})$ is satisfied. Thus $\underline{x}^{(0)}$ becomes the optimal vertex of the initial simplex, and we do not worry about the possibility that this simplex may not be "acceptable".

The vector $\underline{x}^{(*)}$ is not calculated on every iteration, because it is clear sometimes that priority should be given to trying to satisfy the conditions (14). Specifically, this priority is imposed if and only if the previous iteration would have reduced the current value of ρ if its simplex were "acceptable", the priority being initiated by the "Set Branch $= \Delta$" box of Figure 1. In the next paragraph we define a vector $\underline{x}^{(\Delta)}$ that is an alternative new vector of variables that is chosen to improve acceptability. Therefore the current iteration calculates $\underline{x}^{(*)}$ instead of $\underline{x}^{(\Delta)}$ if and only if at least one of the following five conditions holds. (C1) There is no previous iteration. (C2) The previous iteration reduced ρ. (C3) The previous iteration calculated $\underline{x}^{(\Delta)}$. (C4) The previous iteration calculated $\underline{x}^{(*)}$ and reduced the merit function by at least one tenth of the predicted reduction. (C5) The current simplex is "acceptable". The first four conditions are mutually exclusive, but usually condition (C3) or (C4) holds when (C5) is achieved. Exceptions can occur, however, because sometimes the previous iteration will have replaced a vertex of the simplex by a vector $\underline{x}^{(*)}$ that does not satisfy condition (10).

When none of the above conditions holds, the vector $\underline{x}^{(\Delta)}$ is defined as follows. If any of the numbers $\{\eta^{(j)}: j=1,2,\ldots,n\}$ of expression (14) is greater than $\beta\rho$, we let ℓ be the least integer from $[1,n]$ that satisfies the equation

$$\eta^{(\ell)} = \max\{\eta^{(j)}: j=1,2,\ldots,n\}. \tag{15}$$

Otherwise we obtain ℓ from the formula

$$\sigma^{(\ell)} = \min\{\sigma^{(j)}: j=1,2,\ldots,n\}, \tag{16}$$

the inequality $\sigma^{(\ell)} < \alpha\rho$ being implied by the failure of condition (C5). The iteration is going to replace the vertex $\underline{x}^{(\ell)}$ by $\underline{x}^{(\Delta)}$, so we require $\underline{x}^{(\Delta)}$ to be well away from the face of the simplex that is opposite the vertex $\underline{x}^{(\ell)}$. Therefore we let $\underline{v}^{(\ell)}$ be the vector of unit length that is perpendicular to this face, and we make the choice

$$\underline{x}^{(\Delta)} = \underline{x}^{(0)} \pm \gamma\rho\,\underline{v}^{(\ell)}, \tag{17}$$

where the $\pm$ sign is chosen to minimize the approximation $\hat{\Phi}(\underline{x}^{(\Delta)})$ to the new value of the merit function, and where γ is a constant from the interval $(\alpha,1)$ that is set to $\gamma=\frac{1}{2}$ by the software. Then the next iteration is given the simplex that has the vertices $\{\underline{x}^{(j)}: j=0,1,\ldots,n,\ j\neq\ell\}$ and $\underline{x}^{(\Delta)}$. The description of an iteration that calculates $\underline{x}^{(\Delta)}$ is complete.

Alternatively, when an iteration forms $\underline{x}^{(*)}$, we have to choose between three options, namely reducing ρ, or preserving ρ for another iteration that will calculate $\underline{x}^{(*)}$, or preserving ρ for another iteration that will give priority to improving the "acceptability" of the simplex. The rules that govern the choice have been specified already. Because we employ the test (10) if and only if $\underline{x}^{(*)}$ satisfies the condition

$$\|\underline{x}^{(*)}-\underline{x}^{(0)}\|_2 \geq \tfrac{1}{2}\rho, \tag{18}$$

the function values $F(\underline{x}^{(*)})$ and $\{c_i(\underline{x}^{(*)}): i=1,2,\ldots,m\}$ are calculated only when inequality (18) holds. Then we may include these function values in future linear approximations by letting $\underline{x}^{(*)}$ replace one of the vertices $\{\underline{x}^{(j)}: j=1,2,\ldots,n\}$ of the current simplex, any change to the simplex being made in the following way.

The numbers $\{\bar{\sigma}^{(j)}: j=1,2,\ldots,n\}$ are found, where $\bar{\sigma}^{(j)}$ is defined to be the distance from $\underline{x}^{(*)}$ to the face of the current simplex that is opposite $\underline{x}^{(j)}$. These numbers are useful, because some elementary geometry shows that, if $\underline{x}^{(j)}$ is replaced by $\underline{x}^{(*)}$, then the volume of the simplex is multiplied by the factor $\bar{\sigma}^{(j)}/\sigma^{(j)}$. Therefore we take the view that the nonsingularity of the interpolation conditions will not be damaged if j is in the set

$$\mathcal{J} = \{j: \bar{\sigma}^{(j)} \geq \sigma^{(j)}\} \cup \{j: \bar{\sigma}^{(j)} \geq \alpha\rho\}, \tag{19}$$

where α is introduced in expression (14). Moreover, we expect the optimal vertex of the next iteration to be the point

$$\bar{\underline{x}}^{(0)} = \begin{cases} \underline{x}^{(*)}, & \Phi(\underline{x}^{(*)}) < \Phi(\underline{x}^{(0)}) \\ \underline{x}^{(0)}, & \Phi(\underline{x}^{(*)}) \geq \Phi(\underline{x}^{(0)}). \end{cases} \tag{20}$$

Therefore, if $\mathcal{J}$ is nonempty, we let ℓ be the least element of $\mathcal{J}$ that has the property

$$\|\underline{x}^{(\ell)} - \bar{\underline{x}}^{(0)}\|_2 = \max\{\|\underline{x}^{(j)} - \bar{\underline{x}}^{(0)}\|_2 : j \in \mathcal{J}\}. \tag{21}$$

Then the algorithm gives attention to the second of the acceptability requirements (14) by replacing $\underline{x}^{(\ell)}$ by $\underline{x}^{(*)}$ if we have the inequality $\|\underline{x}^{(\ell)} - \bar{\underline{x}}^{(0)}\|_2 > \delta\rho$, where δ is a constant satisfying $1 < \delta \leq \beta$ that is set to $\delta = 1.1$ by the software. Otherwise the new simplex is determined by the rule that its volume is to be maximized, subject to the condition that updating is mandatory when $\Phi(\underline{x}^{(*)})$ is less than $\Phi(\underline{x}^{(0)})$. In other words, if one (or both) of the conditions $\Phi(\underline{x}^{(*)}) < \Phi(\underline{x}^{(0)})$ and $\bar{\sigma}^{(\ell)} > \sigma^{(\ell)}$ holds, then $\underline{x}^{(\ell)}$ is replaced by $\underline{x}^{(*)}$, where now the integer ℓ is derived from the equation

$$\bar{\sigma}^{(\ell)} / \sigma^{(\ell)} = \max\{\bar{\sigma}^{(j)} / \sigma^{(j)} : j = 1, 2, \ldots, n\}. \tag{22}$$

Thus the simplex is revised by most iterations that calculate the objective and constraint functions at $\underline{x}^{(*)}$, the only exception being when, in addition to all the inequalities $\{\bar{\sigma}^{(j)} \leq \sigma^{(j)} : j = 1, 2, \ldots, n\}$ and $\Phi(\underline{x}^{(*)}) \geq \Phi(\underline{x}^{(0)})$, we find that the distance $\|\underline{x}^{(j)} - \bar{\underline{x}}^{(0)}\|_2$ is bounded above by $\delta\rho$ for every j in $\mathcal{J}$.

The description of our algorithm is now complete. Further details of the implementation are available in the Fortran listing that is mentioned at the end of Section 1.

3. Convergence Properties

Our knowledge of the convergence properties of the algorithm is slight. If we tried to establish a global convergence theorem by standard methods, then we would address the following four assertions. (A1) The parameter μ of the merit function (5) remains finite. (A2) The number of reductions in the trust region radius ρ is also finite. (A3) If μ and ρ remain constant, then any optimal vertex $\underline{x}^{(0)}$ cannot be retained for an infinite number of iterations. (A4) If μ and ρ remain constant, then the number of replacements of the optimal vertex is finite. These assertions would imply termination, because, for each ρ, the algorithm has the property that every change to the merit function parameter multiplies μ by at least the factor 4/3.

The following simple example suggests, however, that it would be difficult to make assumptions that provide assertion (A1) without ruling out some optimization calculations that the algorithm should solve. Let $n = 1$ and let the calculation be the problem

$$\left.\begin{array}{ll} \text{minimize} & F(x) = -|x-3|, \quad x \in \mathcal{R}, \\ \text{subject to} & c(x) = \tfrac{1}{4} - |x| \geq 0 \end{array}\right\}, \tag{23}$$

whose solution is $x = -\frac{1}{4}$. Further, let x and ρ satisfy $x > 3$ and $\rho = 1$ initially. Then on the first iteration the approximation (9) to the merit function is the expression

$$\hat{\Phi}(x) = -x + 3 + \mu\,[\,x - \tfrac{1}{4}\,]_+, \quad x \in \mathcal{R}, \tag{24}$$

so the algorithm sets the first value of the merit function parameter to $\mu=2$. It is now straightforward to deduce the action of an iteration if the optimal vertex of the current simplex satisfies $x^{(0)}>\frac{1}{4}$. Specifically, one can show by induction that $x^{(1)}$ and $x^{(*)}$ have the values $x^{(0)}+\rho$ and $x^{(0)}-\rho$ respectively, that $\mu=2$ is preserved, and that the inequality $\Phi(x^{(*)})<\Phi(x^{(0)})$ holds, which causes the next simplex to have the vertices $x^{(0)}-\rho$ and $x^{(0)}$. Further, the trust region radius is not reduced until inequality (10) is satisfied, which requires the condition $x^{(0)}<0.325$, mainly because $\rho=1$, $\mu=2$ and $0.325\le x^{(0)}\le 3$ imply the relation

$$\begin{aligned}\Phi(x^{(0)})-\Phi(x^{(*)}) &= [\,F(x^{(0)})-F(x^{(*)})\,]+2\,\{-c(x^{(0)})-[-c(x^{(0)}-1)\,]_+\}\\ &=1+2\,\{x^{(0)}-\tfrac{1}{4}-[\,|x^{(0)}-1|-\tfrac{1}{4}\,]_+\}\\ &\ge 1+2\,\{0.325-\tfrac{1}{4}-[\,0.675-\tfrac{1}{4}\,]\,\}=0.3\\ &\ge 0.1\,[\,\hat{\Phi}(x^{(0)})-\hat{\Phi}(x^{(*)})\,].\end{aligned} \tag{25}$$

Therefore an iteration can begin with $x^{(0)}=\frac{1}{2}-\epsilon$, and can generate a simplex with the vertices $x^{(0)}=-\frac{1}{2}-\epsilon$ and $x^{(1)}=\frac{1}{2}-\epsilon$, where ϵ is a very small positive number. In this case the approximation to the merit function on the next iteration has the form

$$\hat{\Phi}(x)=x-3+\mu\,[\,\tfrac{1}{4}-2\epsilon x-2\epsilon^2\,]_+,\quad x\in\mathcal{R}. \tag{26}$$

It now follows that we decrease the infeasibility of the linear approximation to the constraint function by *increasing* x. Further, we see that this change to x provides the required reduction in $\hat{\Phi}$ only if μ is made larger than $1/(2\epsilon)$.

Hence, even in the simple case (23), there is no *a priori* upper bound on μ. The difficulty is that the values of a function at the vertices of a simplex can be tiny perturbations of a constant, although the function itself varies substantially for most changes to the variables. Thus the linear approximations that are made by the algorithm may be highly misleading. On the other hand, the numerical results of the next section show that the given algorithm solves a range of nonlinear optimization calculations fairly well. It therefore seems futile to impose restrictions that would allow assertion (A1) to be established analytically. Moreover, a proof of assertion (A4) may be even more elusive, so we ignore this challenge too, at least in the case when the precision of the computer arithmetic is infinite.

On the other hand, the validity of assertion (A2) is an easy consequence of formula (11) and the condition $\rho_{\text{end}}>0$. Further, the following argument establishes that assertion (A3) is also true.

Lemma 1. *Any sequence of iterations of the given algorithm that does not change either ρ or the optimal vertex $\underline{x}^{(0)}$ is finite.*

Proof: We recall that, if a simplex is acceptable at the beginning of an iteration, and if the operations of the iteration do not alter the optimal vertex, then either a reduction in ρ or termination occurs at the end of the iteration. Therefore it is sufficient to prove that the conditions (14) are satisfied after a finite number of iterations of the given sequence.

Now Figure 1 shows that an iteration would change $\underline{x}^{(0)}$ if the previous iteration had calculated an $\underline{x}^{(*)}$ that satisfied $\Phi(\underline{x}^{(*)})<\Phi(\underline{x}^{(0)})$. Therefore alternate iterations in the sequence revise the simplex by replacing one of the vertices

$\{\underline{x}^{(j)} : j=1,2,\ldots,n\}$ by the point (17). Further, if the set $\{j : \|\underline{x}^{(j)}-\underline{x}^{(0)}\|_2>\beta\rho\}$ is nonempty, then the replacement reduces the number of elements in this set by one. Further, none of the other changes to the simplex can increase the number of elements in this set, because every $\underline{x}^{(*)}$ satisfies inequality (7). It follows that the second of the conditions (14) holds for every j after at most n iterations of the given sequence have used formula (17) to introduce a new vertex, which happens on alternate iterations.

The conditions $\|\underline{x}^{(*)}-\underline{x}^{(0)}\|_2\leq \rho$ and $\|\underline{x}^{(\Delta)}-\underline{x}^{(0)}\|_2\leq\rho$ also imply that the given iterations never add any elements to the set $\{j : \|\underline{x}^{(j)}-\underline{x}^{(0)}\|_2 > \delta\rho\}$, where δ is introduced soon after equation (21). Therefore this set does not change after a finite number of iterations. It follows from some details in Section 2 that eventually none of the iterations under consideration reduces the volume of the current simplex. Further, the alternate iterations that replace a vertex $\underline{x}^{(\ell)}$ by the point (17) multiply the volume by a factor that exceeds γ/α. Thus, if the lemma were false, the volume would become unbounded, which would contradict the second of the conditions (14). The proof is complete. ■

The formula (17) that defines $\underline{x}^{(\Delta)}$ also has the following interesting property.

Lemma 2. *Let an iteration replace the vertex $\underline{x}^{(\ell)}$ of the current simplex by $\underline{x}^{(\Delta)}$, and let the distances from the vertices (excluding $\underline{x}^{(0)}$) to their opposite faces in the old and new simplices be $\{\sigma_{\text{old}}^{(j)} : j = 1,2,\ldots,n\}$ and $\{\sigma_{\text{new}}^{(j)} : j = 1,2,\ldots,n\}$ respectively. Then, in addition to the equation $\sigma_{\text{new}}^{(\ell)}=\gamma\rho$, we have the inequalities*

$$\sigma_{\text{new}}^{(j)} \geq \sigma_{\text{old}}^{(j)}, \quad j=1,2,\ldots,n,\ j\neq\ell. \tag{27}$$

Proof: The equation $\sigma_{\text{new}}^{(\ell)} = \gamma\rho$ follows from the choice (17) and the definition of $\underline{v}^{(\ell)}$. Further, this definition implies the relation

$$(\underline{x}^{(\Delta)}-\underline{x}^{(0)},\ \underline{x}^{(i)}-\underline{x}^{(0)}) = 0, \quad i=1,2,\ldots,n,\ i\neq\ell, \tag{28}$$

the left hand side being a scalar product. Let j be any integer from $[1,n]$ that is different from ℓ. Then the closest point to $\underline{x}^{(j)}$ in the opposite face of the new simplex can be expressed in the form

$$\underline{x}^{(0)} + \sum_{\substack{i=1\\ i\neq j,\ell}}^{n} \theta_i(\underline{x}^{(i)}-\underline{x}^{(0)}) + \theta_\ell(\underline{x}^{(\Delta)}-\underline{x}^{(0)}) \tag{29}$$

for some multipliers $\{\theta_i : i=1,2,\ldots,n,\ i\neq j\}$, which gives the bound

$$\begin{aligned}[\sigma_{\text{new}}^{(j)}]^2 &= \|(\underline{x}^{(j)}-\underline{x}^{(0)}) - \sum_{\substack{i=1\\ i\neq j,\ell}}^{n} \theta_i(\underline{x}^{(i)}-\underline{x}^{(0)}) - \theta_\ell(\underline{x}^{(\Delta)}-\underline{x}^{(0)})\|_2^2 \\ &\geq \|(\underline{x}^{(j)}-\underline{x}^{(0)}) - \sum_{\substack{i=1\\ i\neq j,\ell}}^{n} \theta_i(\underline{x}^{(i)}-\underline{x}^{(0)})\|_2^2, \end{aligned} \tag{30}$$

where the last line depends on all the equations (28). We see that this right hand side is the square of the distance from $\underline{x}^{(j)}$ to a point in the opposite face of the *old* simplex. Thus the required inequality (27) is a consequence of the definition of $\sigma_{\text{old}}^{(j)}$. ■

We complete this section by establishing termination under the crude assumption that, due to the limited precision of computer arithmetic, only a finite number of different values of the functions F and $\{c_i : i=1,2,\ldots,m\}$ can occur. In this case assertion (A4) of the opening paragraph of this section holds, because every change to the optimal vertex has to provide a strict reduction in the merit function, which now cannot happen infinitely often. Further, we have noted already that assertions (A2) and (A3) are valid. Therefore it is straightforward to deduce termination if the number of changes to μ is finite, which should be another consequence of the computer arithmetic. Nevertheless, we will respond to the challenge of allowing μ to be any nonnegative real number.

Since every increase in μ is by at least the factor 4/3, and since we have already treated the case when μ is changed finitely often, we can assume without loss of generality that μ is greater than any positive constant Ω. We set $\Omega = 1$ if all calculable values of F are the same, or if $m = 0$, or if all calculable values of the constraint violation function

$$\Gamma(\underline{x}) = [\max\{-c_i(\underline{x}) : i=1,2,\ldots,m\}]_+, \quad \underline{x}\in\mathcal{R}^n, \tag{31}$$

are equal. Otherwise, we let Ω be the greatest change that can occur in F divided by the smallest positive change that can occur in Γ, this ratio being well-defined because of the crude assumption in the previous paragraph. Then, if μ exceeds Ω, the reduction $\Phi(\underline{x}^{(*)}) < \Phi(\underline{x}^{(0)})$ is found in practice if and only if we have either $\Gamma(\underline{x}^{(*)}) < \Gamma(\underline{x}^{(0)})$ or $\Gamma(\underline{x}^{(*)}) = \Gamma(\underline{x}^{(0)})$ and $F(\underline{x}^{(*)}) < F(\underline{x}^{(0)})$, the value of F being immaterial in the former case due to the choice of Ω. Thus the conditions that define an optimal vertex become independent of μ when the merit function parameter is sufficiently large. Hence the number of reductions in the merit function that can be achieved by updating the optimal vertex is finite. Therefore assertion (A4) is valid even if the condition that μ remains constant is replaced by the requirement that μ be sufficiently large, so there is no need for assertion (A1). It follows that the presumed finite precision of the computer arithmetic implies termination. Further, one could take advantage of this argument in practice by forcing a coarse discretization on the values of F and Γ when testing whether a vertex is optimal.

4. Discussion and Numerical Results

The main influence on the design of the algorithm was the belief that linear approximations to nonlinear constraints are highly useful. Further, because one can express a general objective function as a linear objective function subject to an inequality constraint by introducing a slack variable, it is consistent to make a linear approximation to the objective function too. We picked the easiest way of defining these linear approximations, namely interpolation at the vertices of a simplex. It was then necessary to impose a trust region bound in order that each linear programming subproblem has a finite solution. The shape of the trust region was chosen to be spherical, in order to preserve rotational symmetry when one takes a geometrical view of the steps of the algorithm, and the changes to the trust region radius are monotonic, in order to avoid all the careful attention to details that would arise from a strategy that allowed ρ to increase. We picked a merit function that employs the greatest constraint violation because this is often what users want. Here we have in mind that 1000 constraint violations of 10^{-6} are usually preferable to a single constraint violation of 10^{-3}, but a 1-norm merit function would not distinguish between these two cases. On the other hand, a smooth merit function would provide so many advantages that this subject deserves some research.

The example in the second paragraph of Section 4 is worrying because of the severe loss of efficiency that can occur if μ becomes huge. Indeed, consider the simple case when $n=2$ and $m=1$, when F and c_1 are the linear functions

$$F(\underline{x}) = x_2, \quad c_1(\underline{x}) = -x_1, \quad \underline{x} \in \mathcal{R}^2, \tag{32}$$

when $\rho = 1$, and when the vertices of the current simplex have the components $\underline{x}^{(0)} = (0,1)$, $\underline{x}^{(1)} = (\epsilon, 0)$ and $\underline{x}^{(2)} = (1,1)$, where the number ϵ is very small and positive. If μ satisfied the condition $\mu > 1/\epsilon$, then $\underline{x}^{(0)}$ would be the optimal vertex and $\underline{x}^{(*)}$ would have the coordinates $(0,0)$. Further, the current iteration of our algorithm would update the simplex by replacing $\underline{x}^{(1)}$ by $\underline{x}^{(*)}$, although this change to the variables is tiny. Instead, therefore, it might be better to let $\underline{x}^{(*)}$ solve the linear programming problem (4) subject to the trust region bound $\|\underline{x}^{(*)} - \underline{x}^{(1)}\|_2 \le \rho$, which would give the trial vector $\underline{x}^{(*)} = (0, -[1-\epsilon^2]^{1/2})$. This choice, however, is also unsatisfactory, because, if c_1 were replaced by a mildly nonlinear function that satisfied $c_1(\underline{x}) = -x_1$ at the vertices of the current simplex, then it is likely that we would find the increase $\Phi(\underline{x}^{(*)}) > \Phi(\underline{x}^{(0)})$ in the merit function. Thus the large value of μ would cause a reduction in ρ, although there is no evidence that the variables are nearly optimal.

It is difficult, however, to devise techniques that increase and decrease a merit function parameter and that are guaranteed not to cycle in general optimization calculations. Therefore we include the partial remedy of only reducing μ if it seems to be too large when ρ is decreased, which happens finitely often. The procedure that is given in Section 2 is derived from the magnitudes of the two terms on the right hand side of the definition (5) of the merit function. Clearly the numerator of expression (13) is approximately a typical change to F, and it is reasonable to exclude from consideration the constraints whose indices are not in $\mathcal{I}$. Further, the term $[c_i^{(\max)}]_+ - c_i^{(\min)}$ is a typical change to the i-th constraint function if $c_i^{(\max)}$ is nonnegative and otherwise it is a change that has to be made to achieve feasibility, but the choice of "min" rather than "max" in the denominator of the ratio (13) is debatable. Here we take the view that, if μ is decreased, then we want each of the relevant constraints to make a contribution to the merit function that is not much less than a typical change to F. It is therefore helpful if the user scales the constraint functions so that they have similar magnitudes.

The values of the parameters α, β, γ and δ that are mentioned in Section 2 were guided by some numerical tests that did not provide clear answers. We picked $\alpha = \frac{1}{4}$

for the first of the acceptability conditions (14), because $\alpha=\frac{1}{8}$ gave relatively poor results in the numerical experiments, and because the proof of Lemma 1 suggests that there should be enough scope for γ/α to be substantially greater than one. There was little difference in practice between $\gamma=\frac{1}{2}$ and $\gamma=1$, so we preferred the smaller value, because formula (17) tends to move $\underline{x}^{(\Delta)}$ away from any subspace that contains the most successful vectors of variables. We require $\beta>1$ in expression (14). Further, the condition $\beta\geq 2$ is advantageous, because otherwise it would be usual for every edge of the current simplex to be too long immediately after halving ρ, and then "acceptability" might demand the updating of all the suboptimal vertices before the next reduction in ρ. On the other hand, an example in the penultimate paragraph of this section will show that too large a value of β can be highly inefficient. Therefore $\beta=2.1$ was selected. The parameter δ that occurs soon after equation (21) should certainly satisfy $1<\delta\leq\beta$. We picked $\delta=1.1$, because a value that is only a little larger than one helps the newly calculated function values at $\underline{x}^{(*)}$ to be included in the linear approximations of the next iteration.

The algorithm was applied to the following ten problems using single precision arithmetic on a Sparc 2 workstation. In every case the initial trust region radius was $\rho=\frac{1}{2}$ and all components of the initial vector of variables were set to 1. Firstly, some easy tests of the effects of nonlinearity were made by the calculations

$$\text{minimize } F(\underline{x}) = 10(x_1+1)^2 + x_2^2, \quad \underline{x}\in\mathcal{R}^2, \tag{A}$$

$$\left.\begin{array}{ll}\text{minimize} & F(\underline{x}) = x_1x_2, \quad \underline{x}\in\mathcal{R}^2,\\ \text{subject to} & x_1^2+x_2^2\leq 1\end{array}\right\}, \tag{B}$$

and

$$\left.\begin{array}{ll}\text{minimize} & F(\underline{x}) = x_1x_2x_3, \quad \underline{x}\in\mathcal{R}^3,\\ \text{subject to} & x_1^2+2x_2^2+3x_3^2\leq 1\end{array}\right\}, \tag{C}$$

there being no constraints in problem (A). The next two calculations were also unconstrained, being the mild versions

$$\text{minimize } F(\underline{x}) = (x_1^2-x_2)^2 + (1+x_1)^2, \quad \underline{x}\in\mathcal{R}^2, \tag{D}$$

and

$$\text{minimize } F(\underline{x}) = 10(x_1^2-x_2)^2 + (1+x_1)^2, \quad \underline{x}\in\mathcal{R}^2, \tag{E}$$

of the well-known problem of Rosenbrock (1960). We took the constrained calculations

$$\left.\begin{array}{ll}\text{minimize} & F(\underline{x}) = -x_1-x_2, \quad \underline{x}\in\mathcal{R}^2,\\ \text{subject to} & x_1^2\leq x_2 \text{ and to } x_1^2+x_2^2\leq 1\end{array}\right\}, \tag{F}$$

and

$$\left.\begin{array}{ll}\text{minimize} & F(\underline{x}) = x_3, \quad \underline{x}\in\mathcal{R}^3,\\ \text{subject to} & 5x_1-x_2+x_3\geq 0, \quad -5x_1-x_2+x_3\geq 0\\ \text{and to} & x_1^2+x_2^2+4x_2\leq x_3\end{array}\right\} \tag{G}$$

from Fletcher's (1987) book. Finally, problems (H)–(J) are the ones with the numbers 43, 100 and 108 in Hock and Schittkowski (1980), so they have 4, 7 and 9

Problem number	Function values	Final $F(\underline{x})$	Final $\Gamma(\underline{x})$	Final $\|\underline{x}-\underline{x}^{(\mathrm{opt})}\|_2$
(A)	37	1.8×10^{-5}	0	3.3×10^{-3}
(B)	37	−0.5000	2.0×10^{-6}	1.3×10^{-3}
(C)	45	−0.0786	4.7×10^{-6}	1.4×10^{-3}
(D)	100	3.1×10^{-5}	0	1.3×10^{-2}
(E)	347	4.0×10^{-3}	0	1.4×10^{-1}
(F)	30	−1.4142	3.0×10^{-6}	1.2×10^{-4}
(G)	29	−3.0000	1.3×10^{-4}	5.9×10^{-5}
(H)	74	−44.0000	2.9×10^{-6}	1.4×10^{-3}
(I)	198	680.6303	5.7×10^{-5}	5.9×10^{-3}
(J)	143	−0.8660	1.0×10^{-6}	8.9×10^{-4}

Table 1: Problems (A)–(J) when $\rho_{\mathrm{end}}=10^{-3}$

Problem number	Function values	Final $F(\underline{x})$	Final $\Gamma(\underline{x})$	Final $\|\underline{x}-\underline{x}^{(\mathrm{opt})}\|_2$
(A)	65	1.2×10^{-7}	0	2.8×10^{-4}
(B)	44	−0.5000	6.0×10^{-8}	6.1×10^{-5}
(C)	60	−0.0786	0	9.2×10^{-6}
(D)	173	6.4×10^{-7}	0	1.7×10^{-3}
(E)	698	9.5×10^{-5}	0	2.2×10^{-2}
(F)	41	−1.4142	1.5×10^{-7}	4.6×10^{-5}
(G)	33	−3.0000	0	2.4×10^{-8}
(H)	87	−44.0000	2.2×10^{-6}	1.2×10^{-3}
(I)	212	680.6303	0	5.3×10^{-3}
(J)	173	−0.8660	1.2×10^{-7}	9.5×10^{-5}

Table 2: Problems (A)–(J) when $\rho_{\mathrm{end}}=10^{-4}$

variables respectively. This last calculation is intended to maximize the area of a hexagon of unit diameter, but it provides the value $\frac{1}{2}\sqrt{3}\approx0.866$, although the area of a circle of unit diameter is only $\frac{1}{4}\pi\approx0.785$. The reason for the incorrect result is that the formulation of the problem allows one circuit of the hexagon to degenerate to two circuits of an equilateral triangle, the area of the triangle being counted twice. This example is also interesting because it includes some local minima where the area has the value 0.5.

The results are presented in Tables 1 and 2, the difference between the tables being that the final trust region radii are 10^{-3} and 10^{-4}, respectively. The columns of each table display the problem number, the number of calculations of F and $\{c_i : i=1,2,\ldots,m\}$, the final value of the objective function, the final value of the maximum constraint violation (31), and the final value of $\|\underline{x}-\underline{x}^{(\mathrm{opt})}\|_2$, which is the

Euclidean distance from the calculated vector of variables to a true solution. We see that the amount of computation is not excessive for an easy-to-use algorithm that does not require any derivatives, although the Nelder and Mead (1965) method is sometimes much more efficient when there are no constraints, because it adapts the shape of the simplex to the curvature of the objective function. Further, a comparison of the two tables shows that the accuracy in the calculated variables can be controlled approximately by the final value of ρ, but some of the entries in the last column of the tables are rather large.

The reason for these large entries is that linear approximations to nonlinear functions are often misleading near a solution to an optimization calculation. For example, we consider the unconstrained minimization of the quadratic function

$$F(\underline{x}) = x_1^2 + Mx_2^2, \quad \underline{x} \in \mathcal{R}^2, \tag{33}$$

where M is a positive constant, and we let θ be a positive parameter. Then the points

$$\underline{x}^{(0)} = \begin{pmatrix} \theta \\ 0 \end{pmatrix}, \quad \underline{x}^{(1)} = \begin{pmatrix} \theta - \frac{1}{4}\rho \\ (\frac{15}{16})^{1/2}\rho \end{pmatrix} \quad \text{and} \quad \underline{x}^{(2)} = \begin{pmatrix} \theta - \frac{1}{4}\rho \\ -(\frac{15}{16})^{1/2}\rho \end{pmatrix} \tag{34}$$

provide an acceptable simplex, and we have the function values

$$F(\underline{x}^{(0)}) = \theta^2, \quad F(\underline{x}^{(1)}) = F(\underline{x}^{(2)}) = \theta^2 - \tfrac{1}{2}\theta\rho + (\tfrac{1}{16} + \tfrac{15}{16}M)\rho^2. \tag{35}$$

Thus $\underline{x}^{(0)}$ is an optimal vertex if the inequality

$$\theta < (\tfrac{1}{8} + \tfrac{15}{8}M)\rho \tag{36}$$

holds. Further, the algorithm makes the linear approximation

$$\hat{F}(\underline{x}) = \theta^2 + 2[\theta - (\tfrac{1}{8} + \tfrac{15}{8}M)\rho](x_1 - \theta), \quad \underline{x} \in \mathcal{R}^2, \tag{37}$$

to the objective function, so $\underline{x}^{(*)}$ has the components $(\theta+\rho, 0)$ when $\underline{x}^{(0)}$ is the optimal vertex. In this case, unfortunately, we find the inequality $F(\underline{x}^{(*)}) > F(\underline{x}^{(0)})$, so either termination occurs or ρ is reduced, although the distance from $\underline{x}^{(0)}$ to the true solution can exceed $\frac{15}{8}M\rho$. Therefore it is possible for large errors to occur in the calculated solution when the second derivative matrix $\nabla^2 F$ is positive definite and only mildly ill-conditioned. Further, if ρ is halved when $\theta = \frac{15}{8}M\rho$, and if the final variables are nearly optimal, then at least another $\frac{15}{4}M$ iterations are needed. In the light of this example, the numerical results of Tables 1 and 2 are as good as can be expected.

Further, this example suggests that it may be possible to develop a very useful new algorithm by using quadratic instead of linear approximations to the objective and constraint functions. Then each simplex of $(n+1)$ points would have to be replaced by a suitable set of $\frac{1}{2}(n+1)(n+2)$ points, in order that the coefficients of the new approximations can be calculated by interpolation. The author intends to investigate this approach, at least in the unconstrained case, because such research should yield another algorithm that is more suitable for the minimization of noisy functions than the usual methods that employ difference approximations to derivatives.

Acknowledgements

This work began when David Ingram of Westland Helicopters asked for my advice on the technique of extending a version of the Nelder and Mead (1965) method to constrained optimization by forcing the objective function to be infinite at infeasible points. Thus he provided the motivation for the given algorithm. I am also very grateful to the Mathematics Department of the University of Canterbury, New Zealand, because it provided excellent facilities for this research while I was there on sabbatical leave. Further, I offer my thanks to a referee who suggested several improvements to the presentation.

References

R. Fletcher (1987), *Practical Methods of Optimization*, John Wiley and Sons (Chichester).

R.E. Griffith and R.A. Stewart (1961), "A nonlinear programming technique for the optimization of continuous processing systems", *Management Sci.*, Vol. 7, pp. 379–392.

D.M. Himmelblau (1972), *Applied Nonlinear Programming*, McGraw-Hill (New York).

W. Hock and K. Schittkowski (1980), *Test Examples for Nonlinear Programming Codes, Lecture Notes in Economics and Mathematical Systems 187*, Springer-Verlag (Berlin).

J.A. Nelder and R. Mead (1965), "A simplex method for function minimization", *Comput. J.*, Vol. 7, pp. 308–313.

M.J.D. Powell (1978), "A fast algorithm for nonlinearly constrained optimization calculations", in *Numerical Analysis, Dundee 1977, Lecture Notes in Mathematics 630*, ed. G.A. Watson, Springer-Verlag (Berlin), pp. 144–157.

H.H. Rosenbrock (1960), "An automatic method for finding the greatest or least value of a function", *Comput. J.*, Vol. 3, pp. 175–184.

W. Spendley, G.R. Hext and F.R. Himsworth (1962), "Sequential application of simplex designs in optimisation and evolutionary operation", *Technometrics*, Vol. 4, pp. 441–461.

M.B. Subrahmanyam (1989), "An extension of the simplex method to constrained optimization", *J. Optim. Theory Appl.*, Vol. 62, pp. 311–319.

A TRUNCATED SQP ALGORITHM FOR LARGE SCALE NONLINEAR PROGRAMMING PROBLEMS *

PAUL T. BOGGS
Applied and Computational Mathematics Division, National Institute of Standards and Technology, Gaithersburg, MD 20899

JON W. TOLLE
Mathematics Department, University of North Carolina, Chapel Hill, NC 27599

and

ANTHONY J. KEARSLEY
Department of Computational and Applied Mathematics, Rice University, Houston, Texas 77251

Abstract. We consider the inequality constrained nonlinear programming problem and an SQP algorithm for its solution. We are primarily concerned with two aspects of the general procedure, namely, the approximate solution of the quadratic program, and the need for an appropriate merit function. We first describe an (iterative) interior-point method for the quadratic programming subproblem that, no matter when it is terminated, yields a descent direction for a suggested new merit function. An algorithm based on ideas from trust-region and truncated Newton methods is suggested and some of our preliminary numerical results are discussed.

1. Introduction

Large scale optimization problems are gradually submitting to the power of advanced algorithmic development and of modern computing environments, leading to the formulation of models requiring solutions of these problems in a variety of scientific areas. Two excellent recent reviews are given by Coleman (Coleman, 1992) and Conn, Gould and Toint (Conn *et al.*, 1992), who survey some important applications as well as recent trends in algorithms and consider the impact of parallel computing architectures for large scale optimization.

Following these authors we take the term *large scale* to mean any optimization problem that is large enough so that the exploitation of special structure is important. In this paper we are particularly concerned with sparsity, although, as they point out, other problem structures may be important as well. We assume the general nonlinear programming problem to be of the form

$$\begin{aligned} &\min_x f(x) \\ &\text{subject to:} \quad g(x) \leq 0 \end{aligned} \qquad (NLP)$$

where $f : \Re^n \rightarrow \Re^1$, and $g : \Re^n \rightarrow \Re^m$. We note that we could include nonlinear equality constraints in (NLP) without incurring any analytical difficulties, but at

* Contribution of the National Institute of Standards and Technology and not subject to copyright in the United States.

S. Gomez and J-P. Hennart (eds.), Advances in Optimization and Numerical Analysis, 69–77.
© 1994 Kluwer Academic Publishers.

the expense of distracting technicalities. We thus omit them for the purposes of the exposition here, but they have been included in our program.

Our basic tool for the solution of (NLP) is the sequential quadratic programming (SQP) algorithm in which, given an approximate solution-multiplier pair, (x^k, λ^k), (NLP) is approximated by a quadratic program of the form

$$\begin{aligned} &\min_{\delta} \nabla f(x^k)^T \delta + \tfrac{1}{2}\delta^T B^k \delta \\ &\text{subject to: } \nabla g(x^k)^T \delta + g \leq 0. \end{aligned} \tag{QP}$$

Here B^k is taken to be an approximation to the Hessian of the Lagrangian for (NLP), i.e., for

$$\ell(x, \lambda) = f(x) + g(x)^T \lambda$$

we choose

$$B^k \approx \nabla^2_{xx} \ell(x^k, \lambda^k).$$

In this form the solution to (QP) provides a search direction for improving the current iterate, x^k. A steplength is chosen in this direction so as to reduce a *merit function*. Roughly speaking, a merit function is a scalar valued function with a minimum at the solution to (NLP). Thus reducing this function ensures progress and allows for the construction of a globally convergent scheme. (See e.g., (Boggs and Tolle, 1989) and (Boggs *et al.*, 1991).) In a previous paper, (Boggs *et al.*, 1991), the authors introduced a merit function for (NLP) and showed that it is appropriate for use with the SQP algorithm. In this paper we apply these ideas to the large scale case, solving (QP) only approximately by an iterative interior-point algorithm that we can stop prematurely. Such ideas are in the spirit of *truncated* or *inexact* Newton methods. (See (Dembo *et al.*, 1982) and, for a recent discussion of these methods, (Eisenstat and Walker, 1991).)

To be more specific, we use the interior-point quadratic program solver of (Boggs *et al.*, 1992). At each iteration this method constructs a low-dimensional subspace and solves (QP) restricted to that subspace. We can show that halting this procedure after any number of steps yields a descent direction for the merit function. The details of this solver and its properties relative to its use in an SQP algorithm are discussed in §2.

The actual merit function and a related approximate merit function are reviewed in §3. We then state the results just mentioned, namely that the inexact directions are compatible with these functions. In §4 we give the details of the algorithm. One of the problematic points is how to control the number of iterations on (QP). Here we use some ideas from *trust region* methods. We attempt to assess how well (QP) approximates the behavior of the merit function by maintaining an estimate of a trust region radius. §4 also contains a summary of the results of some numerical experimentation with the algorithm on a few large problems of interest. Our results indicate that our procedure is viable for large scale problems. Suggestions for further research are contained in §5.

2. An Interior-Point QP Solver

Interior-point methods for linear programming have been demonstrated to be very successful, especially on large problems; thus it is natural to consider their extension to quadratic programs (QP). One method that has performed well on linear programs, and has been extended to QP with both good numerical results and particularly interesting properties with respect to the SQP method is the *optimal subspace* method of Boggs, et al. (Boggs *et al.*, 1992). (See also (Domich *et al.*, 1991).) We take the QP of §1 to be of the form

$$\begin{aligned} &\min_{\delta} c^T\delta + \tfrac{1}{2}\delta^T Q\delta \\ &\text{subject to:} \quad A^T\delta + b \leq 0 \end{aligned} \tag{1}$$

where $c, \delta \in \Re^n$, $Q \in \Re^{n\times n}$, $A \in \Re^{n\times m}$, and $b \in \Re^m$.

The assumptions on (1) that are necessary to apply the algorithm are that the problem be bounded; that A have full column rank; that there exist feasible points (i.e., that the constraints be consistent); and that Q be positive semidefinite. Note that a full dimensional interior is not required. We comment further on these assumptions at the end of this section.

Briefly, the general algorithm can be expressed as follows.

O3D Algorithm for Quadratic Programming

1. Given a feasible point, δ^0; set $j := 0$.
2. Generate 3 independent search directions

$$p_i, \qquad i = 1, \ldots, 3.$$

   Let P^j be the matrix whose columns are p_i.
3. Form and solve the restricted quadratic program

$$\begin{aligned} &\min_{\zeta} c^T\tilde{\delta} + \tfrac{1}{2}\tilde{\delta}^T Q\tilde{\delta} \\ &\text{subject to:} \quad A^T\tilde{\delta} + b \leq 0 \end{aligned}$$

   where $\tilde{\delta} = \delta^j + P^j\zeta$, and $\zeta \in \Re^3$. Call the solution ζ^*.
4. Set $\delta^{j+1} := \delta^j + \rho P^j\zeta^*$ for an appropriate value of the steplength ρ.
5. If stopping criteria are met, set $J = j$, $\delta_J = \delta^j$ and exit.
6. Go to 2.

The details of the actual algorithm can be found in (Boggs *et al.*, 1992) and (Domich *et al.*, 1991); here we describe those that are the most important for its application in the SQP setting. One of the three directions is always a descent direction with respect to the objective function, thus assuring descent in the objective value at each step. Specifically, the algorithm uses directions that are solutions to

$$\left[AD^2A^T + Q/\beta\right] p_i = t_i \tag{2}$$

where β is a positive scalar depending on the current iterate,

$$D = \operatorname{diag}\{1/r_k,\ k = 1, \ldots, m\},$$

$r_k = -(A_k^T \delta^j + b_k)$, and t_i is a particular right hand side. The form of the matrix in (2) allows for efficient exploitation of the sparsity. Note that if Q is positive semi-definite, then this matrix is positive definite for all interior points. The steplength ρ (cf. Step 4) is set either to obtain the optimal solution in the given direction or to advance 99% of the distance to the boundary.

An important aspect of the algorithm is the procedure for obtaining an initial feasible point, since we certainly do not require that a feasible point be given. The algorithm uses a "Big M" method to construct the Phase I problem:

$$\begin{aligned} &\min_{\delta,\theta} c^T \delta + \tfrac{1}{2}\delta^T Q \delta + M\theta \\ &\text{subject to:} \quad A^T \delta + b - e\theta \leq 0 \end{aligned} \tag{3}$$

where e is a vector of all ones and θ is the "artificial" variable. Clearly for θ^* large enough the point $(\delta, \theta) = (0, \theta^*)$ is feasible for (3). The above procedure is thus used until the artificial variable is negative, at which point the current value of δ is feasible, and the $M\theta$ and $e\theta$ terms are dropped. If no such value of the artificial variable can be found, then the QP is not consistent, i.e., no feasible point exists, and the algorithm stops. In this case, however, one can show that the optimal solution satisfies

$$\theta = \min_{\delta} \max_{j} \{A_j^T \delta + b_j\},$$

and the resulting δ is a reasonable direction for (NLP).

The criteria for convergence of the algorithm are that at least one of the following hold: (a) the relative change in two successive values of the objective function is small; (b) the relative difference between the primal and the dual objective function values is small; or (c) the relative difference between two successive iterates is small. To this list, we have added the criterion (d) the scaled solution vector exceeds a specified length. This last condition has been implemented to allow the use of trust region strategies to monitor the quality of the (QP) approximation. In particular, this procedure will cause the algorithm to halt if (QP) is unbounded, again with a reasonable direction.

Note that the assumptions set forth above ensure that a solution to (3) exists, but that the quadratic subproblems arising in the SQP algorithm may not have solutions. Nevertheless, the directions calculated by O3D are useful directions in the solution of (NLP).

3. The Merit Function

A merit function for (NLP) is typically a scalar valued function that has an unconstrained minimum at x^*, the solution to (NLP). Thus a reduction in this function implies that progress is being made towards the solution.

In (Boggs *et al.*, 1991) we derived a merit function for (NLP) based on the work in (Boggs and Tolle, 1984) and (Boggs and Tolle, 1989) for equality constrained problems. This was done by considering the slack variable problem (see (Tapia, 1980))

$$\begin{aligned} &\min_{x,s} f(x) \\ &\text{subject to:} \quad g(x) + \tfrac{1}{4} S^2 e = 0 \end{aligned} \tag{4}$$

where

$$S = \text{diag}\{s_1, \ldots, s_m\}.$$

The merit function in (Boggs and Tolle, 1984) was then applied to (4). Since the resulting merit function only contained references to s_i^2, and not to just s_i, it was natural to rephrase this merit function in terms of $z_i = s_i^2$. This led to the rather unusual situation of having a *constrained* merit function, i.e., a merit function whose constrained minimum corresponds to the solution of (NLP). Our merit function is

$$\psi_d(x,z) = f(x) + \bar{\lambda}(x,z)^T \bar{c}(x,z) + \tfrac{1}{d}\bar{c}(x,z)^T \mathcal{A}(x,z)^{-1}\bar{c}(x,z) \tag{5}$$

where d is a scalar,

$$\begin{aligned} \bar{c}(x,z) &= g(x) + Ze \\ \mathcal{A}(x,z) &= \nabla g(x)^T \nabla g(x) + Z \\ \bar{\lambda}(x,z) &= -\mathcal{A}(x,z)^{-1}\nabla g(x)^T \nabla f(x) \end{aligned}$$

and

$$Z = \text{diag}\{z_1, \ldots, z_m\}$$

with the z vector constrained to be nonnegative. Although the merit function is constrained, our algorithm ensures that the z_i always remain positive; thus the bounds present neither a theoretical nor a computational difficulty. For a direction, δ^k, in x obtained as the solution to (QP), we take the direction for the change in z to be

$$q^k = -[\nabla g(x^k)\delta^k + g(x^k) + z^k].$$

Thus the next step is

$$\begin{aligned} x^{k+1} &= x^k + \alpha\delta^k \\ z^{k+1} &= z^k + \alpha q^k \end{aligned}$$

for some value of α. Observe that if $z^k \geq 0$ and δ^k is feasible, then

$$q^k + z^k = -[\nabla g(x^k)^T \delta^k + g(x^k)] \geq 0 \tag{6}$$

and it follows that for $\alpha \in (0,1]$, $z^{k+1} = z^k + \alpha q^k \geq 0$.

We show in (Boggs *et al.*, 1991) that ψ_d has certain desirable properties for d sufficiently small. First, under mild conditions, a constrained minimum of ψ_d corresponds to a solution of (NLP). Furthermore, the directions (δ^k, q^k) are descent directions for ψ_d for (x^k, z^k) sufficiently close to feasibility; a steplength of one is acceptable near the solution if the method is converging q-superlinearly; and the directions are always descent directions for $r(x,z) = \|\bar{c}(x,z)\|^2$.

Despite these useful properties, ψ_d has two deficiencies that preclude using it directly in an algorithm. First, as stated above, δ^k is only a descent direction near feasibility, and second, it requires the evaluation of gradients and other nontrivial computation to assess a prospective value of α. Thus we employ an approximate merit function and a globalization strategy that overcome these deficiencies. We use

$$\psi_d^k(x,z) = f(x) + \bar{c}(x,z)^T \bar{\lambda}^k + \tfrac{1}{d}\bar{c}(x,z)^T (\mathcal{A}^k)^{-1}\bar{c}(x,z)$$

where

$$\mathcal{A}^k = \nabla g(x^k)^T \nabla g(x^k) + Z^k$$
$$\bar{\lambda}^k = -(\mathcal{A}^k)^{-1} \nabla g(x^k)^T \nabla f(x^k).$$

We show in (Boggs *et al.*, 1991) that (δ^k, q^k) is a descent direction for ψ_d^k everywhere, that ψ_d^k will not interfere with rapid local convergence, and that the globalization strategy described in §4 is effective.

The main theoretical result described here is that O3D and the merit function are compatible. Specifically, if δ_J^k is only a partial solution to (QP) obtained by J iterations of O3D (see step 5), the above results continue to hold. We state the assumptions that guarantee this. We use the term ***strong local solution*** to mean an optimal point, together with a multiplier vector, of (NLP) at which the following hold.

A1: The active constraint gradients are linearly independent.

A2: Strict complementary slackness holds.

A3: The second order sufficient conditions hold.

In addition we make the following assumptions on the (QP) subproblems:

A4: The matrices $\{B^k\}$ are uniformly positive definite.

A5: For each k (QP) has a strong local solution.

We also need an assumption that guarantees that the merit function is well defined i.e., that $\mathcal{A}$ is nonsingular. As in (Boggs *et al.*, 1991) we formulate this by partitioning the index set of the constraints into two subsets a and u. We can then write, without loss of generality,

$$g(x) = \begin{pmatrix} g_a(x) \\ g_u(x) \end{pmatrix}$$

and correspondingly,

$$z = \begin{pmatrix} z_a \\ z_u \end{pmatrix}.$$

Usually the index subset a will correspond to the set of active constraints for (NLP) or (QP). The necessary assumption in terms of a particular partition is the following.

A6: The set $\{\nabla g_i(x) \ : \ i \in a\}$ is linearly independent and $z_u > 0$.

A discussion of the implications of these assumptions for SQP algorithms is given in (Boggs *et al.*, 1991). The proofs of the results make use of the techniques in (Boggs *et al.*, 1991) combined with an induction argument.

4. Algorithm and Numerical Results

A brief statement of the final algorithm is as follows. Following the statement, we give a brief discussion of some important points.

SQP Algorithm

1. Given x^0, τ (trust region radius), η (globalization parameter), and d (merit function parameter):
   Set $k := 0$.
2. Using O3D, iterate while $\|\delta\| < \tau$ on

$$\begin{aligned}&\min_{\delta} \nabla f(x^k)^T \delta + \tfrac{1}{2}\delta^T B^k \delta\\ &\text{subject to: } \nabla g(x^k)^T \delta + g(x^k) \le 0\end{aligned}$$

   to obtain δ^k.
3. Set $q^k = -[\nabla g(x^k)^T \delta^k + g(x^k) + z^k]$.
4. (Globalization Step)
   Choose α^k such that ψ_d^k is reduced.
   If $\|\bar{c}(x^k + \alpha^k\delta^k, z^k + \alpha^k q^k)\| \ge \|\bar{c}(x^k, z^k)\|$ and $\|\bar{c}(x^k, z^k)\| > \eta$,
   reduce α^k until $\|\bar{c}(x^k + \alpha^k\delta^k, z^k + \alpha^k q^k)\| < \|\bar{c}(x^k, z^k)\|$.
5. If $\psi_d(x^k + \alpha^k\delta^k, z^k + \alpha^k q^k) > \psi_d(x^k, z^k)$
   set $\eta = \frac{1}{2}\|\bar{c}(x^k, z^k)\|$.
6. Set

$$\begin{aligned}x^{k+1} &:= x^k + \alpha^k \delta^k\\ z^{k+1} &:= z^k + \alpha^k q^k.\end{aligned}$$

7. If convergence criteria are met, quit.
8. Adjust τ.
9. Set $k := k + 1$; goto 2.

A few comments are necessary. First, the globalization step is based on the work in (Boggs and Tolle, 1989). In brief, η is an estimate of the radius of the domain containing the feasible region in which the true merit function, $\psi_d(x, z)$, is reduced in the direction (δ, q). For all iterates, the algorithm first requires that the approximate merit function be reduced. If the current iterate lies outside the η-domain, then the algorithm also requires that the constraint infeasibilities be reduced. If the iterate lies inside the η-domain then the true merit function should also be reduced; if not, then η is reduced. This allows steps that may increase the merit function, but only in a controlled way. The steps that increase the merit function are usually seen only in early iterations or after active set changes. Second, our procedure for updating τ is to compute the predicted relative reduction of the merit function based on the (QP) and compare that with the actual relative reduction. This comparison of predicted and actual reductions is done using the approximate merit function if the current iterate lies outside the η-domain. The true merit function is employed otherwise. We then use standard updating strategies to adjust τ (see e.g. (Dennis and Schnabel, 1983) or (Moré and Sorensen, 1983)). Third, the penalty parameter,

d, is updated in a very straightforward manner. Essentially, an estimate of the condition of the problem is monitored. In the event that this estimate increases significantly, d is decreased. Provided that the initial value of d is reasonable, this updating does not occur often, and is only observed when the iterates are outside the η-domain. The iterates do not 'hug', or stick too closely to the constraint manifold, as is the case when the penalty parameter becomes too small. Computationally, this simple procedure for updating d appears to be effective even in the presence of highly nonlinear constraints.

We have used this procedure to solve several problems in the range of 100–500 variables with up to 500 constraints. Many of these problems have arisen from discretizations of control problems where the Hessian of the Lagrangian and the Jacobian of the constraints have some known sparsity structure. These problems are somewhat special, in that we know that the major expense in the calculation of an iterate comes from the solving of the (QP) . Typically, the constraints are nonlinear inequalities that, in some way, limit the control variables, and the objective function is an energy approximation. The number of constraints is greater than the number of variables in many of the problems we solved. In our testing, we use forward finite-difference approximations to gradients and Hessians, and modify the Hessian of the Lagrangian to be positive semidefinite in cases where it is not (e.g. (Gill *et al.*, 1981)). This latter procedure requires the addition of a non-negative diagonal matrix to the Hessian approximation.

Our observations include the following.

- The number of major iterations is reasonable.
- The globalization procedure remains efficient, i.e. many full steps are accepted.
- Close to the solution, the trust region becomes inactive.
- The trust-region strategy is basically effective, i.e., it prevents long, unprofitable steps from being generated at the beginning and after the active set changes.
- In general a small number of iterations of O3D suffices at each major iteration, and a very small number near the solution.
- As in all of our previous work in this area, the parameter d in the merit function is not critical, i.e., the performance of the algorithm is not changed much by changes in the strategy for adjusting d.

5. Future Research

We have described a preliminary version of an extension of the SQP ideas to the large scale case. In doing so, we have used a combination of an interior-point method for solving (QP) with trust region and truncated Newton methods to create a promising algorithm. There are, however, many computational and theoretical aspects of this algorithm that need further analysis and testing. Computationally, we need to continue to test the procedure to discover its strengths and weaknesses. At the same time, the limitation on the QP solver, O3D, that the Hessian must be positive semidefinite, appears to be surmountable. In particular, we believe that the procedure for solving the reduced quadratic program (step 3 of O3D) can be modified to handle an indefinite (or negative definite) Hessian. This would allow us to avoid the extra work of ensuring that the Hessian is positive definite, and to explore directions of negative curvature.

Our theoretical analysis described in §3 relies on the usual strong assumptions that are typically satisfied in the small scale case. In large problems, some of these assumptions are not as likely to be satisfied. In particular, large problems may be highly degenerate. We know that the interior-point algorithms for LP and QP are not affected by these cases and, although the theory does not apply, our SQP algorithm had no difficulty in solving such problems. Also, in some of the problems, inconsistent quadratic subproblems occurred. This, too, caused no difficulty for the algorithm, but is a problem for the theory. Thus, obtaining good theoretical results under a weakened set of assumptions is an important task for further research.

References

Boggs, P. T., Domich, P. D., Rogers, J. E., Witzgall, C., (1991) "An interior point method for linear and quadratic programming problems." *Mathematical Programming Society COAL Newsletter*, 19:32-40.

Boggs, P. T., Tolle, J. W., (1984) "A family of descent functions for constrained optimization" *SIAM J. Num. Anal.*, 26:1146 - 1161.

Boggs, P. T., Tolle, J. W., (1989) "A strategy for global convergence in a sequential quadratic programming algorithm." *SIAM Journal of Numer. Anal.*, 21:600-623.

Boggs, P. T., Tolle, J. W., Kearsley, A. J., (1991) "A merit function for inequality constrained nonlinear programming problems.", Technical Report 4702, NIST, Gaithersburg, MD, 4702.

Coleman, T. F., (1992) "Large Scale Numerical Optimization: Introduction and overview." In *Encyclopedia of Computer Science and Technology.* Marcel Dekker, Inc., New York (to appear).

Conn, A. R., Gould, N. I. M., Toint, Ph. T., (1992) " Large-scale nonlinear constrained optimization." Technical Report 92-02, Facultés Universitaires de Namur, Département de Mathématique, Namur, Belgium.

Dennis, J. E., Schnabel, R. B., (1983) *Numerical Methods for Unconstrained Optimization and Nonlinear Equations.* Prentice-Hall, Englewood Cliffs, NJ.

Dembo, R. S., Eisenstat, S. C., Steihaug, T., (1982) "Inexact Newton methods." *SIAM J. Num. Anal.*, 19:400-408.

Domich, P. D., Boggs, P. T., Rogers, J. E., Witzgall, C., (1991) "Optimizing over three-dimensional subspaces in an interior point method for linear programming." *Linear Algebra and its Applications*,152

Eisenstat, S. C., Walker, H. F., (1991) "Globally convergent inexact Newton methods." Technical Report, Utah State University, Department of Mathematics, Logan, UT.

Gill, P., Murray, W., and Wright, M., (1981) *Practical Optimization.* Academic Press, New York.

Moré, J. J., Sorensen, D. C., (1983) "Computing a trust region step." *SIAM J. Sci. Stat. Comput.*, 4:553 - 572.

Tapia, R. A., (1980) "On the role of slack variables in quasi-Newton methods for unconstrained optimization." In *Numerical Optimization of Dynamical Systems*, L. C. W. Dixon and G. P. Szegö, eds., North-Holland, Amsterdam: 235-246.

PERFORMANCE OF A MULTIFRONTAL SCHEME FOR PARTIALLY SEPARABLE OPTIMIZATION

A. R. CONN
IBM T.J. Watson Research Center, Yorktown Heights, USA

NICK GOULD
Rutherford Appleton Laboratory, Chilton, GB, EC

M. LESCRENIER
ARBED, Luxemburg, Luxemburg, EC

and

PH. L. TOINT
Department of Mathematics, F.U.N.D.P. Namur, Belgium, EC

Abstract. We consider the solution of partially separable minimization problems subject to simple bounds constraints. At each iteration, a quadratic model is used to approximate the objective function within a trust region. To minimize this model, the iterative method of conjugate gradients has usually been used. The aim of this paper is to compare the performance of a direct method, a multifrontal scheme, with the conjugate gradient method (with and without preconditioning). To assess our conclusions, a set of numerical experiments, including large dimensional problems, is presented.

Key words: partially separable optimization, multifrontal Gaussian elimination.

1. Introduction

This paper is concerned with the solution of partially separable optimization problems (defined in Section 2). Such problems appear in a large majority of nonlinear minimization applications, for example finite-elements, network problems and others. The formalism was first introduced by (Griewank and Toint, 1982) and methods using this particular structure have proved to be extremely successful for large dimensional problems (see (Griewank and Toint, 1984) for instance).

To solve these problems, a trust region type algorithm may be applied, which requires the partial solution of a quadratic minimization problem at each step of an iterative scheme. Up to now, only iterative methods, specifically (preconditioned and truncated) conjugate gradient schemes, have been used in practice to solve the quadratic minimization problem. The aim of this paper is to test the use of direct methods, particularly multifrontal schemes, for this purpose.

The authors are aware that this type of direct method can only be used when the solution of the quadratic problem can be found by solving a linear system, that is, when the quadratic has a finite solution. Such a situation normally arises when the Hessian matrix of the quadratic is positive definite. In the indefinite case, the authors use a simple strategy which consists of computing directions of negative curvature. However, they realize that a more sophisticated strategy like the Levenberg-Marquardt algorithm (see (Moré, 1978)) or an attempt to solve the trust region problem by another method may be more appropriate.

The paper is organized as follows. Section 2 defines the concept of partial separability. Section 3 describes the trust region algorithm used to solve partially separable problems. Iterative and direct methods to solve the quadratic minimization problem are proposed in Sections 4 and 5 respectively. In

S. Gomez and J-P. Hennart (eds.), Advances in Optimization and Numerical Analysis, 79–96.
© 1994 *Kluwer Academic Publishers.*

Section 6 we discuss the numerical experiments and conclusions are drawn in Section 7. We end with an appendix that describes some additional test functions not available in the literature.

2. Partial separability

We consider the simple bound constrained minimization problem

$$\min \ f(x) \tag{1}$$

subject to the bounds

$$a_j \leq x_j \leq b_j \tag{2}$$

where x is a vector of $\mathbf{R}^n$ and f is a so called *partially separable function*, that is a function of the form

$$f(x) = \sum_{i=1}^{m} f_i(x) \tag{3}$$

where the *element functions* $f_i(x)$ have Hessian matrices of low rank compared with n, the dimension of the problem.

A typical case is when each element function only depends on a small subset of the variables called the *elemental variables* of that element. It is also frequently the case that, for some elements, the number of elemental variables can be further reduced by applying, for each one of these elements, a linear transformation of its elemental variables in its *internal variables*. For every element, that is for $i = 1, \ldots, m$, the complete transformation from the original variables of the problem (the vector x) to elemental or (when applicable) to internal variables is then given by

$$y_i = U_i x, \tag{4}$$

where the matrix U_i has fewer than n rows. For instance, given the partially separable function ($n = 3$)

$$f(x) = x_1^2 + (x_1 - x_2)^2 + (x_2 - x_3)^2, \tag{5}$$

we have that

$$f_1(x) = x_1^2, \quad f_2(x) = (x_1 - x_2)^2 \text{ and } f_3(x) = (x_2 - x_3)^2, \tag{6}$$

where the sets of elemental variables corresponding to the elements are given by $\{x_1\}$, $\{x_1, x_2\}$ and $\{x_2, x_3\}$. These sets can be further reduced for elements 2 and 3 by defining two 1-component vectors y_2 and y_3 of internal variables by

$$y_2 = \begin{pmatrix} 1 & -1 & 0 \end{pmatrix} x \text{ and } y_3 = \begin{pmatrix} 0 & 1 & -1 \end{pmatrix} x. \tag{7}$$

For the first element, the elemental and internal variables coincide, and we have that

$$y_1 = \begin{pmatrix} 1 & 0 & 0 \end{pmatrix} x. \tag{8}$$

Equations (7) and (8) in our example correspond to (4), and hence

$$U_1 = \begin{pmatrix} 1 & 0 & 0 \end{pmatrix} \quad U_2 = \begin{pmatrix} 1 & -1 & 0 \end{pmatrix} \text{ and } U_3 = \begin{pmatrix} 0 & 1 & -1 \end{pmatrix}. \tag{9}$$

The adaptability of partially separable methods to large problems comes mainly from a compact storage scheme for the Hessian approximation and the corresponding updating technique. The change of variables (4) allows us to consider new element functions $\hat{f}_i$ such that

$$f_i(x) = \hat{f}_i(y_i). \tag{10}$$

The gradients of f_i and $\hat{f}_i$ satisfy

$$g_i(x) = U_i^T \hat{g}_i(y_i), \tag{11}$$

while the Hessian approximations at x satisfy

$$H_i(x) = U_i^T \hat{H}_i(y_i) U_i. \tag{12}$$

The so called *partitioned updating technique* consists in storing and accessing only the gradients and Hessian approximations in internal variables, that is, the $\hat{g}_i(y_i)$ and $\hat{H}_i(y_i)$ from (11) and (12) respectively. The advantage clearly comes from the fact that the number of internal variables is much smaller than the total dimension of the problem and that the matrix approximating the Hessian of f

$$H(x) = \sum_{i=1}^{m} H_i(x) \tag{13}$$

is never explicitly assembled.

Finally, only the non empty columns of U_i (and pointers to the variables relevant to each column) are needed. For many practical problems, sets of the U_i differ *only* in the elemental variables that they select, see for instance U_2 and U_3 in (9). We would not therefore envisage storing each U_i, merely a set of variable-to-column pointers and require that commonly occuring operations, such as the products

$$u = U_i v \quad \text{and} \quad u = U_i^T v \tag{14}$$

be performed by a user provided subroutine.

3. A trust region algorithm for partially separable problems

The algorithm we propose for solving problem (1)–(2) is of trust region type and belongs to the class of methods described by (Conn *et al.*, 1989).

At each iteration, we suppose that we have a feasible point $x^{(k)}$, the gradient $g^{(k)}$ of the objective function (3) at this point and a suitable symmetric approximation $H^{(k)}$ to the Hessian of f at $x^{(k)}$. In the remainder of this paper, by "Hessian" we will always mean an approximation of the true Hessian. This approximation may be computed by a secant updating formula such as the Broyden-Fletcher-Goldfarb-Shanno or the symmetric rank-one update for instance (see (Dennis and Schnabel, 1983)). The gradient and Hessian of the partially separable function f are stored as described in Section 2. We approximate the objective function by a quadratic model

$$m^{(k)}(x^{(k)} + s) = f(x^{(k)}) + s^T g^{(k)} + \frac{1}{2} s^T H^{(k)} s, \tag{15}$$

in a region surrounding the current iterate defined by its radius $\Delta^{(k)}$. This region, called the trust region, is of the form

$$\|x - x^{(k)}\| \leq \Delta^{(k)}. \tag{16}$$

It is convenient to choose the infinity norm, for then the shape of the trust region is aligned with the simple bounds of the problem (1)–(2). If we define

$$l_i^{(k)} = \max[a_i, x_i^{(k)} - \Delta^{(k)}], \quad u_i^{(k)} = \min[b_i, x_i^{(k)} + \Delta^{(k)}], \tag{17}$$

a trial point $x^{(k)} + s^{(k)}$ is constructed by finding an approximation to the solution of the trust region problem

$$\min\ m^{(k)}(x^{(k)} + s) \tag{18}$$

subject to the bounds

$$l_i^{(k)} \le x_i^{(k)} + s_i \le u_i^{(k)} \tag{19}$$

If the function value calculated at this new point is well approximated by its predicted value (the one given by the model), the point is accepted as the next iterate and the trust region is possibly enlarged; otherwise, the point is rejected and the trust region size decreased. More precisely, we compute the ratio of the achieved to the predicted reduction of the objective function,

$$\rho^{(k)} = \frac{f(x^{(k)}) - f(x^{(k)} + s^{(k)})}{f(x^{(k)}) - m^{(k)}(x^{(k)} + s^{(k)})} \tag{20}$$

and set

$$x^{(k+1)} = \begin{cases} x^{(k)} + s^{(k)} & \text{if } \rho^{(k)} > \mu, \\ x^{(k)} & \text{if } \rho^{(k)} \le \mu, \end{cases} \tag{21}$$

and

$$\Delta^{(k+1)} = \begin{cases} \gamma_0 \Delta^{(k)} & \text{if } \rho^{(k)} \le \mu, \\ \Delta^{(k)} & \text{if } \mu < \rho^{(k)} < \eta, \\ \gamma_2 \Delta^{(k)} & \text{if } \rho^{(k)} \ge \eta, \end{cases} \tag{22}$$

where $\gamma_0 < 1 < \gamma_2$, μ and η are appropriate numbers.

It remains now to describe our approximate solution of (18)–(19). We first compute the Generalized Cauchy Point $x_C^{(k)}$, which is defined as the first local minimizer of the univariate function

$$m^{(k)}(P[x^{(k)} - tg^{(k)}]) \tag{23}$$

with respect to $t \in \mathbf{R}$, where $P[\cdot]$ is the projection operator computed componentwise as

$$(P[x])_j = \begin{cases} l_j^{(k)} & \text{if } x_j \le l_j^{(k)}, \\ u_j^{(k)} & \text{if } x_j \ge u_j^{(k)}, \\ x_j & \text{otherwise.} \end{cases} \tag{24}$$

The Generalized Cauchy Point (GCP) is therefore the first local minimizer of the model, along the piecewise linear arc defined by projecting the steepest descent direction into the feasible domain of problem (1)–(2). We then choose $x^{(k)} + s^{(k)} = x_C^{(k)}$ if the reduced model's gradient at this point is smaller in norm than a fraction of the norm of the reduced model's gradient at $x^{(k)}$, that is if

$$\|Z(x_C^{(k)})^T \nabla m^{(k)}(x_C^{(k)})\| \le \eta^{(k)} \stackrel{\text{def}}{=} \min\left[0.1, \sqrt{\|Z(x^{(k)})^T g^{(k)}\|}\right] \|Z(x^{(k)})^T g^{(k)}\|, \tag{25}$$

where $Z(x)^T$ is the orthogonal projector onto the linear subspace corresponding to variables that are not at their bound at x. If the test (25) fails, we define $I(x_C^{(k)}, l^{(k)}, u^{(k)})$ as the active set of $x_C^{(k)}$ with respect to the bounds $l^{(k)}$ and $u^{(k)}$, that is the set of indices of the variables at the GCP violating or lying on a constraint of (19). We also define $C(x_C^{(k)})$ as the linear subspace of the variables that are free at the GCP, that is

$$C(x_C^{(k)}) = \operatorname{span}\{e_i \mid i \notin I(x_C^{(k)}, l^{(k)}, u^{(k)})\} \tag{26}$$

where e_i is the i-th vector of the canonical basis of $\mathbf{R}^n$. In order to find $x^{(k)} + s^{(k)}$, we then compute a better approximation than the GCP to the solution of sub-problem (18)–(19) with the restriction that $s \in C(x_C^{(k)})$, so that the variables in the active set remain fixed at their value at the GCP. This gives us the trial point and can be calculated using an iterative or a direct method. This is the subject of the next two Sections.

4. Minimization of the quadratic model by iterative methods

If the Hessian $H^{(k)}$, restricted to the subspace $C(x_C^{(k)})$, is positive definite and the bounds of sub-problem (18)–(19) that are inactive at the Cauchy point remain inactive, the solution of (18)–(19) may be obtained as the solution of a system of linear equations. However, this may be prohibitively expensive in the context of large scale optimization unless care is taken to solve the resulting linear system as efficiently as possible. To date, iterative schemes (in particular, truncated, preconditioned conjugate gradient methods) seem to have been the most popular approach for approximately solving the sub-problem, as reflected in (Toint, 1981), (Steihaug, 1983), or (Stoer, 1983).

Two main reasons explain this interest. The first one is that a truncated strategy that asymptotically takes the exact quasi-Newton step can save a significant amount of computation during the early iterations, when we are still far from the optimum. The second reason, and probably the main one, is that these types of methods do not involve operations on the matrix entries but only require matrix-vector products. This calculation, which represents the major cost of the algorithm, can be efficiently performed when the sparsity pattern of the matrix is taken into account. In the context of partially separable optimization, the matrix is the Hessian of the objective function, and the matrix-vector product does not require the assembly of the Hessian since it can be computed as

$$(\sum_{i=1}^{m} H_i)d = \sum_{i=1}^{m}(U_i^T \hat{H}_i U_i)d. \tag{27}$$

The conjugate gradient algorithm is applied, starting from $x = x_C^{(k)}$, to the sub-problem (18)–(19) with the restriction that the variables in the set $I(x_C^{(k)}, l^{(k)}, u^{(k)})$ are kept fixed throughout the process. Our algorithm optionally uses a preconditioner given by the Moore-Penrose pseudo-inverse of the diagonal matrix whose entries are the diagonal entries of the restricted Hessian. The algorithm is terminated at the point $\bar{x}$ if

1. the norm of the restricted gradient of the model, that is $||Z(\bar{x})^T \nabla m^{(k)}(\bar{x})||$, is less than $\eta^{(k)}$, for some $\eta^{(k)}$;
2. one or more of the unrestricted variables violate one of the bounds (19). The point $\bar{x}$ is then the point at which the first offending bound(s) is (are) encountered;
3. a direction of negative curvature is found, in which case $\bar{x}$ is chosen as the last point along this direction that still satisfies the bounds.

We refer the reader to (Conn *et al.*, 1988) for a detailed description of this procedure.

A superlinear rate of convergence can be assured provided that the ratio of the norm of the restricted model's gradient at the final point to that at $x^{(k)}$ tends to zero as the iterates approach a Kuhn-Tucker point for the problem and the matrices $H^{(k)}$ restricted to the set of variables active at the solution satisfy the Dennis-Moré condition (see (Dennis and Schnabel, 1983)). A suitable choice for $\eta^{(k)}$ in order to satisfy the first condition is given by the definition of $\eta^{(k)}$ in (25).

5. Minimization of the quadratic model by a direct method

Let us now consider the solution by a direct method of the sub-problem (18)–(19) with the additional constraint that variables that are at their bound at the GCP remain fixed.

We already remarked that the advantage of using the conjugate gradient method to solve (18)–(19) is that there is no need to assemble the Hessian explicitly. Surprisingly, this advantage can be maintained for a class of direct methods called the frontal methods. These methods solve symmetric positive-definite systems of linear equations

$$Ax = b \tag{28}$$

by Gaussian elimination, where the matrix A has a finite-element representation

$$A = \sum_r B^{(r)}, \tag{29}$$

and where each $B^{(r)}$ is zero except in a small number of rows and columns.

In frontal methods, advantage is taken of the fact that elimination steps

$$a_{ij} := a_{ij} - \frac{a_{i\ell}a_{\ell j}}{a_{\ell\ell}} \tag{30}$$

do not have to wait for all the assembly steps

$$a_{ij} := a_{ij} + b_{ij}^{(r)} \tag{31}$$

from (29) to be complete. The operation (30) can be performed as soon as the pivot row (and column) is fully summed, that is, as soon as all the operations (31) have been completed for them. The fully summed rows and columns (not yet eliminated) are stored in a so-called frontal matrix whose size can be maintained sufficiently small.

Frontal methods are particularly well suited to our framework, because of the obvious similarity between (29) and (13). The class of frontal methods includes several algorithms for varying structure in the matrix A (for instance, band, arrowhead, ...). As we were interested in a general purpose solver, we chose to use the approach introduced by (Irons, 1970), allowing for general sparsity pattern of A. We realize however that using this method for simple sparsity patterns may be overly complicated, and simpler direct methods may be more appropriate. We also note that this technique can also exploit sparsity by using several distinct frontal matrices: we will thus refer to it as the "multifrontal" method.

However, this scheme does not apply to indefinite matrices. To overcome this difficulty, (Duff and Reid, 1983) followed (Bunch and Parlett, 1971), using a mixture of 1×1 and 2×2 pivots chosen during the numerical factorization of A, allowing stable symmetric Gaussian elimination for indefinite systems. The method has been implemented by (Duff and Reid, 1982) as a set of Fortran subroutines, currently available through the Harwell library under the name MA27, and is the one we used for our implementation.

During the assembly steps of the frontal method, one needs to access the entries of each element Hessian $H_i^{(k)}$. Since the element Hessians are stored in internal variables, each time we need one of them we must restore it in elemental variables. The user is then asked to write a code to perform operation (12), in addition to the operations (14). Numerical experience shows that this can often be done efficiently since (12) requires few floating point operations. Furthermore, we maintain the advantages of a reduction of storage for the Hessian $H^{(k)}$ and of an efficient updating technique.

One must point out however, that the current version of MA27 does not assume a finite-element representation for the matrix A but requires instead its storage in compact form. Consequently, for our experimental code, one had to assemble the Hessian, even though this type of method does not require it. This does not affect the viability of the frontal approach, or alter our conclusions. Moreover there are plans to introduce a new version of MA27 that will allow an elemental representation of the coefficient matrix (Iain Duff, private communication).

Given the Bunch-Parlett factorization of the Hessian (restricted to the free variables)

$$Z_C^{(k)T} H^{(k)} Z_C^{(k)} = L^{(k)} D^{(k)} L^{(k)T}, \tag{32}$$

where $Z_C^{(k)T} = Z(x_C^{(k)})^T$ is the orthogonal projector onto $C(x_C^{(k)})$, we check the 1x1 and 2x2 pivots stored in $D^{(k)}$ to decide whether it is positive definite, indefinite, or singular. Different strategies will be applied in these different cases.

If the restricted Hessian is positive definite, we can solve the Newton equations

$$Z_C^{(k)T} H^{(k)} Z_C^{(k)} z = -Z_C^{(k)T} \left[g^{(k)} + H^{(k)}(x_C^{(k)} - x^{(k)}) \right] = -Z_C^{(k)T} \nabla m^{(k)}(x_C^{(k)}) \tag{33}$$

for the direction $z \in C(x_C^{(k)})$, using the MA27 solver.

If $Z_C^{(k)T} H^{(k)} Z_C^{(k)}$ is indefinite, the solution of (33) is no longer that of the sub-problem (18)–(19) restricted to $C(x_C^{(k)})$. Fortunately, the decomposition (32) allows us to compute a direction of negative curvature for this sub-problem.

To compute such a direction, we first chose λ, the most negative eigenvalue of $D^{(k)}$ and computed the corresponding eigenvector $v \in C(x_C^{(k)})$. The direction z would then be given at the cost of the backward substitution

$$z = L^{(k)^{-T}} v \tag{34}$$

and the corresponding curvature would be

$$z^T Z_C^{(k)T} H^{(k)} Z_C^{(k)} z = \lambda \|v\|^2. \tag{35}$$

Numerical experiments indicate a drawback of this method. When the restricted Hessian remains indefinite over a number of successive iterations, the directions z often lie in the same subspace and the number of iterations required to reach optimality is unacceptably high. To avoid this defect, when successive directions of negative curvature are encountered, instead of choosing the most negative eigenvalue of $D^{(k)}$, we cycle through its negative eigenvalues. By cycling we mean that we choose the negative eigenvalue ordered next to the one used at the previous iteration until we reach the last, in which case we repeat the cycle starting with the most negative again.

The last case to consider is when the restricted Hessian is singular and positive semidefinite, although we noticed that it occurs very rarely in practice. The strategy used is the one described by (Conn and Gould, 1984), which consists of solving the linear system (33) if it is consistent, or finding a descent direction z otherwise, satisfying

$$Z_C^{(k)T} H^{(k)} Z_C^{(k)} z = 0 \quad \text{and} \quad z^T Z_C^{(k)T} \nabla m^{(k)}(x_C^{(k)}) < 0. \tag{36}$$

Once z is known, the step $s^{(k)}$ is finally computed as

$$s^{(k)} = x_C^{(k)} - x^{(k)} + \min[1, \alpha^{(k)}][Z_C^{(k)T}]^+ z, \tag{37}$$

where $[Z_C^{(k)T}]^+$ is the Moore-Penrose generalized inverse of $Z_C^{(k)T}$ (that is the operator that completes a vector in $C(x_C^{(k)})$ with zeros to obtain a vector in $\mathbf{R}^n$), and where $\alpha^{(k)}$ is the largest admissible steplength for which the bounds (19) are satisfied at the point $x^{(k)} + s^{(k)}$.

6. Numerical experiments

The test problems we used for our experiments come mainly from the set of functions used by Toint for testing partially separable codes. They are fully described in (Toint, 1983) and the numbering we use here refers to that paper. We considered the problems 10 (Rosenbrock function), 11 (Linear minimum surface), 16 (Boundary value problem), 17 (Broyden tridiagonal), 22 (Diagonal quadratic), 31 (Extended ENGVL1), 33 (Extended Freudenstein) and 36 (Cube problem). For these problems, we used the starting points as defined in (Toint, 1983). We also considered five additional problems, described in the appendix to this paper, so as to allow for other sparsity structures in our test set. These problems are chosen as representative of a larger set used by the authors.

All the experiments were performed on the CRAY 2 supercomputer of Harwell Laboratory. Our code is written in Fortran 77 and compiled using the CFT77 Fortran compiler. All timings reported are CPU seconds.

The initial trust region radius was $\Delta^{(0)} = 0.1\|g^{(0)}\|_2$, and we chose $\mu = 0.25$, $\eta = 0.75$, $\gamma_0 = 1/\sqrt{10}$ and $\gamma_2 = \sqrt{10}$. The stopping criteria we used was based on the order of magnitude of the gradient (projected on the feasible domain); we required its norm to be smaller than 10^{-6}. The algorithm was also stopped if the CPU time exceeded 1200 seconds or if more than 10000 iterations were required.

In the tables, the symbol * indicates that the trust region radius has become too small and that the routine has consequently decided to stop. One must point out, however, that in those cases, the projected gradient norm was of order 10^{-4}, or even 10^{-5}, and the failure should be attributed to numerical rounding errors preventing the accurate calculation of the ratio ρ (equation (20)).

For each test problem, we consider three different methods, conjugate gradients (cg), diagonally preconditioned conjugate gradients (pcg) and multifrontal (multif), to obtain an approximate solution to the sub-problem (18)–(19). We also consider three ways of computing the element Hessians: exact derivatives (exact), the Broyden-Fletcher-Goldfarb-Shanno update (BFGS) and the symmetric rank-one update (rk1). Specifically, in the latter two cases the i-th element Hessian, $H_i^{(k)}$, stored in terms of its internal variables, is updated from the BFGS formula

$$\hat{H}_i^{(k+1)} = \hat{H}_i^{(k)} + \frac{\hat{y}_i^{(k)}\hat{y}_i^{(k)T}}{\hat{y}_i^{(k)T}\hat{s}_i^{(k)}} - \frac{\hat{H}_i^{(k)}\hat{s}_i^{(k)}\hat{s}_i^{(k)T}\hat{H}_i^{(k)}}{\hat{s}_i^{(k)T}\hat{H}_i^{(k)}\hat{s}_i^{(k)}}, \tag{38}$$

or from the rank-one formula

$$\hat{H}_i^{(k+1)} = \hat{H}_i^{(k)} + \frac{\hat{r}_i^{(k)}\hat{r}_i^{(k)T}}{\hat{r}_i^{(k)T}\hat{s}_i^{(k)}}. \tag{39}$$

Here $\hat{s}_i^{(k)}$ and $\hat{y}_i^{(k)}$ are, respectively, the change in the internal variables $U_i(x^{(k+1)} - x^{(k)})$, and the change in the elemental gradient $\hat{g}_i^{(k+1)} - \hat{g}_i^{(k)}$, and $\hat{r}_i^{(k)}$ is defined as $\hat{y}_i^{(k)} - \hat{H}_i^{(k)}\hat{s}_i^{(k)}$. The BFGS update is only performed for a given element if the new approximation can be ensured to be positive definite, and this is implemented by only allowing an update if the condition

$$\|\hat{y}_i^{(k)}\|^2 \leq 10^8 \hat{y}_i^{(k)T}\hat{s}_i^{(k)}$$

is satisfied. The rank-one update is only made when the correction has norm smaller than 10^8, i.e. when

$$\|\hat{r}_i^{(k)}\|^2 \leq 10^8 |\hat{r}_i^{(k)T}\hat{s}_i^{(k)}|.$$

The initial estimate of each element Hessian $\hat{H}_i^{(0)}$ is set to the identity matrix when updating schemes are used. This choice is considered satisfactory as the test problems are reasonably well scaled.

The figures we report in each Table are the number of function evaluations (f calls), the number of gradient evaluations (g calls), and the overall CPU-time (total cpu). Under the label "linear system stats" we give the number of conjugate gradient iterations in the case of the iterative method or information of the type *pd nc sc (ratio)* for the direct methods. Here *pd* is the number of positive definite linear systems solved, *nc* is the number of directions of negative curvature taken, *sc* is the number of singular (but consistent) linear systems solved and *ratio* gives an idea of the fill-in during the Gaussian elimination and is equal to the storage space needed to store the factors of the Gaussian decomposition divided by the space needed for the original matrix.

Table 1 presents the performance of the different methods on a set of 13 problems. Each problem was specified with 100 variables.

pb	Hessian	method	f calls	g calls	linear system stats	total cpu
10	exact	cg	441	289	2064	5.74
		pcg	492	259	2133	6.27
		multif	527	272	512 11 0 (1.00)	12.11
	BFGS	cg	617	380	2883	9.44
		pcg	603	368	2016	7.65
		multif	348	279	342 0 0 (1.00)	8.70
	rk1	cg	1047	617	3037	12.18
		pcg	1089	642	1755	11.27
		multif	1747	1021	562 1054 102 (1.00)	44.62
11	exact	cg	35	29	80	0.62
		pcg	38	31	71	0.85
		multif	15	14	15 0 0 (1.89)	0.44
	BFGS	cg	17	18	154	0.65
		pcg	21	19	135	0.80
		multif	17	18	17 0 0 (1.89)	0.56
	rk1	cg	65	35	178	1.35
		pcg	53	36	85	1.32
		multif	96	61	18 78 0 (1.89)	2.96
16	exact	cg	3	4	254	0.94
		pcg	4	5	377	1.45
		multif	3	4	3 0 0 (1.00)	0.11
	BFGS	cg	20	20	2604	9.80
		pcg	17	17	2189	8.40
		multif	21	21	20 0 0 (1.00)	0.78
	rk1	cg	14	11	836	3.21
		pcg	8	7	329	1.38
		multif	6	6	4 1 0 (1.00)	0.20
17	exact	cg	7	8	29	0.17
		pcg	6	7	28	0.21
		multif	5	6	5 0 0 (1.00)	0.17
	BFGS	cg	11	9	30	0.24
		pcg	11	9	32	0.32
		multif	11	9	9 0 0 (1.00)	0.35
	rk1	cg	15	9	40	0.32
		pcg	11	9	35	0.33
		multif	26	9	8 16 0 (1.00)	0.83

pb	Hessian	method	f calls	g calls	linear system stats	total cpu
22	exact	cg	6	7	8	0.05
		pcg	2	3	1	0.02
		multif	3	4	1 0 0 (1.00)	0.04
	BFGS	cg	22	15	76	0.37
		pcg	33	18	77	0.49
		multif	25	16	19 0 0 (1.00)	0.69
	rk1	cg	10	8	10	0.11
		pcg	10	8	4	0.10
		multif	6	4	1 0 0 (1.00)	0.08
31	exact	cg	8	9	21	0.07
		pcg	8	9	13	0.06
		multif	7	8	5 0 0 (1.00)	0.13
	BFGS	cg	13	11	31	0.15
		pcg	13	11	25	0.15
		multif	11	9	7 0 0 (1.00)	0.22
	rk1	cg	13	10	27	0.14
		pcg	25	10	25	0.20
		multif	15	10	6 5 0 (1.00)	0.31
33	exact	cg	11	12	25	0.10
		pcg	7	8	18	0.07
		multif	10	11	4 0 0 (1.00)	0.13
	BFGS	cg	19	15	26	0.19
		pcg	17	13	23	0.18
		multif	17	13	6 0 0 (1.00)	0.24
	rk1	cg	17	13	28	0.17
		pcg	18	13	27	0.18
		multif	37	14	7 20 0 (1.00)	0.73
36	exact	cg	1029	618	7827	17.81
		pcg	1023	594	4030	11.74
		multif	944	562	932 7 0 (1.00)	22.42
	BFGS	cg	1515	963	9143	25.48
		pcg	1410	928	4449	17.00
		multif	1189	812	1176 1 0 (1.00)	29.75
	rk1	cg	3152	1824	9037	35.79
		pcg	3087	1804	4616	28.52
		multif	>10000	–	–	–

pb	Hessian	method	f calls	g calls	linear system stats	total cpu
55	exact	cg	5	6	4	0.03
		pcg	5	6	4	0.04
		multif	5	6	2 0 0 (1.00)	0.08
	BFGS	cg	12	9	2	0.09
		pcg	13	10	9	0.12
		multif	13	10	3 0 0 (1.00)	0.17
	rk1	cg	22	10	14	0.15
		pcg	15	10	11	0.13
		multif	22	10	12 0 0 (1.00)	0.20
56	exact	cg	12	13	14	0.16
		pcg	12	13	0	0.22
		multif	12	13	11 0 0 (1.00)	0.39
	BFGS	cg	19	20	430	1.80
		pcg	19	20	452	2.06
		multif	19	20	13 0 0 (1.00)	0.59
	rk1	cg	39	25	17	0.69
		pcg	32	22	26	0.70
		multif	33	22	1 22 0 (1.01)	0.98
57	exact	cg	107	68	1380	5.83
		pcg	94	57	741	3.94
		multif	15	16	11 0 0 (1.00)	0.47
	BFGS	cg	75	53	942	4.56
		pcg	332	230	741	9.34
		multif	24	22	16 0 0 (1.00)	0.82
	rk1	cg	81	58	1382	6.28
		pcg	238	168	605	6.99
		multif	24	22	16 0 0 (1.00)	0.83
59	exact	cg	8	7	16	0.12
		pcg	10	9	18	0.16
		multif	12	9	4 5 0 (2.68)	0.45
	BFGS	cg	13	12	21	0.29
		pcg	13	12	14	0.28
		multif	13	12	12 0 0 (2.68)	0.70
	rk1	cg	18	8	31	0.33
		pcg	13	10	7	0.24
		multif	21	10	6 11 0 (2.75)	0.94

pb	Hessian	method	f calls	g calls	linear system stats	total cpu
61	exact	cg	13	14	56	0.35
		pcg	11	12	25	0.21
		multif	11	12	10 0 0 (1.00)	0.54
	BFGS	cg	33	25	136	1.05
		pcg	26	20	66	0.70
		multif	26	20	18 0 0 (1.00)	1.19
	rk1	cg	45	22	131	1.23
		pcg	56	25	45	1.08
		multif	116	69	20 88 0 (1.00)	6.18

Table 1 : Performance of the methods on test problems with 100 variables.

This table indicates the viability of the multifrontal method in the context of partially separable optimization and more than that, we already see that the method seems to be really competitive with the iterative schemes in some cases. It is not rare that the number of function and gradient evaluations is significantly reduced and this can lead to significant improvements in terms of computation time.

Two special points in these results are worth more comment.

1. We first observe that, on problem 22 with the multifrontal solver, BFGS requires the solution of 19 linear systems while rk1 only requires 1! This behaviour is explained by the conjunction of the quadratic termination properties of the rk1 update and the particular structure of problem 22. This structure is such that all element Hessians are constant diagonal 3×3 matrices. Because rk1 needs at most p steps to obtain an exact $p \times p$ constant Hessian, the exact Hessian is obtained after 3 steps in problem 22. These first 3 steps did not require the complete solution of (33) because the test (25) was satisfied. A fourth iteration and a single linear system solution are then all that is needed to minimize the quadratic objective exactly. Since the BFGS update does not enjoy similar quadratic termination properties, more iterations are required to form a good approximation of the Hessian and to converge.

2. We also note that, for problem 36 using BFGS, the multifrontal method finds one indefinite system. This is very surprising, as the BFGS update ensures positive definiteness of the Hessian approximations. This last property, although true in exact arithmetic, can however be violated due to rounding errors in the case where the matrix to update has a condition number of the order of the inverse of machine precision. This is what happens in problem 36: the conditioning of some element Hessian matrices gradually builds up and finally exceeds 10^{17}!

   When this situation occurs, it seems inadvisable to keep on updating an indefinite matrix with the BFGS update. We therefore decided to reinitialize the element Hessian approximations to the identity matrix, in effect restarting the algorithm. We are well aware that this technique is merely a "quick fix", and that a more sophisticate procedure is desirable. Finding such a procedure might however be difficult, because one may wish to detect the bad conditioning of the element Hessians before negative curvature is actually produced, while maintaining an unfactored element Hessian representation, as needed in the partitioned updating framework.

In order to investigate more carefully the relative performance of the methods, we increased the dimension n up to a maximum of 5000. The results of those experiments are given in Tables 2 to 7 for a subset of our test problems. The problems were chosen as being fairly representative of the larger set.

pb	Hessian	method	f calls	g calls	linear system stats	total cpu
961	exact	cg	507	453	860	84.45
		pcg	179	130	327	42.39
		multif	26	20	26 0 0 (4.17)	12.53
	BFGS	cg	52	38	565	25.30
		pcg	117	80	550	41.01
		multif	26	21	26 0 0 (4.17)	13.37
	rk1	cg	613	379	771	156.38
		pcg	450	280	349	104.35
		multif	–	–	–	>1200
4900	exact	cg	–	–	–	>1200
		pcg	574	435	848	713.70
		multif	36	29	36 0 0 (6.43)	108.26
	BFGS	cg	133	97	1422	352.69
		pcg	–	–	–	>1200
		multif	32	25	32 0 0 (6.43)	104.66
	rk1	cg	–	–	–	>1200
		pcg	–	–	–	>1200
		multif	–	–	–	>1200

Table 2 : Performance of the methods on the test problem 11 for increasing dimensions.

pb	Hessian	method	f calls	g calls	linear system stats	total cpu
1000	exact	cg	2	3	256	9.18
		pcg	2	3	612	22.31
		multif	5	6	5 0 0 (1.00)	1.75
	BFGS	cg	15	15	10770	1938.77
		pcg	–	–	–	>1200
		multif	37	32	35 1 0 (1.00)	14.53
	rk1	cg	12	8	823	31.97
		pcg	7	6	3085	117.50
		multif	11	11	3 7 0 (1.00)	4.07
5000	exact	cg	2	3	1151	204.75
		pcg	2	3	1830	332.50
		multif	7	8	7 0 0 (1.00)	12.19
	BFGS	cg	–	–	–	>1200
		pcg	–	–	–	>1200
		multif	14	14	14 0 0 (1.01)	27.68
	rk1	cg	14	9	184	43.07
		pcg	18	10	124	39.07
		multif	58	34	11 46 1 (1.01)	105.62

Table 3 : Performance of the methods on the test problem 16 for increasing dimensions.

pb	Hessian	method	f calls	g calls	linear system stats	total cpu
1000	exact	cg	13	14	0	1.23
		pcg	13	14	0	2.33
		multif	13	14	0 0 0 (–)	1.23
	BFGS	cg	21	22	1453	54.41
		pcg	21	22	2628	99.46
		multif	21	22	12 0 0 (1.00)	5.97
	rk1	cg	36	24	17	6.38
		pcg	35	26	30	7.81
		multif	42	27	0 30 0 (1.00)	21.10
5000	exact	cg	14	15	0	6.62
		pcg	14	15	0	12.49
		multif	14	15	0 0 0 (–)	6.56
	BFGS	cg	22	23	2170	396.23
		pcg	–	–	–	>1200
		multif	22	23	11 0 0 (1.00)	28.81
	rk1	cg	36	25	13	29.79
		pcg	38	28	35	42.60
		multif	–	–	–	>1200

Table 4 : Performance of the methods on the test problem 56 for increasing dimensions.

pb	Hessian	method	f calls	g calls	linear system stats	total cpu
1000	exact	cg	143	93	1964	81.67
		pcg	206	128	955	64.91
		multif	17	18	10 0 0 (1.00)	9.45
	BFGS	cg	210	147	2177	102.33
		pcg	560	410	860	134.44
		multif	19	17	15 0 0 (1.00)	14.28
	rk1	cg	218	152	2103	100.51
		pcg	673	476	911	152.97
		multif	19	17	15 0 0 (1.00)	14.18
5000	exact	cg	146	94	1774	382.91
		pcg	154	99	735	248.88
		multif	18	19	10 0 0 (1.00)	158.42
	BFGS	cg	289	205	3235	751.54
		pcg	–	–	–	>1200
		multif	20	18	16 0 0 (1.00)	259.90
	rk1	cg	222	158	2329	559.15
		pcg	–	–	–	>1200
		multif	20	18	16 0 0 (1.00)	252.50

Table 5 : Performance of the methods on the test problem 57 for increasing dimensions.

pb	Hessian	method	f calls	g calls	linear system stats	total cpu
1000	exact	cg	10	7	23	1.59
		pcg	14	11	31	2.31
		multif	40	24	6 31 0 (12.13)	52.28
	BFGS	cg	14	13	44	3.51
		pcg	12	11	10	2.33
		multif	13	12	11 0 0 (12.13)	15.97
	rk1	cg	24	17	12	5.02
		pcg	16	13	6	2.82
		multif	36	17	2 24 0 (12.04)	36.00
5000	exact	cg	19	12	44	15.62
		pcg	10	8	16	7.10
		multif	16	10	4 10 0 (52.39)	488.78
	BFGS	cg	15	14	49	19.93
		pcg	13	12	15	14.05
		multif	15	14	14 0 0 (51.98)	447.67
	rk1	cg	38	19	17	51.98
		pcg	19	13	6	16.35
		multif	40	21	2 29 0 (52.34)	1070.84

Table 6 : Performance of the methods on the test problem 59 for increasing dimensions.

pb	Hessian	method	f calls	g calls	linear system stats	total cpu
1000	exact	cg	14	15	64	4.10
		pcg	11	12	24	2.32
		multif	12	13	10 0 0 (1.00)	10.81
	BFGS	cg	36	25	151	12.19
		pcg	29	22	61	* 7.80
		multif	28	22	18 0 0 (1.00)	22.37
	rk1	cg	57	27	175	15.50
		pcg	46	25	44	* 9.42
		multif	–	–	–	>1200
5000	exact	cg	14	15	56	* 23.58
		pcg	11	12	23	16.03
		multif	12	13	10 0 0 (1.00)	171.92
	BFGS	cg	41	29	119	* 62.12
		pcg	28	21	67	42.90
		multif	30	22	20 0 0 (1.00)	388.19
	rk1	cg	40	24	93	53.55
		pcg	75	34	100	82.58
		multif	–	–	–	>1200

Table 7 : Performance of the methods on the test problem 61 for increasing dimensions.

When exact derivatives are available, the multifrontal method seems competitive with respect to the conjugate gradient type algorithms in many cases. It appears to be also the case when the Broyden-Fletcher-Goldfarb-Shanno update is applied. If the symmetric rank-one update is used however, conjugate gradient methods are preferable.

We observe the excellent performances of the multifrontal method when the quadratic model is convex (see Table 5 for instance). However, if the fill-in of the factorization is too high, the number of operations for the Gaussian elimination is dominant and the multifrontal scheme is no longer competitive. A good example of this is shown in Table 6 where the sparsity pattern of the Hessian is randomly generated.

We observe particularly poor performances of the direct method when directions of negative curvature are taken. This happens, obviously, more often with the symmetric rank-one update of the Hessian. Illustration of this behaviour can be found in Tables 3 and 6. The proposed strategy seems to be clearly inefficient and other strategies must definitely be found to handle such cases, if direct methods are to be used.

We note that the negative curvature directions generated by the rank-one update are always taken into account when using our multifrontal approach, in contrast with the (preconditioned) conjugate gradient technique. This last calculation only considers the first few Krylov subspaces spanned by the gradient and its successive images by the Hessian approximation, especially in the early iterations when this approximation may be quite poor and the trust region radius small. The comparison of the performance of the rank-one update with the (preconditioned) conjugate gradient and multifrontal schemes, in cases where negative curvature is encountered, tends to show that this possible neglect of negative curvature in the early stages of the computation might be advantageous.

We also wanted our algorithm to converge to a local minimum from any starting point. We therefore ran several additional tests using the same objective functions as before, but with different starting points. These experiments did not affect the conclusions above.

Finally, in (Conn *et al.*, 1988), the simplest of the updating schemes, the symmetric rank-one method, appeared to perform better than the BFGS method for many of the problems tested. In the context of partial separability, this conclusion does not appear to apply. One could attempt to explain this phenomenon by the fact that it is not uncommon for the projection of successive search directions into the range space of certain elements to be close to linear dependence, despite the independence of the overall search directions. Linear dependence of the search directions can be highly undesirable for the rank-one update.

7. Conclusions

The numerical experiments show that the use of a direct method instead of an iterative one can sometimes lead to very significant improvements in terms of computation time. They also clearly demonstrate that the improvements can be achieved only in cases where the quadratic model of the function at the current iterate is convex (or at least not too often non-convex) and the structure of the Hessian is sufficiently regular to avoid high fill-in during the factorization.

When the approximation of the function Hessian is indefinite, the use of directions of negative curvature inhibits fast convergence of direct methods and can even lead to a dramatic increase in computation time. This conclusion convinced the authors that in this particular case, other strategies must definitely be used.

The main conclusion is that an efficient code for partially separable optimization must provide a choice of methods more adapted to the specific problem being solved. Such strategies are incorporated in the LANCELOT package (see (Conn *et al.*, 1992)).

Acknowledgements

The project has been financially supported by the Belgian National Fund for Scientific Research, Harwell Laboratory (U.K.A.E.A.) and by grant A8639 from the Natural Sciences and Engineering Research Council of Canada. This support is gratefully acknowledged.

Appendix

The five problems that we added to introduce other types of sparsity structure are now given. For each of them we mention (a) the element functions, (b) any bounds on the variables and (c) the starting point.

Test problem 55

(a) $f_i(x) = (x_i^2 + x_n^2)^2 - 4x_i + 3, \quad (i = 1, ..., n-1).$

(c) $x = (1, 1, \ldots, 1).$

Test problem 56

(a) $f_i(x) = (x_i + x_{i+1})e^{-x_{i+2}(x_i + x_{i+1})}, \quad (i = 1, \ldots, n-2).$

(b) $x_i \geq 0, \quad (i = 1, \ldots, n)$

(c) $x = (1, 1, \ldots, 1)$

Test problem 57

(a) $f_i(x) = (x_i + x_{i+1} + x_n)^4, \quad (i = 1, \ldots, n-2)$

$f_{n-1}(x) = (x_1 - x_2)^2,$

$f_n(x) = (x_{n-1} - x_n)^2.$

(c) $x = (1, -1, 1, -1, \ldots).$

Test problem 59

(a) $f_i(x) = x_i^2 e^{-x_{j_i}} \quad (i = 1, \ldots, n),$

$f_i(x) = x_{i-n}^2 e^{-x_{j_i}} \quad (i = n+1, \ldots, 2n),$

where the indices j_i are randomly generated between 1 and n in order by subroutine FA04BS of the Harwell Subroutine Library, starting with the default seed, but with the provision that any j_i equal to i is rejected and the next random number in the sequence taken.

(c) $x = (1, -1, 1, -1, \ldots).$

Test problem 61

(a) $f_i(x) = (x_i^2 + 2x_{i+1}^2 + 3x_{i+2}^2 + 4x_{i+3}^2 + 5x_n^2)^2 - 4x_i + 3, \quad (i = 1, \ldots, n-4).$

(c) $x = (1, 1, \ldots, 1).$

References

J.R. Bunch and B.N. Parlett. Direct methods for solving symmetric indefinite systems of linear equations. *SIAM Journal on Numerical Analysis*, 8:639–655., 1971.

A. R. Conn and N. Gould. On the location of directions of infinite descent for nonlinear programming algorithms. *SIAM Journal on Numerical Analysis*, 21(6):302–325, 1984.

A. R. Conn, N. I. M. Gould, and Ph. L. Toint. Global convergence of a class of trust region algorithms for optimization with simple bounds. *SIAM Journal on Numerical Analysis*, 25:433–460, 1988. See also same journal 26:64–767, 1989.

A. R. Conn, N. I. M. Gould, and Ph. L. Toint. Testing a class of methods for solving minimization problems with simple bounds on the variables. *Mathematics of Computation*, 50:399–430, 1988.

A. R. Conn, N. I. M. Gould, and Ph. L. Toint. LANCELOT*: a Fortran package for large-scale nonlinear optimization (Release A)*. Springer Verlag, Heidelberg, Berlin, New York,1992.

J. E. Dennis and R. B. Schnabel. *Numerical methods for unconstrained optimization and nonlinear equations.* Prentice-Hall, Englewood Cliffs, USA, 1983.

I. S. Duff and J. K. Reid. MA27: A set of Fortran subroutines for solving sparse symmetric sets of linear equations. Report R-10533, AERE Harwell Laboratory, Harwell, UK, 1982.

I. S. Duff and J. K. Reid. The multifrontal solution of indefinite sparse symmetric linear equations. *ACM Transactions on Mathematical Software*, 9(3):302–325, 1983.

A. Griewank and Ph. L. Toint. Partitioned variable metric updates for large structured optimization problems. *Numerische Mathematik*, 39:429–448, 1982.

A. Griewank and Ph. L. Toint. Numerical experiments with partially separable optimization problems. In D. F. Griffiths, editor, *Numerical Analysis: Proceedings Dundee 1983*, pages 203–220, Berlin, 1984. Springer Verlag. Lecture Notes in Mathematics 1066.

B.M. Irons. A frontal solution program for finite-element analysis. *Int. J. Numer. Meth. Eng.*, 2:5–32, 1970.

J. J. Moré. The Levenberg-Marquardt algorithm: implementation and theory. In G. A. Watson, editor, *Proceedings Dundee 1977*, Berlin, 1978. Springer Verlag. Lecture Notes in Mathematics.

T. Steihaug. The conjugate gradient method and trust regions in large scale optimization. *SIAM Journal on Numerical Analysis*, 20(3):626–637, 1983.

J. Stoer. Solution of large linear systems of equations by conjugate gradient type methods. In A. Bachem, M. Grötschel, and B. Korte, editors, *Mathematical Programming: The State of the Art*, pages 540–565, Berlin, 1983. Springer Verlag.

Ph. L. Toint. Towards an efficient sparsity exploiting Newton method for minimization. In I. S. Duff, editor, *Sparse Matrices and Their Uses*, London, 1981. Academic Press.

Ph. L. Toint. Test problems for partially separable optimization and results for the routine pspmin. Technical Report 83/4, Department of Mathematics, FUNDP, Namur, Belgium, 1983.

TOWARDS SECOND-ORDER METHODS FOR STRUCTURED NONSMOOTH OPTIMIZATION

MICHAEL L. OVERTON and XIANJIAN YE
Computer Science Department
Courant Institute of Mathematical Sciences
New York University

Abstract. Structured nonsmooth optimization objectives often arise in a composite form $f = h \circ a$, where h is convex (but not necessarily polyhedral) and a is smooth. We consider the case where the structure of the nonsmooth convex function h is known. Specifically, we assume that, for any given point in the domain of h, a parameterization of a manifold Ω, on which h reduces locally to a smooth function, is given. We discuss two linear spaces: the tangent space to the manifold Ω at a point, and the subspace parallel to the affine hull of the subdifferential of h at the same point, and explain that these are typically orthogonal complements. We indicate how the construction of locally second-order methods is possible, even when h is nonpolyhedral, provided the appropriate Lagrangian, modeling the structure, is used. We illustrate our ideas with two important convex functions: the ordinary max function, and the max eigenvalue function for symmetric matrices, and we solicit other interesting examples with genuinely different structure from the community.

Key words: nonsmooth optimization, convex composite optimization, second derivatives

A minimization objective which often arises in practice is the composite function $f = h \circ a$ where

$$h: \quad \Re^n \to \Re \qquad \text{is convex}$$

and

$$a: \quad \Re^m \to \Re^n \qquad \text{is } C^1$$

(continuously differentiable). Thus

$$f: \quad \Re^m \to \Re$$

with

$$f(x) = h(a(x)).$$

In practice, the convex function h often has a rather special structure. The simplest example is

$$h_1(a) = \text{ the maximum element of the vector } a.$$

In this case, h_1 is said to be *polyhedral,* since the epigraph of h_1 (the set of points $(a, \eta) \in \Re^{n+1}$ satisfying $\eta \geq h_1(a)$) is a polyhedron.

S. Gomez and J-P. Hennart (eds.), Advances in Optimization and Numerical Analysis, 97–109.
© 1994 *Kluwer Academic Publishers.*

Some more interesting examples are obtained by considering convex functions whose argument, while conveniently denoted as a vector $a \in \Re^n$, is really a matrix associated with that vector. If $n = N(N+1)/2$ for some integer N, define

$$A = \mathbf{Sym}\ a$$

to be the N by N symmetric matrix defined by copying the elements of a consecutively into its upper triangle, multiplying the off-diagonal elements by the factor $1/\sqrt{2}$ for convenience, and let

$$a = \mathbf{vecsym}\ A$$

denote the inverse operation, defining the vector a in terms of the elements of A. Let

$$h_2(a) = \text{ the maximum eigenvalue of the symmetric matrix } \mathbf{Sym}\ a.$$

By Rayleigh's variational principle, h_2 is convex, but it is not polyhedral.

First-order optimality conditions for a convex function h are conveniently described in terms of its *subdifferential* [17], defined by

$$\partial h(\widehat{a}) = \{d \in \Re^n : \ h(a) \geq h(\widehat{a}) + \langle a - \widehat{a}, d \rangle, \ \ \forall a \in \Re^n\}.$$

A necessary condition for $\widehat{a}$ to minimize h is then

$$0 \in \partial h(\widehat{a}).$$

The subdifferential $\partial h(\widehat{a})$ is compact and reduces to a single point if and only if h is differentiable at $\widehat{a}$, in which case $\partial h(\widehat{a}) = \nabla h(\widehat{a})$. We shall need to refer to the *affine hull* of the subdifferential,

$$\mathbf{aff}\ \partial h(\widehat{a}),$$

the affine space of smallest dimension containing $\partial h(\widehat{a})$. We have $\mathbf{aff}\ \partial h(\widehat{a}) = \partial h(\widehat{a})$ if and only if h is differentiable at $\widehat{a}$.

Duality principles which give a concise representation of the subdifferential of h are known for the functions h_1 and h_2. In the first case, let $\widehat{a}$ be a given point and assume without loss of generality that

$$\widehat{a}_1 = \cdots = \widehat{a}_t > \widehat{a}_{t+1} \geq \cdots \geq \widehat{a}_n, \tag{1}$$

for some integer t. Then, as is well known [8],

$$\partial h_1(\widehat{a}) = \{\begin{bmatrix} u \\ 0 \end{bmatrix} : \ \ u \in \Re^t, \ \ e^T u = 1, \ \ u \geq 0\}, \tag{2}$$

where $e = [1 \ldots 1]^T$. Consequently

$$\mathbf{aff}\ \partial h_1(\widehat{a}) = \{\begin{bmatrix} u \\ 0 \end{bmatrix} : \ \ u \in \Re^t, \ \ e^T u = 1\}. \tag{3}$$

In the case of h_2, suppose without loss of generality that the eigenvalues of $\widehat{A} = \mathbf{Sym}\ \widehat{a}$ are

$$\widehat{\lambda}_1 = \cdots = \widehat{\lambda}_t > \widehat{\lambda}_{t+1} > \cdots > \widehat{\lambda}_N \tag{4}$$

with $\widehat{Q}$ a matrix whose columns are a corresponding orthonormal set of eigenvectors, i.e.

$$\widehat{A} = \widehat{Q}\widehat{\Lambda}\widehat{Q}^T, \tag{5}$$

where $\widehat{\Lambda} = \mathrm{Diag}(\widehat{\lambda}_1, \ldots, \widehat{\lambda}_N)$. It follows from a simple argument given in [12] that

$$\partial h_2(\widehat{a}) = \{\mathbf{vecsym}\ \widehat{Q}U\widehat{Q}^T : \ U \in S\Re^{N\times N}, \ U = \begin{bmatrix} U_{11} & 0 \\ 0 & 0 \end{bmatrix},$$

$$U_{11} \in S\Re^{t\times t}, \ \mathbf{tr}\, U_{11} = 1, \ U_{11} \geq 0\}. \tag{6}$$

Here $S\Re^{N\times N}$ denotes the set of real symmetric matrices, **tr** denotes trace and $U_{11} \geq 0$ means that U_{11} is positive semi-definite. It follows that

$$\mathbf{aff}\ \partial h_2(\widehat{a}) = \{\mathbf{vecsym}\ \widehat{Q}U\widehat{Q}^T : \ U \in S\Re^{N\times N}, \ U = \begin{bmatrix} U_{11} & 0 \\ 0 & 0 \end{bmatrix},$$

$$U_{11} \in S\Re^{t\times t}, \ \mathbf{tr}\, U_{11} = 1\}. \tag{7}$$

The subdifferential gives all the relevant *first-order* variational information about h, but does not, of course, give second-order information. In the case of h_1, there are no second-order effects to consider, since h_1 is polyhedral. But in the case of h_2, we would like to use second-order information as well; the subdifferential does *not* contain this information.

For the composite function $f = h \circ a$, it is well known that a generalized subdifferential, more often known as a generalized gradient, can be obtained by means of a chain rule [6, 8]. Composing the subdifferential of the convex function h with the gradient of the smooth function a, we have

$$\partial f(x) = \partial h(a) \circ \nabla a(x)$$

$$= \{v \in \Re^m : \ v_j = \langle d, \frac{\partial a}{\partial x_j}(x)\rangle, j = 1, \ldots, m, \text{ for some } d \in \partial h(a)\}.$$

A necessary condition for optimality is then

$$0 \in \partial f(x).$$

Let $\widehat{x}$ be a given point, defining $\widehat{a} = a(\widehat{x})$, and consider $f_1 = h_1 \circ a$, $f_2 = h_2 \circ a$. We have

$$\partial f_1(\widehat{x}) = \{v : \ v_k = \sum_{i=1}^{t} u_i \frac{\partial a_i}{\partial x_k}, \ u \in \Re^t, \ e^T u = 1, \ u \geq 0\}$$

and

$$\partial f_2(\widehat{x}) = \{v : \ v_k = \mathbf{tr}\ U\widehat{Q}^T(\mathbf{Sym}\ \frac{\partial a}{\partial x_k})\widehat{Q}, \ U = \begin{bmatrix} U_{11} & 0 \\ 0 & 0 \end{bmatrix},$$

$$U_{11} \in S\Re^{t\times t}, \ \mathbf{tr}\, U_{11} = 1, \ U_{11} \geq 0\}.$$

In the case where h is polyhedral, it is straightforward to further obtain the second-order information in f from the composition of the subdifferential of h with the Hessian of a. But this is *not* possible in the case of f_2, since the second-order information in h_2 is not contained in the subdifferential.

There has been a great deal of recent work concerning *second-order analytical tools* for *general* convex functions. An excellent overview is given by Burke[3]; other references include [4, 5, 9, 10, 16, 18, 20]. We take a very different approach here, assuming specifically that a lot is known about the *structure* of the convex function. Work on *numerical methods* for convex composite optimization has, on the other hand, been mostly limited to the case where the convex function h is *polyhedral*: see [8, 23, 24, 26]. These methods do not generalize to the nonpolyhedral case in a satisfactory way since the subproblems which need to be solved are not tractable in general. However, see Burke [2] for a more general treatment, including an interesting historical account of convex composite optimization. Another general treatment is given by [25].

Our *key assumption* is that, given any point $\widehat{a}$, the *local structure* of the convex function h is known. By this we mean that a manifold $\Omega(\widehat{a})$, containing the point $\widehat{a}$ and contained in $\Re^n$, on which h reduces to a smooth function near $\widehat{a}$, is known. More specifically, suppose that $\Omega(\widehat{a})$ is the solution set of a known equation

$$\Phi(a,b)=0, \tag{8}$$

projected into $\Re^n$. Here the *structure function*

$$\Phi: \quad \Re^n \times \Re^r \rightarrow \Re^s \quad \text{ is } C^1,$$

with $\Phi(\widehat{a},\widehat{b})=0$ for some $\widehat{b}\in\Re^r$, and with, for some positive ϵ,

$$\Phi(a,b)=0 \text{ and } \|a-\widehat{a}\|\le\epsilon \quad \Rightarrow \quad h(a)=g(b)$$

where

$$g: \quad \Re^r \rightarrow \Re \quad \text{ is } C^1.$$

We have introduced the additional variables $b\in\Re^r$ in order to facilitate the description of Ω by the structure function Φ. Our use of the structure function Φ is partly inspired by the notion of *structure functionals* for polyhedral convex functions introduced by Osborne[11] and may be viewed as a generalization of this concept to the nonpolyhedral case.

Before applying these ideas to h_1 and h_2, let $n=2$ and consider the simple convex function

$$h_0(a)=\frac{1}{2}a_1^2+|a_2|.$$

To avoid the trivial case, assume $\widehat{a}_2=0$. The subdifferential is then easily seen to be

$$\{a: \quad a_1=\widehat{a}_1; \quad a_2\in[-1,1]\},$$

so

$$\textbf{aff }\partial h_0(\widehat{a})=\{a: \quad a_1=\widehat{a}_1\}.$$

Define the manifold $\Omega(\widehat{a})$ to be the axis $\{a : \ a_2 = 0\}$. Note particularly that $\Omega(\widehat{a})$ and the subspace parallel to **aff** $\partial h_0(\widehat{a})$ are *orthogonal complements.* To parameterize Ω in the structure function notation, let $r = 1$, and write $b = \beta$ to emphasize that $b \in \Re$. Let $s = 2$,

$$g(b) = \frac{1}{2}\beta^2 \quad \text{and} \quad \Phi(a, \beta) = \begin{bmatrix} a_1 - \beta \\ a_2 \end{bmatrix}.$$

Clearly, the conditions on the structure function are satisfied. Note that, in this case, $n + r - s$, the number of variables in a and b reduced by the number of equations in Φ, equals 1, the dimension of Ω.

Now consider h_1, the ordinary max function, with t defined in (1). Define the manifold $\Omega(\widehat{a})$ by constraining $a_1 = \cdots = a_t$, giving a linear space with dimension $n + 1 - t$ (codimension $t - 1$). As was the case with h_0, we see from (3) that $\Omega(\widehat{a})$ and the subspace parallel to **aff** $\partial h_1(\widehat{a})$ are orthogonal complements. Let $r = 1$, $b = \beta \in \Re$, $s = t$,

$$g(\beta) = \beta \quad \text{and} \quad \Phi(a, \beta) = \begin{bmatrix} a_1 - \beta \\ \vdots \\ a_t - \beta \end{bmatrix}.$$

Letting $\widehat{\beta} = h_1(\widehat{a})$, we have $\Phi(\widehat{a}, \widehat{\beta}) = 0$. The equation $\Phi(a, \beta) = 0$ implies that $a_1 = \cdots = a_t = \beta$, and hence, if a is close enough to $\widehat{a}$, the max element equals β. Thus, the conditions on the structure function are satisfied. We have $n + r - s = n + 1 - t$, the dimension of $\Omega(\widehat{a})$.

Now consider $h_2(a)$, the maximum eigenvalue of $A = \textbf{Sym}\ a$. We have $n = N(N+1)/2$, where N is the dimension of A. Consider a given point $\widehat{a}$, with $\widehat{A} = \textbf{Sym}\ \widehat{a}$, $\widehat{a} = \textbf{vecsym}\ \widehat{A}$, and t equal to the multiplicity of the largest eigenvalue of $\widehat{A}$ (see (4)). Define the manifold $\Omega(\widehat{a})$ by constraining the t largest eigenvalues of $\textbf{Sym}\ a$ to be equal. This manifold is *nonlinear*, with two components: one linear, and one nonlinear. The linear component corresponds exactly to the manifold Ω for the ordinary max function h_1, consisting of $\textbf{vecsym}\ \widehat{Q}D\widehat{Q}^T$, where D is diagonal with the first t entries equal to an unspecified common value, and $\widehat{Q}$ is defined by (5). This component has dimension $N - t + 1$. The nonlinear component is the set of vectors corresponding to the *orbit* of $\widehat{A}$, which is the set of symmetric matrices similar to $\widehat{A}$, i.e. having the same eigenvalues. Let $\mathbf{T}\Omega(\widehat{a})$ be the linear space which is *tangent* to $\Omega(\widehat{a})$ at $\widehat{a}$. (By convention, the tangent space is shifted to include the origin and so does not generally contain $\widehat{a}$.) The tangent space $\mathbf{T}\Omega(\widehat{a})$ has two components: the linear component of $\Omega(\widehat{a})$, and the tangent space to the orbit of $\widehat{A}$. It can be shown, using elementary techniques from differential geometry described in [1, Sec. 2] or [7, Prop. 2.1.1], that the orthogonal complement of the tangent space to the orbit at $\widehat{A}$ is precisely the space of matrices which commute with $\widehat{A}$. Using (4)–(5), this latter subspace is clearly the set of matrices of the form

$$\widehat{Q}\begin{bmatrix} U_{11} & 0 \\ 0 & U_{22} \end{bmatrix}\widehat{Q}^T,$$

where U_{11} is any symmetric t by t matrix and U_{22} is any diagonal matrix of order $N - t$. This set has dimension $t(t+1)/2 + N - t$. The orthogonal complement of the tangent space $\mathbf{T}\Omega(\widehat{a})$ is a smaller space, since its elements must also be orthogonal to the linear component of $\Omega(\widehat{a})$. This condition imposes the constraints that $U_{22} = 0$ and $\mathbf{tr}\ U_{11} = 0$, resulting in a linear space of dimension $t(t+1)/2 - 1$. Applying the **vecsym** operator to this space gives the unique subspace of $\Re^n$ parallel to **aff** $\partial h_2(\widehat{a})$, defined in (7). We therefore see that the *tangent space* $\mathbf{T}\Omega(\widehat{a})$ *and the subspace parallel to* **aff** $\partial h_2(\widehat{a})$ *are orthogonal complements.* It follows that $\Omega(\widehat{a})$ has codimension $t(t+1)/2 - 1$ (dimension $n + 1 - t(t+1)/2$)).

The best way to parameterize $\Omega(\widehat{a})$ using a structure function is not obvious, but one convenient way is as follows. Let

$$b = (\beta, \mathbf{vecskew}\ B, \mathbf{vecdiag}\ D),$$

where $\beta \in \Re$, B is a *skew-symmetric* matrix of order N, i.e. $B = -B^T$, with **vecskew** B the corresponding vector in $\Re^{N(N-1)/2}$, and D is a *diagonal* matrix of order $N - t$, with **vecdiag** D the corresponding vector in $\Re^{N-t}$. Thus, $b \in \Re^r$, where $r = N(N+1)/2 - t + 1$. Define $g(b) = \beta$ and

$$\Phi(a, b) = \mathbf{vecsym}\left(e^{-B}\widehat{Q}^T(\mathbf{Sym}\ a)\widehat{Q}e^B - \begin{bmatrix} \beta I & 0 \\ 0 & D \end{bmatrix}\right),$$

where $\widehat{Q}$ is defined in (4) and I denotes the identity matrix of order t. We have $s = N(N+1)/2$. Since B is skew-symmetric, e^B is orthogonal. Clearly, Φ is smooth, and letting

$$\widehat{b} = (h_2(\widehat{a}), \mathbf{vecskew}\ 0, (\widehat{\lambda}_{t+1}, \ldots, \widehat{\lambda}_N)),$$

we have $\Phi(\widehat{a}, \widehat{b}) = 0$. Furthermore, $\Phi(a, b) = 0$ implies that $A = \mathbf{Sym}\ a$ is similar to a diagonal matrix whose first t elements are β. Since the eigenvalues of a matrix are a continuous function of the matrix elements, this implies that the maximum eigenvalue of A equals β if a is sufficiently close to $\widehat{a}$.

This choice of structure function needs modification, however, since $\widehat{b}$ is not a unique solution of $\Phi(\widehat{a}, b) = 0$. The general solution is

$$\widehat{b} = (h_2(\widehat{a}), \mathbf{vecskew}\ B, (\widehat{\lambda}_{t+1}, \ldots, \widehat{\lambda}_N)),$$

where

$$B = \begin{bmatrix} B_{11} & 0 \\ 0 & 0 \end{bmatrix}$$

and the leading t by t block B_{11} is any skew-symmetric matrix. Without loss of generality, therefore, we can constrain B to have leading t by t block equal to zero. This reduces r by $t(t-1)/2$, so that $n + r - s = n + 1 - t(t+1)/2$, the dimension of $\Omega(\widehat{a})$.

An alternative approach to parameterizing $\Omega(\widehat{a})$ for the max eigenvalue function has recently been proposed by [19].

It is no surprise to find that $\mathbf{T}\Omega(\widehat{a})$ and the subspace parallel to **aff** $\partial h(\widehat{a})$ are orthogonal complements, since h is smooth when restricted to Ω and, on the other hand, the subdifferential describes exactly directions in which h is nonsmooth. Theorem 13.1 of [17] is highly relevant in this regard. More worthy of note, perhaps, is that for the interesting function h_2, beautifully simple arguments from convex analysis show that $\partial h_2(\widehat{a})$ has the form (6), while equally beautiful and simple arguments from differential geometry show that the tangent space $\mathbf{T}\Omega(\widehat{a})$ is orthogonal to the affine hull of (6). We must, however, be cautious in making general statements. In particular, nothing has yet been said about the uniqueness of Ω. Clearly, one should define Ω to have dimension as large as possible, but even when this is done, the choice of Ω is generally not unique. For example, in the case of $h_0(a) = \frac{1}{2}a_1^2 + |a_2|$, for $\widehat{a} = 0$, we chose $\Omega(\widehat{a})$ to be the axis defined by $a_2 = 0$, since this reflects the global structure of h, but from a local point of view we could just as well have chosen the parabola $a_2 = a_1^2$, which is tangent to the axis at the origin. However, this alternative choice of Ω is less convenient when it comes to defining the structure function Φ and the associated objective function g. Also, observe that even though $\Omega(\widehat{a})$ is not unique, $\mathbf{T}\Omega(\widehat{a})$ is unique. Another interesting example is $h(a) = \max(a_1^2 + a_2^2, |a_1|)$. In this case, the global structure of h is understood by considering the two one-dimensional manifolds M_1, M_2 which form the solution set of $a_1^2 + a_2^2 = |a_1|$, and which are tangent to each other at the origin. For any point $\widehat{a} \in M_1$, the manifold $\Omega(\widehat{a})$ can be taken to be M_1 itself or any manifold tangent to M_1 at $\widehat{a}$. This is true even in the case $\widehat{a} = 0$. In all these cases, $\mathbf{T}\Omega(\widehat{a})$ is unique and is the orthogonal complement of the subspace parallel to **aff** $\partial h(\widehat{a})$. The choice $\Omega = M_1$ (or M_2) is by far the most convenient from the point of view of defining Φ and g.

One must be concerned about exceptional cases where the orthogonal complement property does not hold, as illustrated by the example[1]

$$h(a) = \max(2a_1^2 + a_2^2, a_1^2 + 2a_2^2).$$

At all points satisfying $\widehat{a}_1 = \pm\widehat{a}_2$, except $\widehat{a} = 0$, the manifold $\Omega(\widehat{a})$ on which h is locally smooth can be taken to be one of the lines $a_1 = \pm a_2$. The subdifferential $\partial h(\widehat{a})$ has dimension one at any of these points, and the orthogonal complement property holds. At the point where the two lines cross, namely $\widehat{a} = 0$, $\partial h(\widehat{a})$ reduces to the single point 0. For the orthogonal complement property to hold at this point, therefore, we would need $\Omega(0) = \Re^2$, but this is not possible. On the contrary, the maximal-dimension choice for $\Omega(0)$ is *any* one-dimensional manifold passing through 0 (hence $\mathbf{T}\Omega(0)$ is not unique). The difficulty here is that h is differentiable at the point 0, but not in any neighborhood of 0. A related example is

$$h(a) = \max(|a_1|, |a_2|),$$

since, for all $\widehat{a}_1 = \pm\widehat{a}_2$, except $\widehat{a} = 0$, $\Omega(\widehat{a})$ can again be taken to be one of the lines $a_1 = \pm a_2$. However, in this case $\Omega(0) = 0$ and **aff** $\partial h(0) = \Re^2$, so the orthogonal complement property holds everywhere.

When all these considerations are taken properly into account, it seems that it should be possible to establish a rather general result on the orthogonal complementarity of $\mathbf{T}\Omega(\widehat{a})$ and the subspace parallel to **aff** $\partial h(\widehat{a})$.

We now explain the purpose of the structure function Φ. Instead of attempting to minimize the nonsmooth function h directly, we may consider minimizing the smooth function g subject to the condition that its argument lie on the manifold Ω,

[1] This example was suggested by J.V. Burke.

where h agrees with g. Since Ω is parameterized by (8), let us introduce a Lagrangian associated with g and Φ, namely

$$L(a,b,u) = g(b) + \langle u, \Phi(a,b)\rangle.$$

Because the definition of the structure function Φ depends on $\widehat{a}$, so does the definition of the Lagrangian. Let $\nabla_a L$ denote the gradient of L with respect to a, etc., with $\nabla_a \Phi$, for example, being the matrix with i,j element equal to $\partial\Phi_j/\partial a_i$, evaluated at (a,b). Then

$$\nabla_a L(a,b,u) = \nabla_a \Phi\ u,$$

$$\nabla_b L(a,b,u) = \nabla g + \nabla_b \Phi\ u,$$

and, of course,

$$\nabla_u L(a,b,u) = \Phi.$$

Define the set

$$\Psi(a) = \{d:\ \ \exists b,\ u \text{ s.t. } \nabla_a L(a,b,u) = d;\ \ \nabla_b L(a,b,u) = 0;\ \ \Phi(a,b) = 0\}.$$

Since the definition of L depends on $\widehat{a}$, so does the definition of Ψ. In the case of the ordinary max function h_1, we have $b = \beta$,

$$\nabla_b L(a,b,u) = \frac{\partial L(a,b,u)}{\partial \beta} = 1 - e^T u$$

and

$$\nabla_a L(a,b,u) = \begin{bmatrix} I \\ 0 \end{bmatrix} u = \begin{bmatrix} u \\ 0 \end{bmatrix}$$

so we see from (3) that, if a is near enough to $\widehat{a}$, the set $\Psi(a)$ is *precisely* **aff** $\partial h_1(a)$.

Now consider the max eigenvalue function h_2. It is straightforward to show that

$$\nabla_a L(\widehat{a},\widehat{b},u) = \textbf{vecsym}\ \widehat{Q} U \widehat{Q}^T,$$

where $U = \textbf{Sym}\ u$. (This equation does not hold in a neighborhood of $(\widehat{a},\widehat{b})$, but only at the point.) The first component of $\nabla_b L(a,b,u)$ is

$$\frac{\partial L(a,b,u)}{\partial \beta} = 1 - \textbf{tr}\ U_{11},$$

where U_{11} is the leading t by t block of U. Furthermore, it can be shown [15, Thm 5.2] that setting the other components of $\nabla_b L(\widehat{a},\widehat{b},u)$ to zero gives the condition that all elements of U outside the U_{11} block must be zero. Therefore, we see from (6) that $\Psi(\widehat{a})$ is precisely **aff** $\partial h_2(\widehat{a})$.

We conjecture that this result, $\Psi(\widehat{a}) = $ **aff** $\partial h(\widehat{a})$, can be stated and proved in a fairly general setting.

Now recall that our objective is to minimize the composite function $f = h \circ a$. Suppose that a point $\widehat{x}$ is given: this defines $\widehat{a} = a(\widehat{x})$, which in turn defines a composite structure function Φ and the associated smooth function g and point $\widehat{b}$. Given a regularity condition on the function $a(x)$, we can expect to find the same orthogonal decomposition of the variable space into the tangent space to the manifold on which the composite function is smooth and the affine hull of the subdifferential (generalized gradient) of the composite function. Consider the equality-constrained nonlinear program

$$\min_{x \in \Re^m,\ b \in \Re^r} \quad g(b) \tag{9}$$

$$\text{s.t.} \quad \Phi(a(x), b) = 0. \tag{10}$$

The corresponding Lagrangian is

$$\widetilde{L}(x, b, u) = g(b) + \langle u, \Phi(a(x), b) \rangle.$$

We have $\nabla_b \widetilde{L} = \nabla_b L$ and

$$\nabla_x \widetilde{L}(x, b, u) = \nabla_x a \nabla_a L(a(x), b, u) = \nabla_x a \nabla_a \Phi(a(x), b)\ u.$$

Let

$$\widetilde{\Psi}(x) = \{v : \ \exists b,\ u \ \text{s.t.} \ \nabla_x \widetilde{L}(x, b, u) = v; \ \ \nabla_b \widetilde{L}(x, b, u) = 0; \ \ \Phi(a(x), b) = 0\}.$$

Then if $\Psi(\widehat{a}) = \textbf{aff}\ \partial h(\widehat{a})$, we also have $\widetilde{\Psi}(\widehat{x}) = \textbf{aff}\ \partial f(\widehat{x})$. Given a regularity condition on Φ, a necessary condition for $(\widehat{x}, \widehat{b}, u)$ to locally solve the nonlinear program (9)–(10) is that $\nabla_x \widetilde{L}(\widehat{x}, \widehat{b}, u) = 0$, $\nabla_b \widetilde{L}(\widehat{x}, \widehat{b}, u) = 0$, $\Phi(a(\widehat{x}), \widehat{b}) = 0$. If $\widetilde{\Psi}(\widehat{x}) = \textbf{aff}\ \partial f(\widehat{x})$, this means that $0 \in \textbf{aff}\ \partial f(\widehat{x})$. We therefore have the standard necessary condition for $\widehat{x}$ to minimize the composite function $f = h \circ a$, *with the exception of the inequality conditions on the dual variables defining* $\partial f(\widehat{x})$. [2] The presence or absence of these inequality conditions precisely define the difference between ∂f and **aff** ∂f.

These ideas suggest that the *Newton step for the nonlinear program* (9)–(10) is the main ingredient needed to construct a local second-order minimization algorithm for $h \circ a$. The immediate difficulty with this is that the choice of appropriate structure function Φ depends on $\widehat{x}$, the unknown desired solution. Let us postpone discussion of this difficulty for the moment. Also, we wish to ensure that not only is $0 \in \textbf{aff}\ \partial f(\widehat{x})$, but also that $0 \in \partial f(\widehat{x})$. The best way to ensure this seems to be to compute approximate dual variable information at every step, check the necessary inequality condition on these quantities ($u \geq 0$ in the case of h_1, $U_{11} \succeq 0$ in the case of h_2), and, if this condition does *not* hold, conclude that the current manifold (defined by $\widehat{x}$) is *not* optimal, and take a step *away* from the manifold instead of *towards* it. It may not be obvious how to do this: in the case of h_2, see [12, Sec. 3].

Suppose that convergence does take place towards a point $\widehat{x}$ satisfying $0 \in \partial f(\widehat{x})$. In order to guarantee that the convergence rate is second-order, we need the manifold

[2] It is instructive to compare these formulas and conclusions with the results for the general polyhedral case given by Osborne [11, pp.191-193].

Ω and the associated structure function to be more than just C^1, but rather C^2 with a Lipschitz condition on the second derivative. (If we insist that Ω be C^2 in its original definition, we cannot claim that $\mathbf{T}\Omega(\widehat{a})$ and **aff** ∂h are, in general, orthogonal complements. A counterexample is obtained by considering a function which is C^1 but not C^2 in a certain direction.)

A Newton step for (9)–(10) linearizes Φ to make first-order approximations to the constraint manifold, and uses a quadratic objective based on the Hessian of $\widetilde{L}(x, b, u)$. Considering again the orthogonal decomposition of the variable space into the tangent space to the constraint manifold and the affine hull of the generalized gradient, we see that *first-order* approximation is used in the *latter* direction while *second-order* information is needed in the *former* direction. Given regularity conditions on the functions $a(x)$ and $\Phi(a, b)$, together with positive definiteness of the Hessian of the Lagrangian function restricted to the tangent space, quadratic convergence of such a Newton method can be expected.

Now let us address the fact that the definition of the structure function itself depends on the unknown point $\widehat{x}$. In the case of the function h_1, this simply means that an "active set" of indices must be identified. This is a standard approach in nonlinear and minmax programming and is quite acceptable, at least for nondegenerate problems, since it only requires estimating which elements of $a(x)$ are coalescing to the same maximum value. In the case of the function h_2, the multiplicity t of the maximum eigenvalue must be identified: this is also acceptable for the same reason. But more is needed in the case of h_2: the definition of the structure function also requires $\widehat{Q}$, which is assumed to be a set of eigenvectors corresponding to an *exactly* multiple eigenvalue. In practice, however, the eigenvalue will be multiple only at the limit point of the iteration. The best we can do is to define Φ in terms of the eigenvectors of the current matrix iterate. This amounts to using *a different structure function for each Newton step*, with the structure functions themselves converging to the optimal structure function. It turns out that, in the case of h_2, *quadratic convergence is still obtained.* However, the convergence analysis is substantially complicated by this difficulty: see [15] for details.

We conclude by soliciting examples of other interesting convex functions. We are aware of many examples related to those discussed here, such as other polyhedral functions [8, 11], the maximum singular value of a nonsymmetric matrix (see Appendix), other norm functions [22], diagonal scaling problems [13, 21], sums of eigenvalues [14], etc. The ideas discussed in this paper are apparently applicable to all these functions. We would be very interested to hear of other structured convex functions that arise in applications and have a *genuinely different structure* from those discussed here.

Acknowledgements

Helpful conversations with Jim Burke, Jean-Pierre Haeberly, Mike Osborne and Rob Womersley are gratefully acknowledged. Both authors were supported in part by National Science Foundation Grant CCR-9101649.

References

1. V.I. Arnold. On matrices depending on parameters. *Russian Mathematical Surveys*, 26:29–43, 1971.
2. J.V. Burke. Descent methods for composite nondifferentiable optimization problems. *Mathematical Programming*, 33:260–279, 1985.
3. J.V. Burke. Second order necessary and sufficient conditions for convex composite NDO. *Mathematical Programming*, 38:287–302, 1987.
4. J.V. Burke and R. A. Poliquin. Optimality conditions for non-finite valued convex composite functions, 1992.
5. R.W. Chaney. On second derivatives for nonsmooth functions. *Nonlinear Analysis: Theory, Methods and Applications*, 9:1189–1209, 1985.
6. F. H. Clarke. *Optimization and Nonsmooth Analysis*. John Wiley, New York, 1983.
7. T.F. Fairgrieve. The application of singularity theory to the computation of the Jordan canonical form. Master's thesis, University of Toronto, 1986.
8. R. Fletcher. *Practical Methods of Optimization*. John Wiley, Chichester and New York, second edition, 1987.
9. J.B. Hirriart-Urruty. A new set-valued second order derivative for convex functions. In J.B. Hirriart-Urruty, editor, *Fermat Days 85: Mathematics for Optimization*, pages 157–182, Amsterdam, 1986. North-Holland.
10. A.D. Ioffe. On some recent developments in the theory of second order optimality conditions. In S.Dolecki, editor, *Optimization*, Lecture Notes in Math. 1405, pages 55–68. Springer-Verlag, 1989.
11. M.R. Osborne. *Finite Algorithms in Optimization and Data Analysis*. John Wiley, Chichester and New York, 1985.
12. M.L. Overton. Large-scale optimization of eigenvalues. *SIAM Journal on Optimization*, 2:88–120, 1992.
13. M.L. Overton and R.S. Sezginer. The largest singular value of $e^X A_0 e^{-X}$ is convex on convex sets of commuting matrices. *IEEE Transactions on Automatic Control*, 35:229–230, 1990.
14. M.L. Overton and R.S. Womersley. Optimality conditions and duality theory for minimizing sums of the largest eigenvalues of symmetric matrices. *Mathematical Programming*, 1993. To appear.
15. M.L. Overton and R.S. Womersley. Second derivatives for optimizing eigenvalues of symmetric matrices. Computer Science Dept. Report 626, Courant Institute of Mathematical Sciences, NYU, 1993. Submitted to *SIAM J. Matrix Anal. Appl.*
16. R. Poliquin and R.T. Rockafellar. A calculus of epi-derivatives applicable to optimization. *Canadian Journal of Mathematics*. To appear.
17. R.T. Rockafellar. *Convex Analysis*. Princeton University Press, Princeton, N.J., 1970.
18. R.T. Rockafellar. Second-order optimality conditions in nonlinear programming obtained by way of epiderivatives. *Mathematics of Operations Research*, 14:462–484, 1989.
19. A. Shapiro and M.K.H. Fan. On eigenvalue optimization, 1993. Manuscript.
20. D. Ward. Calculus for parabolic second-order derivatives. Submitted to *Canadian J. Math.*
21. G.A. Watson. An algorithm for optimal l_2 scaling of matrices. *IMA Journal on Numerical Analysis*, 1991. To appear.
22. G.A. Watson. Characterization of the subdifferential of some matrix norms. *Linear Algebra and its Applications*, 170:33, 1992.
23. R.S. Womersley. Local properties of algorithms for minimizing nonsmooth composite functions. *Mathematical Programming*, 32:69–89, 1985.
24. E. Yamakawa, M. Fukushima, and T. Ibaraki. An efficient trust region algorithm for minimizing nondifferentiable composite functions. *SIAM Journal on Scientific and Statistical Computing*, 10:562–580, 1989.
25. X.-J. Ye. On the local properties of algorithms for minimizing nonsmooth composite functions, 1991. Manuscript, Nanjing University.
26. Y. Yuan. On the superlinear convergence of a trust region algorithm for nonsmooth optimization. *Mathematical Programming*, 31:269–285, 1985.

Appendix

Suppose that $n = N^2$, define

$$A = \textbf{Mat } a$$

to be the N by N matrix defined by the elements of a, and let

$$a = \textbf{vecmat } A.$$

Let

$$h_3(a) = \text{ the maximum singular value of the matrix } \textbf{Mat } a.$$

Let $\widehat{a}$ be given, with $\widehat{A} = \textbf{Mat } \widehat{a}$. Suppose the singular values of $\widehat{A}$ are

$$\widehat{\sigma}_1 = \cdots = \widehat{\sigma}_t > \widehat{\sigma}_{t+1} \geq \cdots \geq \widehat{\sigma}_N,$$

with $\widehat{P}$ and $\widehat{Q}$ orthogonal matrices whose columns are respectively the left and right singular vectors of $\widehat{A}$, i.e.

$$\widehat{A}\widehat{Q} = \widehat{P}\widehat{\Sigma}$$

where $\widehat{\Sigma} = \text{Diag}(\widehat{\sigma}_i)$. The following result is already known [22], but the simple derivation may be of interest.

Theorem.

$$\partial h_3(\widehat{a}) = \{\textbf{vecsym } \widehat{Q}U\widehat{P}^T : \ U \in S\Re^{N\times N}, \ U = \begin{bmatrix} U_{11} & 0 \\ 0 & 0 \end{bmatrix},$$

$$U_{11} \in S\Re^{t\times t}, \ \textbf{tr } U_{11} = 1, \ U_{11} \geq 0\}.$$

Proof: A standard variational result is

$$\begin{aligned} h_3(\widehat{a}) &= \max \ \{y^T\widehat{A}z : \ y \in \Re^N, \ z \in \Re^N, \ \|y\| = \|z\| = 1\} \\ &= \max \ \{\textbf{tr } \widehat{A}U \ : \ U = zy^T, \ y \in \Re^N, \ z \in \Re^N, \ \|y\| = \|z\| = 1\}. \end{aligned}$$

This maximum is achieved by matrices of the form

$$\{U = \widehat{Q}ww^T\widehat{P}^T : \ w = [u \ \ 0]; \ u \in \Re^t, \ \|u\| = 1\}, \tag{11}$$

since then

$$\textbf{tr } \widehat{A}U = w^T\widehat{\Sigma}w = \widehat{\sigma}_1.$$

Notice the appearance of the *symmetric* rank-one matrix ww^T. It follows from standard results in convex analysis [17] that $\partial h_3(a)$ is the *convex hull* of (11). The proof is completed by noting that the convex hull of

$$\{uu^T : \ u \in \Re^t, \ \|u\| = 1\}$$

is [12, Lemma 1]

$$\{ U_{11} \in S\Re^{t\times t}, \ \textbf{tr } U_{11} = 1, \ U_{11} \geq 0\}.$$

Alternatively, the theorem may be proved less directly by applying [12, Thm 2] to the maximum eigenvalue of

$$\begin{bmatrix} 0 & A \\ A^T & 0 \end{bmatrix}$$

noting that the negative copies of the singular values are of no interest.

It is interesting to note that [12, Thm 1] does *not* seem to generalize nicely to the singular value case, because there does not seem to be any simple formula for the convex hull of

$$\{zy^T : \ y, \ z \in \Re^N, \ \|y\| = \|z\| = 1\}.$$

Constructing a structure function for h_3 is similar to the process for h_2. Let

$$b = (\beta, \textbf{vecskew } B, \textbf{vecskew } C, \textbf{vecdiag } D)$$

where B and C are skew-symmetric matrices of order N, and D is a diagonal matrix of order $N - t$. Define $g(b) = \beta$ and

$$\Phi(a, b) = \textbf{vecsym } (e^{-B}\widehat{P}^T(\textbf{Sym } a)\widehat{Q}e^C - \begin{bmatrix} \beta I & 0 \\ 0 & D \end{bmatrix}).$$

As in the case of h_2, it is necessary to introduce further restrictions on b in order to obtain the right dimension count and a regularity condition on Φ. In this case, the appropriate restriction seems to be that the leading t by t blocks of B and C should be equal.

HOMOTOPY METHODS IN CONTROL SYSTEM DESIGN AND ANALYSIS

LAYNE T. WATSON
Department of Computer Science
Virginia Polytechnic Institute & State University
Blacksburg, VA 24061-0106 USA

STEPHEN RICHTER
Harris Corporation, MS 22-4848
Government Aerospace Systems Division
Melbourne, FL 32902 USA

and

DRAGAN ŽIGIĆ
Department of Computer Science
Virginia Polytechnic Institute & State University
Blacksburg, VA 24061-0106 USA

Abstract. Recent technologies have led to stringent control system requirements. This has increased the importance and complexity of the analysis and design of control systems, which often require the solution of systems of nonlinear equations of high order. Some challenging computational problems in control design include model order reduction, high dimensional Riccati equations, fixed-structure optimization, robust analysis and feedback synthesis, sensor/actuator placement, and simultaneous controller/structure design. This paper describes these problems, and the directions in which globally convergent homotopy methods must be extented in order to be applicable to computational problems in control. By way of illustration, a computationally effective probability-one homotopy algorithm is presented for the optimal projection formulation of the reduced order model problem, together with some numerical results.

Key words: control system design, controller design, fixed-structure optimization, homotopy, globally convergent

1. Introduction

Over the past several decades considerable effort has been devoted to developing homotopic continuation methods for numerically solving nonlinear algebraic equations. References [1]–[13] provide a representative, although by no means exhaustive, historical overview of this development. Homotopy algorithms are now available for widespread usage due to the recent completion

S. Gomez and J-P. Hennart (eds.), Advances in Optimization and Numerical Analysis, 111–143.
© 1994 *Kluwer Academic Publishers.*

of a self-contained software package known as HOMPACK ([14]). Homotopy methods have been applied to problems in a variety of engineering disciplines, including solid geometric modeling [15], finite element analysis [16]–[20] and control theory [21]–[30].

In practice, homotopy methods are of both computational and theoretical interest. Intuitively, the idea behind these methods is to replace a difficult computational problem by an easier problem and follow a continuous path connecting the solutions of the two problems. We briefly describe such methods in Section 2. Computationally, homotopy methods are globally convergent for a broad class of problems even in the presence of high dimensionality and small domain of attraction. Furthermore, homotopy methods can provide theoretical knowledge of the number of solutions and properties of the solution space.

The primary objective of this paper is to discuss the potential applicability of homotopy algorithms for the computational solution of a variety of problems in control system design and analysis. Control design is an active area of research and has proven historically to be a rich source of computationally challenging problems. In this regard, we describe a broad spectrum of computational control problems. Roughly speaking, these problems fall into one of three classes, namely, well researched problems, partially researched problems, and highly challenging problems.

By well researched problems we mean those problems which have undergone considerable computational development and for which there exist highly developed software packages. A representative example in this class is the numerical solution of algebraic Riccati equations, discussed in Section 3.2. The principal goal of new research in this regard is to investigate improvements for problems which remain difficult because of high dimensionality.

By partially researched problems we mean those problems which have undergone considerable theoretical development but have received only ad hoc computational attention. These problems are often approached numerically using Newton's method, gradient search, or methods devised specifically for the problem at hand. However, because of nonlinearity and nonconvexity, these methods encounter difficulties with convergence and multiple solutions. Such difficulties are also exacerbated as dimensionality increases. In Sections 3.3 and 3.4 we focus on two examples within this class of problems, namely, fixed-structure optimization and robust analysis and feedback synthesis.

By highly challenging problems we mean those problems in control design which have been discussed conceptually but, due to their high complexity, have only recently received serious computational attention. Hence new research explores the utility of homotopy methods in rendering such problems tractable. In this class we mention two such problems in Section 3.5, specifically, sensor and actuator placement and simultaneous controller/structure design.

Within each class of problems the goal is to exploit the structure of the problem so as to efficiently adapt homotopy algorithms. To illustrate the necessity of careful utilization of homotopy algorithms, consider the straightforward application of a homotopy algorithm to the n-dimensional symmetric Riccati equation. Viewing the Riccati equation as $n(n+1)/2$ equations in an equal number of unknowns leads to as many as 2^n continuation paths. However, exploiting our interest in positive definite solutions leads to the consideration of only a single path. In Section 3.2 we describe a novel homotopy based Riccati solver with this feature. Section 3.1 discuses the H_2 optimal reduced order model problem, for which a homotopy is described in detail in Section 4. Section 5 gives some numerical results.

2. Probability-one Homotopy Algorithms

Homotopy methods are globally convergent, which distinguishes them from most iterative methods, which are only locally convergent. The general idea of homotopy methods is to make a continuous transformation from an initial problem, which can be solved trivially, to the target problem.

Following [95], the theoretical foundation of all probability-one globally convergent homotopy methods is given in the differential geometry theorem below, which requires the following concept:

DEFINITION. Let $U \subset \mathbb{R}^m$ and $V \subset \mathbb{R}^p$ be open sets, and let $\rho : U \times [0,1) \times V \to \mathbb{R}^p$ be a C^2 map. ρ is said to be transversal to zero if the Jacobian matrix $D\rho$ has full rank on $\rho^{-1}(0)$.

THEOREM 1. *If $\rho(a, \lambda, x)$ is transversal to zero, then for almost all $a \in U$ the map*

$$\rho_a(\lambda, x) = \rho(a, \lambda, x)$$

is also transversal to zero; i.e, with probability one the Jacobian matrix $D\rho_a(\lambda, x)$ has full rank on $\rho_a^{-1}(0)$.

The recipe for constructing a globally convergent homotopy algorithm to solve the nonlinear system of equations

$$f(x) = 0,$$

where $f : \mathbb{R}^p \to \mathbb{R}^p$ is a C^2 map, is as follows: For an open set $U \subset \mathbb{R}^m$ construct a C^2 homotopy map $\rho : U \times [0,1) \times \mathbb{R}^p \to \mathbb{R}^p$ such that

1) $\rho(a, \lambda, x)$ is transversal to zero,
2) $\rho_a(0, x) = \rho(a, 0, x) = 0$ is trivial to solve and has a unique solution $x_0 = x_0(a)$ for each $a \in U$,
3) $\rho_a(1, x) = f(x)$,
4) $\rho_a^{-1}(0)$ is bounded.

Then for almost all $a \in U$ there exists a zero curve γ of ρ_a, along which the Jacobian matrix $D\rho_a$ has rank p, emanating from $(0, x_0)$ and reaching a zero $\bar{x}$ of f at $\lambda = 1$. This zero curve γ does not intersect itself, is disjoint from any other zeros of ρ_a, and has finite arc length in every compact subset of $[0,1) \times \mathbb{R}^p$. Furthermore, if $Df(\bar{x})$ is nonsingular, then γ has finite arc length. The general idea of the algorithm is to follow the zero curve γ emanating from $(0, x_0)$ until a zero $\bar{x}$ of $f(x)$ is reached (at $\lambda = 1$).

The zero curve γ is tracked by the normal flow algorithm [14], a predictor-corrector scheme. In the predictor phase, the next point is produced using Hermite cubic interpolation. Starting at the predicted point, the corrector iteration involves computing (implicitly) the Moore-Penrose pseudo-inverse of the Jacobian matrix at each point. The most complex part of the homotopy algorithm is the computation of the tangent vectors to γ, which involves the computation of the kernel of the $p \times (p+1)$ Jacobian matrix $D\rho_a$. The kernel is found by computing a QR factorization of $D\rho_a$, and then using back substitution. This strategy is implemented in the mathematical software package HOMPACK [14], which was used for the curve tracking here.

Two different homotopy maps are used for solving the optimal projection equations. When the initial problem, $g(x; a) = 0$, can be solved, then the homotopy map is [96]

$$\rho_a(\lambda, x) = F(a, \lambda, x) \equiv \lambda f(x) + (1 - \lambda) g(x; a), \tag{2.1}$$

where $f(x) = 0$ is the final problem, and a is a parameter vector used in defining the function g.

When the initial problem is not solved exactly, i.e., $g(x_0; b) \neq 0$, then the map is a Newton homotopy [97]

$$\rho_a(\lambda, x) = F(b, \lambda, x) - (1 - \lambda) F(b, 0, x_0), \tag{2.2}$$

where $a = (b, x_0)$. For $\lambda = 0$, $\rho_a(0, x_0) = F(b, 0, x_0) - F(b, 0, x_0) = 0$, and for $\lambda = 1$, $\rho_a(1, x) = F(b, 1, x) = f(x) = 0$.

For most of the homotopies actually used for control problems, the theoretical verification of properties 1) and 4) is highly technical and has not been done.

3. Computational Problems in Control Design

As discussed in Section 1, a variety of computational problems will be considered ranging from well-researched problems to highly challenging problems. In the following subsections we briefly discuss representative problems from each area.

3.1. Reduced Order Model Problem

We begin the list of problems with a somewhat detailed description of the H^2 optimal reduced order model problem, for which a homotopy map is developed in Section 4 and for which some computational results are presented in Section 5. The order of topics is a statement of the problem, the necessary optimality conditions (which are computationally intractable), some linear algebra background material, and finally a computationally tractable version of the optimality conditions.

Given the controllable and observable, time invariant, continuous time system

$$\dot{x}(t) = A\,x(t) + B\,u(t), \tag{3.1}$$

$$y(t) = C\,x(t), \tag{3.2}$$

where $A \in \mathbb{R}^{n \times n}$, $B \in \mathbb{R}^{n \times m}$, $C \in \mathbb{R}^{l \times n}$, the goal is to find, for given $n_m < n$, a reduced order model

$$\dot{x}_m(t) = A_m\,x_m(t) + B_m\,u(t),$$

$$y_m(t) = C_m\,x_m(t),$$

where $A_m \in \mathbb{R}^{n_m \times n_m}$, $B_m \in \mathbb{R}^{n_m \times m}$, $C_m \in \mathbb{R}^{l \times n_m}$, which minimizes the quadratic model-reduction criterion

$$J(A_m, B_m, C_m) \equiv \lim_{t \to \infty} \mathbb{E}\left[(y - y_m)^t\,R\,(y - y_m)\right],$$

where the input $u(t)$ is white noise with positive definite intensity V and R is a positive definite weighting matrix. $x(t)$ is the state vector and $y(t)$ is the (measured or observed) output vector.

It is assumed that A is asymptotically stable and diagonalizable, and a solution (A_m, B_m, C_m) is sought in the set

$$A_+ = \big\{(A_m, B_m, C_m) : A_m \text{ is stable, } (A_m, B_m) \text{ is controllable and } (A_m, C_m) \text{ is observable}\big\}.$$

DEFINITION. Given symmetric positive semidefinite matrices $\hat{Q}$, $\hat{P} \in \mathbb{R}^{n\times n}$ such that rank $(\hat{Q})$ = rank $(\hat{P})$ = rank $(\hat{Q}\hat{P})$ = n_m, matrices $G, \Gamma \in \mathbb{R}^{n_m\times n}$ and positive semisimple $M \in \mathbb{R}^{n_m\times n_m}$ are called a (G, M, Γ)-factorization (projective factorization) of $\hat{Q}\hat{P}$ if

$$\hat{Q}\,\hat{P} = G^t\,M\,\Gamma,$$
$$\Gamma\,G^t = I_{n_m}.$$

Positive semisimple means similar to a symmetric positive definite matrix.

The following theorem from [92] gives necessary conditions for the optimal solution to the reduced order model problem.

THEOREM 2. *Suppose $(A_m, B_m, C_m) \in A_+$ solves the optimal model-reduction problem. Then there exist symmetric positive semidefinite matrices $\hat{Q}$, $\hat{P} \in \mathbb{R}^{n\times n}$ such that for some (G, M, Γ)-factorization of $\hat{Q}\hat{P}$, A_m, B_m and C_m are given by*

$$A_m = \Gamma\,A\,G^t, \tag{3.3}$$
$$B_m = \Gamma\,B, \tag{3.4}$$
$$C_m = C\,G^t, \tag{3.5}$$

and such that, with $\tau \equiv G^t\Gamma$, the following conditions are satisfied:

$$0 = \tau[A\,\hat{Q} + \hat{Q}\,A^t + B\,V\,B^t], \tag{3.6}$$
$$0 = [A^t\,\hat{P} + \hat{P}\,A + C^t\,R\,C]\,\tau, \tag{3.7}$$
$$\text{rank }(\hat{Q}) = \text{rank }(\hat{P}) = \text{rank }(\hat{Q}\,\hat{P}) = n_m. \tag{3.8}$$

The equations (3.6)–(3.7) can be written in an equivalent form

$$A\,\hat{Q} + \hat{Q}\,A^t + \tau\,B\,V\,B^t + B\,V\,B^t\,\tau^t - \tau\,B\,V\,B^t\,\tau^t = 0,$$
$$A^t\,\hat{P} + \hat{P}\,A + \tau^t\,C^t\,R\,C + C^t\,R\,C\,\tau - \tau^t\,C^t\,R\,C\,\tau = 0.$$

The matrices $\hat{Q}$ and $\hat{P}$ are called the *controllability* and *observability pseudogramians* [92], respectively, since they are analogous to the Gramians G_c and G_o which satisfy the dual Lyapunov equations

$$A\,G_c + G_c\,A^t + B\,V\,B^t = 0,$$
$$A^t\,G_o + G_o\,A + C^t\,R\,C = 0.$$

τ is an oblique projection (idempotent) operator since $\tau^2 = \tau$. The projection matrix τ can be expressed as

$$\tau = (\hat{Q}\,\hat{P})\,(\hat{Q}\,\hat{P})^{\sharp},$$

where $(\hat{Q}\,\hat{P})^{\sharp}$ is the Drazin inverse [94], which is different from the more widely known Moore-Penrose inverse.

The necessary optimality conditions (3.6)–(3.8) are elegant but computationally impractical for several reasons. The rank conditions (3.8) are difficult to enforce numerically. Furthermore, even though τ is differentiable, neither the (G, M, Γ) nor Drazin inverse representations of τ can be easily differentiated [100]. A more computationally tractable version of (3.6)–(3.8) can be derived using the contragredient transformation. Since this transformation is not widely known, we give next some basic results about it.

The following theorem from [92] gives a sufficient condition for simultaneous reduction of two symmetric positive semidefinite matrices to a diagonal form using a contragredient transformation.

THEOREM 3. [92] *Let symmetric positive semidefinite* $Q,\, P \in I\!R^{n\times n}$ *satisfy*

$$\text{rank}\;(Q) = \text{rank}\;(P) = \text{rank}\;(Q\;P) = n_m, \tag{3.9}$$

where $n_m \le n$. *Then, there exists a nonsingular* $W \in I\!R^{n\times n}$ *(contragredient transformation) and positive definite diagonal* $\Sigma,\, \Omega \in I\!R^{n_m\times n_m}$ *such that*

$$Q = W\begin{pmatrix}\Sigma & 0\\ 0 & 0\end{pmatrix}W^t, \quad P = W^{-t}\begin{pmatrix}\Omega & 0\\ 0 & 0\end{pmatrix}W^{-1}. \tag{3.10}$$

REMARK 4. [92] *Let* Q *and* P *be as in Theorem 3. Then there exists a nonsingular* $U \in I\!R^{n\times n}$ *and positive definite diagonal* $\Lambda \in I\!R^{n_m\times n_m}$ *such that*

$$Q = U\begin{pmatrix}\Lambda & 0\\ 0 & 0\end{pmatrix}U^t, \quad P = U^{-t}\begin{pmatrix}\Lambda & 0\\ 0 & 0\end{pmatrix}U^{-1}.$$

The following lemma defines a factorization which plays an important role in the optimal projection approach for solving the reduced order model problem.

LEMMA 5. [92] *Let symmetric positive semidefinite* $\hat{Q},\, \hat{P} \in I\!R^{n\times n}$ *satisfy the rank conditions* (3.9). *Then, there exist* $G,\, \Gamma \in I\!R^{n_m\times n}$ *and positive semisimple* (positive semisimple *means similar to a symmetric positive definite matrix) $M \in \mathbb{R}^{n_m \times n_m}$ such that*

$$\hat{Q}\,\hat{P} = G^t\,M\,\Gamma, \tag{3.11}$$

$$\Gamma\,G^t = I_{n_m}. \tag{3.12}$$

Proof. Due to Remark 4 there exist nonsingular $W \in \mathbb{R}^{n \times n}$ and positive definite diagonal $\Sigma \in \mathbb{R}^{n_m \times n_m}$ such that

$$\hat{Q} = W \begin{pmatrix} \Sigma & 0 \\ 0 & 0 \end{pmatrix} W^t, \quad \hat{P} = W^{-t} \begin{pmatrix} \Sigma & 0 \\ 0 & 0 \end{pmatrix} W^{-1}. \tag{3.13}$$

The equations (3.13) can be expressed in the equivalent form

$$\hat{Q} = W_1\,\Sigma\,W_1^t, \quad \hat{P} = U_1^t\,\Sigma\,U_1, \tag{3.14}$$

where

$$W = (\overbrace{W_1}^{n_m} \quad W_2), \quad W^{-1} = U = {}^{n_m}\{ \begin{pmatrix} U_1 \\ U_2 \end{pmatrix}. \tag{3.15}$$

¿From (3.14)–(3.15) with $G \equiv W_1^t$ and $\Gamma \equiv U_1$ follow (3.11) and (3.12).

Q. E. D.

Matrices G, M and Γ from Lemma 5 are a (G, M, Γ)-*factorization* of $\hat{Q}\hat{P}$, and (3.14) shows how to implicitly enforce the rank conditions (3.9). Note also that $\tau = G^t\Gamma = W_1 U_1$ gives an explicit differentiable expression for the skew projection operator τ in terms of the components of W_1 and U_1. Using this contragredient machinery yields a computationally tractable form of the necessary optimality conditions (3.6)–(3.8), which is written out in detail in Section 4.

3.2. Riccati Equations

One of the most fundamental computational tasks arising in control theory, as well as related areas, is the numerical solution of Riccati equations. Since Riccati equations are central to modern linear-quadratic estimation and control design, their theoretical properties have been thoroughly studied (see, e.g., [31]–[33]). Several numerical solution techniques have been developed for Riccati equations including eigenvalue methods ([34]–[40]), the Chandrasekhar algorithm ([41]–[43]), and the matrix sign function technique

([44]). The software implementation of Riccati solvers is widely available and is included in numerous control-design packages.

In spite of the advanced state of development of Riccati solvers, there remains a need for numerical methods which operate efficiently and reliably on large dimensional problems. To this end we have developed a prototypical homotopy-based Riccati solver which has demonstrated potential savings on large scale problems. To describe the underlying idea (for details, see [45]) consider the Riccati equation

$$A^t X + XA - X\Sigma X + Q = 0 \tag{3.16}$$

where Σ and Q are symmetric, positive definite matrices. Let $\lambda \in [0,1]$ be the continuation parameter and replace (3.16) by

$$[A-(1-\lambda)\beta I]^t X(\lambda) + X(\lambda)[A-(1-\lambda)\beta I] - X(\lambda)\Sigma X(\lambda) + \lambda Q = 0, \tag{3.17}$$

where A is $n \times n, I$ denotes the $n \times n$ identity, and $\beta > 0$ is a constant. For $\lambda << 1$, the solution $X(\lambda)$ is of order λ so that the quadratic term can be neglected. The resulting Lyapunov equation

$$[A-(1-\lambda)\beta I]^t X(\lambda) + X(\lambda)[A-(1-\lambda)\beta I] + \lambda Q = 0 \tag{3.18}$$

is solvable so long as β is large enough to render $A-(1-\lambda)\beta I$ stable. To advance along the trajectory, note that the derivative $X'(\lambda)$ is given by

$$\begin{aligned} &[A-(1-\lambda)\beta I - \Sigma X(\lambda)]^t X'(\lambda) \\ &\quad + X'(\lambda)[A-(1-\lambda)\beta I - \Sigma X(\lambda)] + Q + 2\beta X(\lambda) = 0 \end{aligned} \tag{3.19}$$

so that X can be updated according to

$$X(\lambda + \Delta\lambda) = X(\lambda) + \Delta\lambda X'(\lambda). \tag{3.20}$$

It can be shown (see [45]) that the coefficient of $X'(\lambda)$ in (3.19) remains stable.

The above algorithm has been implemented in prototype software and appears to possess several features which are potentially advantageous in practice. First note that the eigenproblem required for solving (3.19) is of dimension n, while the Hamiltonian approach involves a $2n$-dimensional eigenproblem needing eight times as much computation. It is also possible in solving (3.19) to exploit the structure of A (e.g., for modal systems) and the fact that $\Sigma X(\lambda)$ is a low-rank perturbation of A. We have also found that much of the computation can be performed with single precision arithmetic using double precision only as λ approaches 1. Overall, the operation count appears comparable to Hamiltonian methods. It is also interesting to note as an additional benefit that if β can be set to zero at some point along the trajectory, then the remaining trajectory consists of useful solutions corresponding to increasing controller authority. As shown in [45] using topological degree theory, the above algorithm is guaranteed to converge. Also, it is important to point out that the above homotopy scheme is fundamentally different from the Newton method based iterative scheme of Kleinman ([35]) which requires an initial stabilizing gain.

3.3. Fixed-structure Optimization

In practice, controller design must account for numerous implementation constraints. For example, a decentralized controller architecture may be required when communication between particular sensors and actuators is precluded by physical constraints. As another example, the dynamic order of a feedback controller may be limited by the desire to minimize real time computational requirements for executing the control law. The approach to control design which explicitly recognizes and imposes such implementation constraints at the start of the design process is known as *fixed-structure optimization*. The literature on fixed-structure optimization is extensive. References [46]–[60] constitute only a representative sample, while a comprehensive review appears in [57].

In general, fixed-structure optimization problems are nonlinear and nonconvex. Hence it is not surprising that Newton-Raphson and gradient search methods have been widely applied. In addition, the structure of fixed-structure optimization problems has often been exploited to develop ad hoc algorithms. In spite of these computational developments, three key issues remain unresolved, namely, startup, guaranteed convergence, and global optimality.

Significant progress has recently been achieved in [59] with regard to startup and convergence. Specifically, for a fixed-order dynamic compensation problem, a continuation algorithm was developed which involves the replacement of the dynamic matrix A by a left-shifted matrix $A - \lambda I$, where λ is the continuation parameter. This approach thus overcomes stabilization problems in algorithm startup. The global optimality issue is not addressed in [59], however.

To address the issue of global optimality we have developed a homotopy algorithm in [60] which is based upon the optimal projection formulation of the necessary conditions for fixed-order dynamic compensation ([53], [56]). To illustrate the main ideas, consider the following control problem. Let the plant be given by

$$\dot{x} = Ax + Bu + w_1, \qquad x \in \mathbb{R}^n, \qquad u \in \mathbb{R}^m, \tag{3.21}$$

where x is the state vector, u is the control vector, w_1 is white (Gaussian) noise, let the plant measurements be

$$y = Cx + w_2, \qquad y \in \mathbb{R}^\ell, \tag{3.22}$$

and consider a controller of the form

$$\dot{x}_c = A_c x_c + B_c y, \qquad x_c \in \mathbb{R}^{n_c}, \tag{3.23}$$

$$u = C_c x_c. \tag{3.24}$$

The plant is assumed to be of some high dimension n while the controller is constrained to be of some low dimension $n_c \leq n$. The plant disturbance w_1 and measurement noise w_2 are white noise signals possessing intensities V_1 and V_2, respectively. The goal of the design problem is to determine the gains A_c, B_c and C_c to minimize the cost functional

$$J(A_c, B_c, C_c) = \lim_{t\to\infty} \mathbb{E}\left[x^t R_1 x + u^t R_2 u\right]. \tag{3.25}$$

As shown in [53], the optimality conditions for this problem are given by the following system of four coupled matrix equations in the unknown $n \times n$ matrices Q, P, $\hat{Q}$, $\hat{P}$:

$$0 = AQ + QA^t + V_1 - Q\overline{\Sigma}Q + \tau_\perp Q\overline{\Sigma}Q\tau_\perp^t, \tag{3.26}$$
$$0 = A^tP + PA + R_1 - P\Sigma P + \tau_\perp^t P\Sigma P\tau_\perp, \tag{3.27}$$
$$0 = (A - \Sigma P)\hat{Q} + \hat{Q}(A - \Sigma P)^t + Q\overline{\Sigma}Q - \tau_\perp Q\overline{\Sigma}Q\tau_\perp^t, \tag{3.28}$$
$$0 = (A - Q\overline{\Sigma})^t\hat{P} + \hat{P}(A - Q\overline{\Sigma}) + P\Sigma P - \tau_\perp^t P\Sigma P\tau_\perp, \tag{3.29}$$

$$\operatorname{rank}\hat{Q} = \operatorname{rank}\hat{P} = \operatorname{rank}\hat{Q}\hat{P} = n_c, \tag{3.30}$$

$$\hat{Q}\hat{P} = G^tM\Gamma, \qquad \Gamma G^t = I_{n_c}. \tag{3.31}$$

$$\tau = G^t\Gamma, \qquad \tau_\perp = I_n - \tau, \tag{3.32}$$

$$\Sigma = BR_2^{-1}B^t, \qquad \overline{\Sigma} = C^tV_2^{-1}C, \tag{3.33}$$

while the extremal gains are given by

$$A_c = \Gamma(A - Q\overline{\Sigma} - \Sigma P)G^t, \tag{3.34}$$

$$B_c = \Gamma Q C^t V_2^{-1}, \tag{3.35}$$

$$C_c = -R_2^{-1} B^t P G^t. \tag{3.36}$$

These equations use the (G, M, Γ)-factorization and projection operator τ defined in Section 3.1. Of course, when $n_c = n, \tau$ is the identity and the four equations immediately reduce to

$$0 = AQ + QA^t + V_1 - Q\overline{\Sigma}Q,$$

$$0 = A^t P + PA + R_1 - P\Sigma P,$$

which is the standard LQG result for full-order controllers.

In [60] a homotopy algorithm was developed for solving the optimal projection equations (3.26)–(3.29). This algorithm has been implemented and applied to an 8th-order design problem used in [61] to compare several controller reduction schemes. As reported in [60], the optimal projection homotopy algorithm reliably yielded stabilizing controllers even at high authority levels. The algorithm has proven to be numerically efficient, requiring the solution of four equations each of order $n_c n$ rather than n^2. Hence, for $n_c << n$, the computational burden is eased considerably.

The optimal projection homotopy algorithm is based upon a topological degree analysis [62] of the optimal projection equations. This analysis provides, for the first time, an upper bound on the number of solutions of the optimal projection equations. This result thus provides the means for determining all solutions of the equations including the global minimum of the original optimization problem. The principal result [60] is as follows.

THEOREM 6. *Assume that the plant is stabilizable and detectable, $V_1 > 0$, $R_1 > 0$ and $n_u \le n_c$, where n_u is the dimension of the unstable subspace of A. Then, in the class of nonnegative definite solutions Q, P, $\hat{Q}$, $\hat{P}$, with*

$$\text{rank } \hat{Q} = \text{rank } \hat{P} = \text{rank } \hat{Q}\hat{P} = n_c, \tag{3.37}$$

the optimal projection equations possess at most

$$\begin{cases} \dbinom{\min(n, m, \ell) - n_u}{n_c - n_u}, & n_c \le \min(n, m, \ell), \\ 1, & \text{otherwise}, \end{cases} \tag{3.38}$$

stabilizing solutions. Each such solution is reachable via a homotopic path with starting point corresponding to diagonal initial data. Furthermore, if the plant is stabilizable by means of an n_cth-order dynamic compensator, then there exists at least one stabilizing solution.

For the example considered in [60]–[61], $n = 8$, $n_u = 2$, and $\ell = m = 1$ (i.e., single-input single-output). Thus, $\min(n, m, \ell) = 1$ and, for controllers of order $n_c \geq 2$, there exists at most one solution. Since the plant is stabilizable by controllers of order $n_c \geq 2$, the numerical solution of the optimal projection equations is guaranteed to be the global optimum.

As with the homotopy algorithm for solving the Riccati equation the current homotopy for solving the optimal projection equations (3.26)–(3.29) uses a simplified path following algorithm. The outstanding research question is to devise a homotopy solver for the optimal projection equations that will substantially increase the numerical robustness and reliability of the current solution procedure.

3.4. Robust Analysis and Feedback Synthesis

A central problem in feedback control is to achieve stability and acceptable performance in the presence of uncertain knowledge of plant dynamics and disturbances. There is thus an obvious necessity for effective tools for the robust analysis and synthesis of feedback control systems. Many of the approaches that have been developed or proposed offer significant computational challenges, especially when applied to high dimensional systems or systems with highly structured perturbations. Open research questions concern the application of homotopic continuation methods to three categories of robustness related problems:

(i) robustness analysis via parameter search,
(ii) structured singular value computation,
(iii) solutions of modified and coupled Riccati equations.

3.4.1. Robustness Analysis Via Parameter Search

This class of robustness analysis methodologies considers systems with parametric uncertainties. The stability and worst case performance of an uncertain system is determined by essentially determining all possible parameter variations and computing the system behavior for each of these variations. Homotopy techniques can aid in these problems by allowing an algorithm to move efficiently from point to point in the parameter space. An example of this type of approach is presented in [63].

The obvious advantage of techniques in this category is that they are nonconservative. However, even for relatively few uncertain parameters these techniques have unrealistic computational requirements. Thus, if there are more than a few (say four) uncertain parameters, alternative analysis methodologies must be considered.

3.4.2. Structured Singular Value Computation

Another class of approaches to robustness analysis is based on singular value computations ([64]–[66]). Since there exists reliable software to efficiently and accurately compute the singular values of a matrix, these approaches are relatively easy to implement. Unfortunately, these techniques may be arbitrarily conservative when the uncertainty is structured. This deficiency motivated the development of the structured singular value ([67]–[71]).

Structured singular value analysis considers perturbations with block diagonal structure. For systems which can be transformed to this form, structured singular value analysis eliminates or reduces the conservatism inherent in the standard singular value analysis. The structured singular value is the solution of an optimization problem. If there are three or fewer perturbation blocks, the optimization problem is convex and the solution can always be obtained by standard nonlinear programming methods ([67], [70]). However, for four or more perturbation blocks, the optimization problem is nonconvex and the computation of the structured singular value is a much more difficult problem. Since homotopy methods are effective for nonconvex problems, they are particularly appropriate for the computation of the structured singular value. Some preliminary work on applying a homotopy method to the structured singular value problem is presented in [71]. It should also be mentioned that the structured singular value can also be used for robustness feedback synthesis ([69]).

3.4.3. Modified Riccati Equation Solutions

Over the past decade increasing attention has been given to two related approaches to robust feedback design. The first class of methodologies uses Lyapunov's direct method to synthesize control laws which satisfy sufficient conditions for robust stability. (See [72]–[75] for representative results.) Important recent results ([79]–[81]) have shown that methodologies in this class also yield solutions to H^∞ optimal control problems. The second class of design methodologies depends upon multiplicative white noise to capture the effect of parametric uncertainty. (See e.g., [76]–[78].) Although the two approaches were developed separately, strong unifying links have recently been discovered ([78]). The similarity of the approaches is seen in a heuristic sense by the forms of the equations which arise in the developments. For linear systems with parametric uncertainty both approaches lead to synthesis procedures which require the solution of modified algebraic Riccati equations or coupled, modified Riccati and Lyapunov equations. Homotopy methods can often provide an effective means of solving these systems of equations as illustrated by the results of [45], [60], [82].

To illustrate the types of modified Riccati equations which occur in the above methodologies, consider uncertain systems of the form

$$\dot{x} = \left(A + \sum_{i=1}^{p} r_i(t) A_i \right) x(t) + Bu(t) \tag{3.39}$$

where

$$|r_i(t)| \leq \bar{r}_i, \quad i = 1, \ldots, p. \tag{3.40}$$

The synthesis procedures of Chang and Peng [72], Peterson and Hollot [74] and Kosmidou and Bertrand [75] each use quadratic Lyapunov functions to develop robust state feedback laws for the system described by (3.39)–(3.40). These three procedures each require a positive definite solution P to one of the following modified Riccati equations.
Chang and Peng [72]:

$$0 = A^t P + PA + Q - PBR^{-1}B^t P + \sum_{i=1}^{p} \mathcal{U}(P, A_i). \tag{3.41}$$

Peterson and Hollot [74]:

$$0 = A^t P + PA + Q - PBR^{-1}B^t P + PEP + D, \quad D \geq 0, \ E \geq 0. \tag{3.42}$$

Kosmidou and Bertrand [75]:

$$0 = A^t P + PA + Q - PBR^{-1}B^t P + \sum_{i=1}^{p} A_i^t P A_i. \tag{3.43}$$

In (3.41) the function $\mathcal{U}(P, A_i)$ is defined as follows. For $i = 1, \ldots, p$ let the symmetric matrix $A_i^t P + PA_i^t$ have the modal decomposition

$$A_i^t P + PA_i = V_i \Lambda_i V_i^t \tag{3.44}$$

where V_i is orthogonal and Λ_i is diagonal. The matrix $\mathcal{U}(P, A_i)$ is then defined by

$$\mathcal{U}(P, A_i) \triangleq V_i \left| \Lambda_i \right| V_i^t \tag{3.45}$$

where $|\Lambda_i|$ is the diagonal matrix whose elements are the absolute values of the elements of Λ_i.

If reduced order dynamic compensation is desired, synthesis procedures associated with quadratic Lyapunov functions and multiplicative white noise often require the solution of coupled Riccati and Lyapunov equations. For example, consider the following control problem. Let the plant be given by

$$\dot{x} = \left(A + \sum_{i=1}^{p} v_i A_i\right) x + \left(B + \sum_{i=1}^{p} v_i B_i\right) u + w_1, \qquad x \in \mathbb{R}^n,\ u \in \mathbb{R}^m, \tag{3.46}$$

with measurements

$$y = \left(C + \sum_{i=1}^{p} v_i C_i\right) x + w_2, \qquad y \in \mathbb{R}^\ell, \tag{3.47}$$

and consider a controller of the form

$$\dot{x}_c = A_c x_c + B_c y, \quad x_c \in \mathbb{R}^{n_c}, \tag{3.48}$$

$$u = C_c x_c, \tag{3.49}$$

where $n_c \leq n$. The plant disturbance w_1 and measurement noise w_2 are white noise signals with respective intensities V_1 and V_2 and cross correlation V_{12}, while $v_1, \ldots, v_p$ are unit intensity white noise processes. The goal of the design is to determine the controller parameters A_c, B_c and C_c which minimize the performance criterion

$$J(A_c, B_c, C_c) = \lim_{t \to \infty} \mathbb{E}\left[x^t R_1 x + 2x^t R_{12} u + u^t R_2 u\right]. \tag{3.50}$$

For convenience in stating the optimality conditions define the following notation for Q, P, $\hat{Q}$, $\hat{P} \in \mathbb{R}^{n \times n}$.

$$A_s \triangleq A + \frac{1}{2}\sum_{i=1}^{p} A_i^2,$$

$$B_s \triangleq B + \frac{1}{2}\sum_{i=1}^{p} A_i B_i,$$

$$C_s \triangleq C + \frac{1}{2}\sum_{i=1}^{p} C_i A_i,$$

$$R_{2s} \triangleq R_2 + \sum_{i=1}^{p} B_i^t (P + \hat{P}) B_i,$$

$$V_{2s} \triangleq V_2 + \sum_{i=1}^{p} C_i(Q+\hat{Q})C_i^t,$$

$$Q_s \triangleq QC_s^t + V_{12} + \sum_{i=1}^{p} A_i(Q+\hat{Q})C_i^t,$$

$$P_s \triangleq B_s^t P + R_{12}^t + \sum_{i=1}^{p} B_i^t(P+\hat{P})A_i.$$

$$A_{Qs} \triangleq A_s - Q_s V_{2s}^{-1} C_s,$$

$$A_{Ps} \triangleq A_s - B_s R_{2s}^{-1} P_s,$$

$$\tau \triangleq G^t\Gamma, \quad \tau_\perp = I_n - \tau.$$

As usual $\hat{Q}\,\hat{P} = G^t M\Gamma$ is a (G, M, Γ)-factorization. Then as shown in [76] the optimality conditions require solutions Q, P, $\hat{Q}$, and $\hat{P}$ of the four coupled Maximum Entropy Optimal Projection (MEOP) design equations

$$0 = A_s Q + QA_s^t + V_1 + \sum_{i=1}^{p}\left[A_i Q A_i^t + (A_i - B_i R_{2s}^{-1} P_s)\hat{Q}(A_i - B_i R_{2s}^{-1} P_s)^t\right] - Q_s V_{2s}^{-1} Q_s^t + \tau_\perp Q_s V_{2s}^{-1} Q_s^t \tau_\perp^t, \tag{3.51}$$

$$0 = A_s^t P + PA_s + R_1 + \sum_{i=1}^{p}\left[A_i^t P A_i + (A_i - Q_s V_{2s}^{-1} C_i)^t \hat{P}(A_i - Q_s V_{2s}^{-1} C_i)\right] - P_s^t R_{2s}^{-1} P_s + \tau_\perp^t P_s^t R_{2s}^{-1} P_s \tau_\perp, \tag{3.52}$$

$$0 = A_{Ps}\hat{Q} + \hat{Q}A_{Ps}^t + Q_s V_{2s}^{-1} Q_s^t - \tau_\perp Q_s V_{2s}^{-1} Q_s^t \tau_\perp^t, \tag{3.53}$$

$$0 = A_{Qs}^t \hat{P} + \hat{P} A_{Qs} + P_s^t R_{2s}^{-1} P_s - \tau_\perp^t P_s^t R_{2s} P_s \tau_\perp, \tag{3.54}$$

where

$$\operatorname{rank} \hat{Q} = \operatorname{rank} \hat{P} = \operatorname{rank} \hat{Q}\hat{P} = n_c.$$

The controller parameters are then given by

$$A_c = \Gamma(A_s - B_s R_{2s}^{-1} P_s - Q_s V_{2s}^{-1} C_s)G^t, \tag{3.55}$$

$$B_c = \Gamma Q_s V_{2s}^{-1}, \tag{3.56}$$

$$C_c = -R_{2s}^{-1} P_s G^t. \tag{3.57}$$

It is of interest to note that the coupled equations (3.26)–(3.29) are a special case of the MEOP equations (3.51)–(3.54). The latter can be handled by a homotopy map similar to that given in Section 4. Other examples of coupled modified Riccati and Lyapunov equations are presented in [78].

3.5. Sensor/actuator Placement and Simultaneous Controller/structure Design

The controller design problems discussed in the preceding sections involve primarily the determination of suitable feedback gains. In practice, the control system designer must first select suitable locations for the sensors and actuators. Hence this is a fundamental problem in control design which is of considerable engineering interest. Representative references include [83]–[88]; see in particular the survey paper [88].

One of the primary theoretical approaches to sensor and actuator placement involves optimization in conjunction with a distributed parameter model. As in LQG theory, it is possible to express the solution in terms of suitable Riccati equations ([86]). As with fixed-structure optimization, however, the placement problem leads to local minima. This follows directly from the periodic nature of structural modes. Numerical approaches to optimal sensor/actuator placement have received only limited attention. Thus the goal of current research is to examine aspects of this problem which are amenable to homotopy algorithms.

Perhaps one of the most ambitious goals of control system design for large spacecraft is simultaneous design of both the structure and the controller. Although the literature is fairly limited, references [89]–[91] represent significant attempts at this class of problems. As in the problem of structural optimization alone, difficulties arise due to high dimensionality and local minima. Accordingly, gradient search methods are utilized in [91].

Because of the computational difficulties encountered in these problems, it is important to explore potential advantages of alternative algorithms such as continuation methods.

4. Homotopy for the Reduced Order Model Problem—an Example

Homotopy algorithms for solving the optimal projection equations for the reduced order model problem of Section 3.1 can be designed using decompositions of the pseudogramians based on contragredient transformations.

The equations (3.6)–(3.7) can be considered in another, equivalent form. If (3.6) is multiplied by U_1 from the left, and (3.7) is multiplied by W_1 from the right, using the contragredient transformation

$$\hat{Q} = W_1 \Sigma W_1^t, \quad \hat{P} = U_1^t \Sigma U_1,$$

the following two equations are obtained:

$$U_1\, A\, W_1 \Sigma\, W_1^t + \Sigma\, W_1^t\, A^t + U_1\, B\, V\, B^t = 0, \tag{4.1}$$

$$A^t\, U_1^t\, \Sigma + U_1^t\, \Sigma\, U_1\, A\, W_1 + C^t\, R\, C\, W_1 = 0. \tag{4.2}$$

The third equation

$$U_1\, W_1 - I = 0 \tag{4.3}$$

determines the relationship between W_1 and U_1.

The matrix equations (4.1)–(4.3) contain $2\, n\, n_m + n_m^2$ scalar equations. On the other hand, the only natural unknowns in (4.1)–(4.3) W_1, U_1 and diagonal Σ, contain $2\, n\, n_m + n_m$ variables. Hence, some additional techniques are necessary in order to make an exact match between the number of equations and the number of unknowns.

One approach is to consider Σ to be symmetric and all elements of Σ as unknowns. (Σ at the solution must be symmetric, but along the homotopy zero curve, Σ need not be symmetric, hence all its elements are unknowns.) This is appropriate, since the equations (4.1)–(4.3) with a full symmetric Σ can be transformed into equations of the same form with a diagonal Σ by computing

$$\Sigma = T\, \bar{\Sigma}\, T^t, \quad \bar{W}_1 = W_1\, T, \quad \bar{U}_1 = T^t\, U_1,$$

where $\bar{\Sigma}$ is diagonal and T is orthogonal.

The following is a description of the algorithm for the method determined by the equations (4.1)–(4.3). The algorithm is based on the normal flow algorithm for dense Jacobian matrices described in [14]. Depending on the relative size of $F(a, 0, x_0)$ the algorithm may be modified. If $F(a, 0, x_0)$ is relatively large, computational experience shows that it is desirable (but not theoretically necessary) to enforce the symmetry of Σ along the homotopy path. This is done by observing that a symmetrized Σ corresponds to *some* homotopy map that *could* have been chosen initially. In effect, x_0 is changed in the homotopy map at each step along the homotopy zero curve γ. Obviously, in that case the homotopy map (2.2) must be used.

The algorithm is using the homotopy map (2.1) or (2.2), where $A(\lambda) = \lambda A + (1 - \lambda) D$ for some matrix D and $F(a, \lambda, x)$ is represented by three equations:

$$U_1\, A(\lambda)\, W_1 \Sigma\, W_1^t + \Sigma\, W_1^t\, A^t(\lambda) + U_1\, B\, V\, B^t = 0, \tag{4.4}$$

$$A^t(\lambda)\, U_1^t\, \Sigma + U_1^t\, \Sigma\, U_1\, A(\lambda)\, W_1 + C^t\, R\, C\, W_1 = 0, \tag{4.5}$$

$$U_1\, W_1 - I = 0. \tag{4.6}$$

In summary, the whole algorithm is:

1) Define D using formula (4.7) or (4.8) below.
2) Choose a starting point (Q_0, P_0) using one of the strategies explained in [30]. Compute $(W_1)_0$, $(U_1)_0$ and Σ_0 using a contragredient transformation.
3) Set $\lambda := 0$, $x := x_0 = ((W_1)_0, (U_1)_0, \Sigma_0)$.
4) Evaluate $\rho_a(\lambda, x)$ given by (2.1) or (2.2), and (4.4)–(4.6).
5) Evaluate $D\rho_a(\lambda, x)$.
6) Take a step along the curve and obtain $x_1 = (W_1, U_1, \Sigma)$, $\bar{\lambda}$.
7) Compute $\bar{x}_1 = (W_1, U_1, \bar{\Sigma}) = (W_1, U_1, (\Sigma + \Sigma^t)/2)$.
8) Change the homotopy to

$$F(a, \lambda, x) - (1 - \lambda)v = 0,$$

where $v = F(a, \bar{\lambda}, \bar{x}_1)/(1 - \bar{\lambda})$.

9) If $\bar{\lambda} < 1$, then set $x := \bar{x}_1$, $\lambda := \bar{\lambda}$, and go to Step 4.
10) If $\bar{\lambda} \geq 1$, compute the solution $\bar{x}_1$ at $\bar{\lambda} = 1$. Compute the reduced order model by diagonalizing $\Sigma = T\,\hat{\Sigma}\,T^t$.

Note: if $F(a, 0, x_0)$ is small, Steps 7 and 8 can be omitted without a serious loss of efficiency.

4.1. Choosing an Initial System and the Starting Point

While with homotopy algorithms in general an initial problem can be chosen practically at random, this problem has some special limitations. The reason is that Theorem 2 provides necessary conditions on a solution only under certain assumptions. In other words, every intermediate problem solution satisfies these equations only if the system is asymptotically stable, controllable and observable. While the absence of these features does not automatically mean that the intermediate problem solution will not satisfy the equations, it is clearly better to define a homotopy path in such a way that each problem along it corresponds to an asymptotically stable, controllable and observable system. Existence of a solution to the reduced order problem follows from [98]. Theorem 7, proved in [30], defines a class of initial systems such that these conditions are satisfied.

THEOREM 7. *For the given system* (3.1)–(3.2), *let* $A = X\,\Lambda\,X^{-1}$, *with* Λ *diagonal. Define* $D = X\,\Omega\,X^{-1}$ *for any diagonal matrix* Ω = *diag* $(\omega_1, \ldots, \omega_n)$, *such that all* ω_i, *for* $i = 1, 2, \ldots, n$, *are in the open left half plane. Then for almost all such* D *any convex combination* $(A(\alpha), B, C)$ *of* *the systems (D, B, C) and (A, B, C) will be asymptotically stable, controllable and observable.*

While the random construction of the matrix D given in Theorem 7 is theoretically plausible, in practice it may not be wise. The reason is that the matrix X is complex in general, which for many choices of Ω leads to a complex matrix D, which is undesirable. Hence, it is better to directly construct a matrix D such that Ω satisfies the conditions given in Theorem 7.

One simple choice for D is

$$D \equiv -c_1 I + \text{diag}\ \{\epsilon_1, \ldots, \epsilon_n\}, \tag{4.7}$$

where $c_1 > 0$ and ϵ_i are small random numbers that correspond to the parameter a in the theory. In this case Ω is a small perturbation of $-c_1 I$.

Also, the matrix D can be defined as

$$D \equiv -c_1 I + c_2 A, \tag{4.8}$$

for $c_1, c_2 > 0$. In this case $\Omega = -c_1 I + c_2 \Lambda$.

The starting point $x_0 = (Q_0, P_0)$ of the homotopy algorithm can be chosen using a number of different strategies.

One strategy is to choose Q_0 and P_0 that are positive semidefinite and satisfy the rank conditions, but are otherwise random. This approach may lead to relatively large values of $F(a, 0, x_0) = g(x_0; a)$.

The second strategy, which can be applied for any choice of the matrix D described above, generally leads to relatively small but nonzero $g(x_0; a)$. Since the matrix D is asymptotically stable, the Lyapunov equation

$$D\, Q + Q\, D^t + B\, V\, B^t = 0 \tag{4.9}$$

has a unique solution Q. Let $Q = T\, \Sigma\, T^t$, where T is orthogonal and

$$\Sigma = \text{diag}\ \{\sigma_1, \ldots, \sigma_n\}.$$

Next, define (assuming the σ_j are ordered in decreasing size)

$$\Sigma_1 \equiv \text{diag}\ \{\sigma_1, \ldots, \sigma_{n_m}, 0, \ldots, 0\}, \qquad Q_0 \equiv T\, \Sigma_1\, T^t.$$

If Q_0 is substituted for Q in (4.9), the equation will not be satisfied, but in general, if the dropped σ_i are sufficiently small, it will not be very different from zero. A similar procedure can be applied to compute P_0 that will 'almost' satisfy the equation

$$D^t P + P D + C^t R C = 0.$$

The point $x_0 = (Q_0, P_0)$ chosen in this way may lead to small values of $g(x_0; a)$. Also, this x_0 can be used as the initial guess for a quasi-Newton algorithm which may find a solution to the initial problem

$$\tau[D Q + Q D^t + B V B^t] = 0, \tag{4.10}$$

$$[D^t P + P D + C^t R C]\tau = 0. \tag{4.11}$$

Three other strategies, as well as comparisons between the strategies, are given in [30].

5. Numerical Results

This section contains numerical results and observations for the optimal reduced order model problem of Section 3.1. A number of problems taken from the literature have been solved by the homotopy algorithm of Section 4, but only one realistic problem will be described here. Numerical solutions for other problems are in [30].

EXAMPLE. This is a state space model of the transfer function between a torque activator and an approximately collocated torsional rate sensor for the ACES structure [99], located at NASA Marshall Space Flight Center, Huntsville, AL. The system in this example is of size $n = 17$, $m = 1$, $l = 1$. The nonzero elements of A are

$A(1,1) = A(2,2) = -0.031978272$, $A(1,2) = -A(2,1) = -78.54$,
$A(1,17) = 0.0097138566$, $A(2,17) = -0.0060463517$,
$A(3,3) = A(4,4) = -5.152212$, $A(3,4) = -A(4,3) = -51.457677$,
$A(3,17) = -0.021760771$, $A(4,17) = -0.0054538246$.
$A(5,5) = A(6,6) = -0.1351159$, $A(5,6) = -A(6,5) = -15.417859$,
$A(5,17) = -0.02179972$, $A(6,17) = -0.015063913$,
$A(7,7) = A(8,8) = -0.42811443$, $A(7,8) = -A(8,7) = -14.698408$,
$A(7,17) = -0.01042631$, $A(8,17) = -0.0088479697$,
$A(9,9) = A(10,10) = -0.064896745$, $A(9,10) = -A(10,9) = -12.077045$,
$A(9,17) = -0.030531575$, $A(10,17) = -0.030260987$,
$A(11,11) = A(12,12) = -0.048520356$,
$A(11,12) = -A(12,11) = -8.9654448$,
$A(11,17) = -0.016843335$, $A(12,17) = -0.011449591$,

$A(13,13) = A(14,14) = -0.036781718,$
$A(13,14) = -A(14,13) = -4.9057426,$
$A(13,17) = -0.1248007, \qquad A(14,17) = -0.0005136047,$
$A(15,15) = A(16,16) = -0.025112482,$
$A(15,16) = -A(16,15) = -3.8432892,$
$A(15,17) = -0.035415526, \qquad A(16,17) = -0.028115589,$
$A(17,17) = -92.399784.$

The matrices B and C are

$$B = \begin{pmatrix} 1.8631111 \\ -1.1413786 \\ -1.2105758 \\ 0.31424169 \\ 0.013307797 \\ -0.211128913 \\ 0.19552894 \\ -0.037391511 \\ -0.01049736 \\ -0.011486242 \\ -0.029376402 \\ 0.0082391613 \\ -0.012609562 \\ -0.0022040505 \\ -0.030853234 \\ 0.0011671662 \\ 0 \end{pmatrix}, \quad C^t = \begin{pmatrix} -0.0097138566 \\ 0.0060463517 \\ 0.021760771 \\ -0.0054538246 \\ 0.02179972 \\ 0.015063913 \\ -0.01042631 \\ -0.0088479697 \\ 0.030531575 \\ 0.030260987 \\ 0.016843335 \\ 0.011449591 \\ 0.1248007 \\ -0.0005136047 \\ 0.035415526 \\ 0.028115589 \\ 184.79957 \end{pmatrix}$$

A model of order $n_m = 6$ is

$$A_m = \begin{pmatrix} -0.23442 & -52.59052 & -0.17250 & 0.06085 & 0.000005 & 0.019936 \\ 52.59052 & -10.53247 & -0.91009 & 0.54970 & 0.00004 & 0.13584 \\ 0.17250 & -0.91009 & -0.18410 & 15.43994 & 0.00001 & 0.046293 \\ 0.06085 & -0.54970 & -15.43994 & -0.03904 & -0.000006 & -0.02371 \\ 0.000005 & -0.00004 & -0.00001 & -0.000006 & -0.000000005 & -78.54024 \\ -0.019936 & 0.13584 & 0.046293 & 0.02371 & 78.54024 & -0.06402 \end{pmatrix},$$

$$B_m = \begin{pmatrix} 0.02447 \\ 0.16899 \\ 0.06006 \\ -0.027808 \\ -0.00004 \\ -0.15819 \end{pmatrix}, \quad C_m^t = \begin{pmatrix} 0.02447 \\ -0.16899 \\ -0.06006 \\ -0.027808 \\ -0.00004 \\ 0.15819 \end{pmatrix}.$$

This model yields the cost $J = 4.19165 \cdot 10^{-5}$.

A model of order $n_m = 8$ is given by

$$A_m = \begin{pmatrix} -70.147 & 21.918 & -2.7406 & 2.9917 & -0.3721 & 0.228 & 0.0246 & 0.083 \\ 54.161 & -32.186 & 4.6829 & 9.2995 & -0.4958 & 0.180 & 0.0289 & 0.093 \\ 3.5118 & -4.6512 & -0.2083 & -51.396 & 0.1211 & -0.013 & -0.0049 & -0.0157 \\ -22.253 & 19.045 & 51.852 & -12.043 & 1.0945 & -0.639 & -0.0741 & -0.243 \\ 1.2271 & -1.1976 & -0.2000 & 1.1602 & -0.1936 & 15.44 & 0.0243 & 0.0807 \\ 0.5249 & -0.5415 & -0.0764 & 0.6934 & -15.450 & -0.014 & -0.0125 & 0.041 \\ -0.0705 & 0.0708 & 0.0106 & -0.0770 & 0.0238 & 0.012 & 0.0181 & -78.574 \\ -0.2393 & 0.2397 & 0.0357 & -0.2610 & 0.0803 & 0.042 & 78.508 & -0.082 \end{pmatrix},$$

$$B_m = \begin{pmatrix} -0.05753 \\ -0.06445 \\ 0.01043 \\ 0.16983 \\ -0.05959 \\ 0.02622 \\ 0.04591 \\ 0.15167 \end{pmatrix}, \quad C_m^t = \begin{pmatrix} -0.16432 \\ 0.16512 \\ 0.02442 \\ -0.18165 \\ 0.05966 \\ 0.02629 \\ -0.04472 \\ -0.15162 \end{pmatrix}.$$

This model yields the cost $J = 3.95223 \cdot 10^{-5}$. Obtaining this 8th-order model required approximately 77 CPU hours on a DECstation 3100, with 2900 Jacobian matrix evaluations. The dimension of the nonlinear system is 336, and a single Jacobian matrix evaluation requires 96 sec on the DECstation 3100.

5.1. Choice of the Initial System

Although the methods work successfully with most choices (4.7) and (4.8) of the initial systems, this choice has a significant impact on the performance of the algorithm. Heuristically, it seems that the best choice is $D \equiv -c_1 I$, with c_1 of the same order of magnitude as the spectral radius of the matrix A.

The following example shows that the performance of the algorithm is strongly affected by the choice of the initial system. The system from Example 5 of [100] is considered and the model of order $n_m = 1$ is sought. The data are obtained using the algorithm of Section 4 and Strategy 1 for choosing the starting point.

With the initial system $D = -0.0003\, I$, 3100 steps are necessary to solve the problem. Figure 1 shows the behavior of the variable $x_7 = \Sigma_{11}$.

With the initial system $D = -10\, I$, 21 steps are sufficient. Figure 2 shows Σ_{11} for this case.

A possible reason for the poor performance in the first case may be that the initial system is close to being asymptotically unstable.

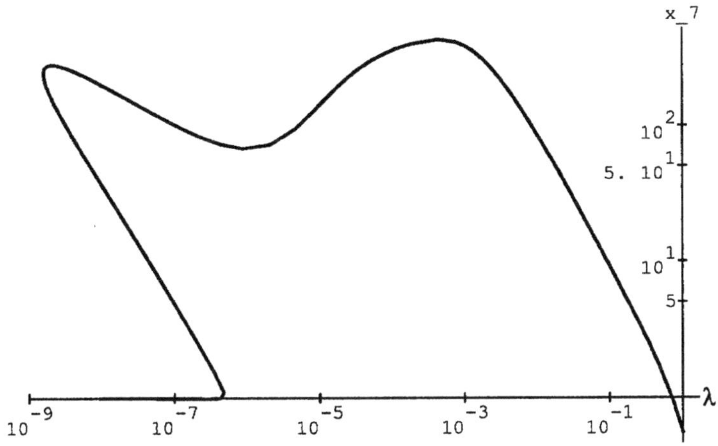

FIGURE 1. *Bad choice of the initial system.*

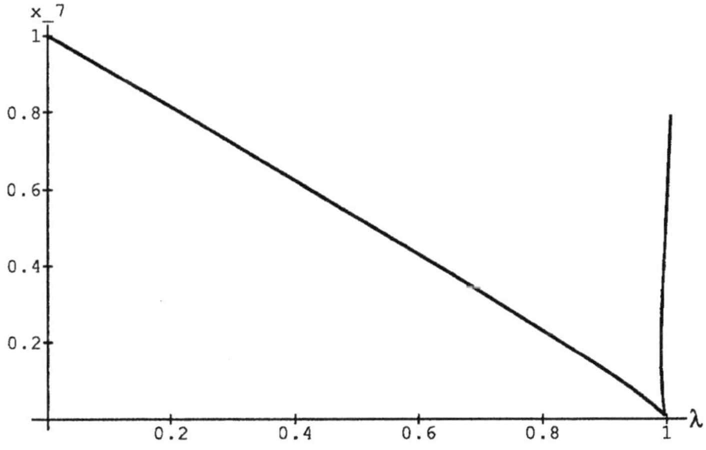

FIGURE 2. *Good choice of the initial system.*

6. Conclusion

This brief list of control problems suggests that even for well researched problems for which good numerical algorithms exist, homotopy methods may have something to contribute for the high dimensional case. For problems that are well understood theoretically but for which no efficient numerical algorithms exist, software such as HOMPACK could be a useful tool. For the class of highly challenging problems, the homotopy approach may provide both existence proofs, information about the nature of the solution, and practical numerical algorithms. While on the order of 100 papers applying homotopy methods in control theory have been published, it is clear that only the surface has been scratched.

7. Acknowledgements

This work was supported in part by AFOSR Grant 89-0497 and DOE Grant DE-FG05-88ER25068. The authors wish to thank the referees for helpful suggestions.

8. References

[1] F. A. Ficken, *The continuation method for functional equations*, Comm. Pure Appl. Math., 4 (1951), pp. 435–456.

[2] G. H. Meyer, *On solving nonlinear equations with a one-parameter operator imbedding*, SIAM J. Numerical Analysis, 5 (1968), pp. 739–752.

[3] P. Laasonen, *An imbedding method of iteration with global convergence*, Computing, 5 (1970), pp. 253–258.

[4] J. M. Ortega and W. C. Rheinboldt, *Iterative solution of nonlinear equations in several variables*, Academic Press, New York, 1970.

[5] E. Wasserstrom, *Numerical solutions by the continuation method*, SIAM Review, 15 (1973), pp. 89–119.

[6] J. H. Avila, *The feasibility of continuation methods for nonlinear equations*, SIAM J. Numer. Anal., 11 (1974), pp. 102–122.

[7] H. Wacker, *Continuation methods*, Academic Press, New York, 1978.

[8] J. C. Alexander and J. A. Yorke, *The homotopy continuation method: Numerically implementable procedures*, Trans. Amer. Math. Soc., 242 (1978), pp. 271–284.

[9] C. B. Garcia and W. I. Zangwill, *Pathways to solutions, fixed points and equilibria*, Prentice-Hall, Englewood Cliffs, NJ, 1981.

[10] B. C. Eaves, F. J. Gould, H. O. Peitgen and M. J. Todd, *Homotopy methods and global convergence*, Plenum Press, New York, 1983.

[11] S. L. Richter and R. A. DeCarlo, *Continuation methods: Theory and applications*, IEEE Trans. Circ. Sys., CAS-30 (1983), pp. 347–352.

[12] L. T. Watson, *Numerical linear algebra aspects of globally convergent homotopy methods*, SIAM Rev., 28 (1986), pp. 529–545.

[13] E. L. Allgower and K. Georg, *Numerical Continuation Methods: An Introduction*, Springer-Verlag, Berlin, 1990.

[14] L. T. Watson, S. C. Billups and A. P. Morgan, *HOMPACK: A suite of codes for globally convergent homotopy algorithms*, ACM Trans. Math. Software, 13 (1987), pp. 281–310.

[15] A. P. Morgan and R. F. Sarraga, *A method for computing three surface intersection points in GMSOLID*, Tech. Rep. GMR-3964, G. M. Research Lab., Warren, MI, 1982.

[16] W. C. Rheinboldt, *Numerical analysis of continuation methods for nonlinear structural problems*, Computers & Structures, 13 (1981), pp. 103–113.

[17] L. T. Watson and W. H. Yang, *Optimal design by a homotopy method*, Applicable Anal., 10 (1980), pp. 275–284.

[18] L. T. Watson, S. M. Holzer, and M. C. Hansen, *Tracking nonlinear equilibrium paths by a homotopy method*, Nonlinear Anal., 7 (1983), pp. 1271–1282.

[19] L. T. Watson, M. P. Kamat, and M. H. Reaser, *A robust hybrid algorithm for computing multiple equilibrium solutions*, Engrg. Comput., 2 (1985), pp. 30–34.

[20] Y. S. Shin, R. T. Haftka, L. T. Watson, and R. H. Plaut, *Tracing structural optima as a function of available resources by a homotopy method*, Comput. Methods Appl. Mech. Engrg., 70 (1988), pp. 151–164.

[21] J. D. Turner and H. M. Chun, *Optimal distributed control of a flexible spacecraft during a large-angle maneuver*, J. Guidance, Control, Dynamics, 7 (1984), pp. 257–264.

[22] J. P. Dunyak, J. L. Junkins and L. T. Watson, *Robust nonlinear least squares estimation using the Chow-Yorke homotopy method*, J. Guidance, Control, Dynamics, 7 (1984), pp. 752–755.

[23] S. Richter and R. DeCarlo, *A homotopy method for eigenvalue assignment using decentralized state feedback*, IEEE Trans. Aut. Control, AC-29 (1984), pp. 148–155.

[24] S. Lefebvre, S. Richter and R. DeCarlo, *A continuation algorithm for eigenvalue assignment by decentralized constant-output feedback*, Int. J. Control, 41 (1985), pp. 1273–1292.

[25] M. MARITON AND R. BERTRAND, *A homotopy algorithm for solving coupled Riccati equations*, Optim. Control Appl. Meth., 6 (1985), pp. 351–357.

[26] D. R. SEBOK, S. RICHTER, AND R. DECARLO, *Feedback gain optimization in decentralized eigenvalue assignment*, Automatica, 22 (1986), pp. 433–447.

[27] L. G. HORTA, J. N. JUANG AND J. L. JUNKINS, *A Sequential linear optimization approach for controller design*, J. Guidance, Control, Dynamics, 9 (1986), pp. 699–703.

[28] S. RICHTER, *A homotopy algorithm for solving the optimal projection equations for fixed-order dynamic compensation: Existence, convergence and global optimality*, Proc. Amer. Control Conf., Minneapolis, MN, June 1987, pp. 1527–1531.

[29] P. T. KABAMBA, R. W. LONGMAN AND S. JIAN-GUO, *A homotopy approach to the feedback stabilization of linear systems*, J. Guidance, Control, Dynamics, 10 (1987), pp. 422–432.

[30] D. ŽIGIĆ, *Homotopy methods for solving the optimal projection equations for the reduced order model problem*, M.S. thesis, Dept. of Computer Sci., Virginia Polytechnic Institute and State Univ., Blacksburg, VA, 1991.

[31] M. JAMSHIDI, *An overview of the solutions of the algebraic matrix Riccati equation and related problems*, Large Scale Systems, 1 (1980), pp. 167–192.

[32] M. A. SHAYMAN, *Geometry of the algebraic Riccati equation*, SIAM J. Control Optim., 21 (1983), pp. 375–409.

[33] I. GOHBERG, P. LANCASTER, AND L. RODMAN, *On Hermitian solutions of the symmetric algebraic Riccati equation*, SIAM J. Control Optim., 24 (1986), pp. 1323–1334.

[34] J. E. POTTER, *Matrix quadratic solutions*, SIAM J. Appl. Math., 14 (1966), pp. 496–501.

[35] D. L. KLEINMAN, *On an iterative technique for Riccati equation computations*, IEEE Trans. Aut. Control, AC-13 (1968), pp. 114–115.

[36] A. J. LAUB, *A Schur method for solving algebraic Riccati equations*, IEEE Trans. Aut. Control, AC-24 (1979), pp. 913–921.

[37] P. VAN DOOREN, *A generalized eigenvalue approach for solving Riccati equations*, SIAM J. Sci. Stat. Comp., 2 (1981), pp. 121–135.

[38] W. F. ARNOLD AND A. J. LAUB, *Generalized eigenproblem algorithms and software for algebraic Riccati equations*, Proc. IEEE, 72 (1984), pp. 1746–1754.

[39] T. Pappas, A. J. Laub and N. R. Sandell, Jr., *On the numerical solution of the discrete-time algebraic Riccati equation*, IEEE Trans. Aut. Control, AC-25 (1980), pp. 631–641.

[40] A. J. Laub, *Numerical linear algebra aspects of control design computations*, IEEE Trans. Aut. Control, AC-30 (1985), pp. 97–108.

[41] J. Casti and L. Ljung, *Some new analytic and computational results for operator Riccati equations*, SIAM J. Control Optim., 13 (1975), pp. 817–826.

[42] T. Kailath, *Some Chandrasekhar-type algorithms for quadratic regulators*, Proc. IEEE Conf. Decis. Control, New Orleans, LA, 1972, pp. 219–223.

[43] K. Ito and R. K. Powers, *Chandrasekhar equations for infinite dimensional systems*, SIAM J. Control Optim., 25 (1987), pp. 596–611.

[44] R. Byers, *Solving the algebraic Riccati equation with the matrix sign function*, preprint.

[45] S. Richter, *Reduced-order control design via the optimal projection approach: A homotopy algorithm for global optimality*, Proc. Sixth VPI&SU Symp. Dyn. Control Large Str., Blacksburg, VA, June 1987.

[46] W. S. Levine and M. Athans, *On the determination of the optimal constant output feedback gains for linear multivariable systems*, IEEE Trans. Aut. Control, AC-15 (1970), pp. 44–48.

[47] W. S. Levine, T. L. Johnson and M. Athans, *Optimal limited state variable feedback controllers for linear systems*, IEEE Trans. Aut. Control, AC-16 (1971), pp. 785–793.

[48] N. R. Sandell, Jr., P. Varaiya, M. Athans and M. G. Safanov, *Survey of decentralized control methods for large scale systems*, IEEE Trans. Aut. Control, AC-23 (1978), pp. 108–128.

[49] C. J. Wenk and C. H. Knapp, *Parameter optimization in linear systems with arbitrarily constrained controller structure*, IEEE Trans. Aut. Control, AC-25 (1980), pp. 496–500.

[50] D. P. Looze and N. R. Sandell, Jr., *Gradient calculations for linear quadratic fixed-control structure problems*, IEEE Trans. Aut. Control, AC-25 (1980), pp. 285–288.

[51] P. T. Kabamba and R. W. Longman, *An integrated approach to reduced-order control theory*, Optimal Control Appl. Methods, 4 (1983), pp. 405–415.

[52] J. R. Broussard and N. Halyo, *Active flutter control using discrete optimal constrained dynamic compensators*, Proc. 1983 Amer. Control Conf., San Francisco, CA, 1983.

[53] D. C. Hyland and D. S. Bernstein, *The optimal projection equations for fixed-order dynamic compensation*, IEEE Trans. Aut. Control, AC-29 (1984), pp. 1034–1037.

[54] D. D. Moerder and A. J. Calise, *Convergence of a numerical algorithm for calculating optimal output feedback gains*, IEEE Trans. Aut. Control, AC-30 (1985), pp. 900–903.

[55] U. L. Ly, A. Bryson, and R. H. Cannon, *Design of low-order compensators using parameter optimization*, Automatica, 21 (1985), pp. 315–318.

[56] D. S. Bernstein and D. C. Hyland, *The optimal projection equations for finite-dimensional fixed-order dynamic compensation of infinite-dimensional systems*, SIAM J. Control Optim., 24 (1986), pp. 122–151.

[57] P. M. Makila and H. T. Toivonen, *Computational methods for parametric LQ problems—a survey*, IEEE Trans. Aut. Control, AC-32 (1987), pp. 658–671.

[58] V. Mukhopadhyay, *Digital robust active control law synthesis for large order systems using constrained optimization*, Proc. AIAA Guid. Nav. Control Conf., pp. 1414–1423, Monterey, CA, August 1987.

[59] P. T. Kabamba, R. W. Longman and S. Jian-Guo, *A homotopy approach to the feedback stabilization of linear systems*, J. Guidance, Control, Dynamics, 10 (1987), pp. 422–432.

[60] S. Richter, *A homotopy algorithm for solving the optimal projection equations for fixed-order dynamic compensation: Existence, convergence and global optimality*, Proc. Amer. Control Conf., Minneapolis, MN, June 1987, pp. 1527–1531.

[61] Y. Liu and B. D. O. Anderson, *Controller reduction via stable factorization and balancing*, Int. J. Control, 44 (1986), pp. 507–531.

[62] N. G. Lloyd, *Degree Theory*, Cambridge University Press, London, 1978.

[63] R. R. E. De Gaston and M. G. Safanov, *A homotopy method for nonconservative stability robustness analysis*, Proc. 24th IEEE Conf. Decis. Control, Fort Lauderdale, FL, 1985, pp. 1294–1301.

[64] J. C. Doyle and G. Stein, *Multivariable feedback design*, IEEE Trans. Aut. Control, AC-26 (1981), pp. 4–16.

[65] M. G. Safanov, A. J. Laub and G. L. Hartman, *Feedback properties of multivariable systems: the role and use of the return difference matrix*, IEEE Trans. Aut. Control, AC-26 (1981), pp. 47–65.

[66] N. A. Lehtomaki, N. R. Sandell, Jr. and M. Athans, *Robustness results in linear-quadratic Gaussian based multivariable control designs*, IEEE Trans. Aut. Control, AC-26 (1981), pp. 75–93.

[67] J. DOYLE, *Analysis of feedback systems with structured uncertainties*, IEE Proc., Pt. D., 129 (1982), pp. 242–250.

[68] J. DOYLE, J. E. WALL AND G. STEIN, *Performance and robustness analysis for structured uncertainty*, Proc. 21st IEEE Conf. Decis. Control, Orlando, FL, 1982, pp. 629–636.

[69] J. C. DOYLE, *Structured uncertainty in control system design*, Proc. 24th IEEE Conf. Decis. Control, Fort Lauderdale, FL, 1985, pp. 260–265.

[70] M. K. H. FAN AND A. L. TITS, *Characterization and efficient computation of the structured singular value*, IEEE Trans. Aut. Control, AC-31 (1986), pp. 734–743.

[71] M. G. SAFANOV, *Exact calculation of the structured-singular-value stability margin*, Proc. 23rd IEEE Conf. Decis. Control, Las Vegas, NV, 1984, pp. 1224–1225.

[72] S. S. L. CHANG AND T. K. C. PENG, *Adaptive guaranteed cost control of systems with uncertain parameters*, IEEE Trans. Aut. Control, AC-17 (1972), pp. 474–483.

[73] E. NOLDUS, *Design of robust state feedback laws*, Int. J. Control, 35 (1982), pp. 935–944.

[74] I. R. PETERSON AND C. V. HOLLOT, *A Riccati equation approach to the stabilization of uncertain linear systems*, Automatica, 22 (1986), pp. 397–411.

[75] O. I. KOSMIDOU AND R. BERTRAND, *Robust-controller design for systems with large parameter variations*, Int. J. Control, 45 (1987), pp. 927–938.

[76] D. S. BERNSTEIN AND D. C. HYLAND, *The optimal projection/maximum entropy approach to designing low-order, robust controllers for flexible structures*, Proc. 24th IEEE Conf. Decis. Control, Fort Lauderdale, FL, 1985, pp. 745–752.

[77] D. S. BERNSTEIN AND S. W. GREELEY, *Robust controller synthesis using the maximum entropy design equations*, IEEE Trans. Aut. Control, AC-31 (1986), pp. 362–364.

[78] D. S. BERNSTEIN, *Robust static and dynamic output-feedback stabilization: deterministic and stochastic perspectives*, IEEE Trans. Aut. Control, AC-32 (1987), pp. 1076–1084.

[79] I. R. PETERSEN, *Disturbance attenuation and H^∞ optimization: a design method based on the algebraic Riccati equation*, IEEE Trans. Aut. Control, AC-32 (1987), pp. 427–429.

[80] P. P. KHARGONEKAR, I. R. PETERSEN AND K. ZHOU, *Robust stabilization of uncertain systems and H^∞ optimal control*, preprint.

[81] D. S. BERNSTEIN AND W. M. HADDAD, *LQG control with an H_∞ performance bound*, preprint.

[82] M. MARITON AND R. BERTRAND, *A homotopy algorithm for solving coupled Riccati equations*, Optimal Control Appl. Meth., 6 (1985), pp. 351–357.

[83] S. KUMAR AND J. H. SEINFELD, *Optimal location of measurements for distributed parameter estimation*, IEEE Trans. Aut. Control, AC-23 (1978), pp. 690–698.

[84] M. I. J. CHANG AND T. T. SOONG, *Optimal controller placement in modal control of complex systems*, J. Math. Anal. Appl., 75 (1980), pp. 340–358.

[85] P. C. HUGHES AND R. E. SKELTON, *Controllability and observability for flexible spacecraft*, J. Guidance, Control, Dynamics, 3 (1980), pp. 452–459.

[86] S. OMATU AND J. H. SEINFELD, *Optimization of sensor and actuator locations in a distributed parameter system*, J. Franklin Inst., 315 (1983), pp. 407–421.

[87] W. E. VANDER VELDE AND C. R. CARIGNON, *Number and placement of control system components considering possible failures*, J. Guidance, Control, Dynamics, 7 (1984), pp. 703–709.

[88] C. S. KUBRUSLY AND H. MALEBRANCHE, *Sensors and controllers location in distributed systems—a survey*, Automatica, 21 (1985), pp. 117–128.

[89] A. L. HALE, R. J. LISOWSKI AND W. E. DAHL, *Optimal simultaneous structural and control design of maneuvering flexible spacecraft*, J. Guidance, Control, Dynamics, 8 (1985), pp. 86–93.

[90] D. S. BODDEN AND J. L. JUNKINS, *Eigenvalue optimization algorithms for structure/controller design iterations*, J. Guidance, Control, Dynamics, 8 (1985), pp. 697–706.

[91] D. F. MILLER AND J. SHIM, *Gradient-based combined structural and control optimization*, J. Guidance, Control, Dynamics, 10 (1987), pp. 291–298.

[92] D. C. HYLAND AND D. S. BERNSTEIN, *The optimal projection equations for model reduction and the relationships among the methods of Wilson, Skelton, and Moore*, IEEE Trans. Aut. Control, AC-30 (1985), pp. 1201–1211.

[93] D. S. BERNSTEIN AND D. C. HYLAND, *The optimal projection equations for reduced-order state estimation*, IEEE Trans. Aut. Control, AC-30 (1985), pp. 583–585.

[94] S. L. CAMPBELL AND C. D. MEYER, JR., *Generalized Inverses of Linear Transformations*, Pitman, London, 1979, Ch. 7, pp. 120–127.

[95] L. T. WATSON, *Globally convergent homotopy methods: A tutorial*, Appl. Math. Comput., 31BK (1989), pp. 369–396.

[96] L. T. WATSON, *Globally convergent homotopy algorithms for nonlinear systems of equations*, Nonlinear Dynamics, 1 (1990), pp. 143–191.

[97] S. SMALE, *A convergent process of price adjustment and global Newton methods*, J. Math. Econ., 3 (1976), pp. 107–120.

[98] J. T. SPANOS, M. H. MILMAN AND D. L. MINGORI, *Optimal model reduction and frequency-weighted extension*, AIAA paper number AIAA–90–3345–CP, 1990, pp. 271–284.

[99] E. G. COLLINS, JR., D. J. PHILLIPS AND D. C. HYLAND, *Robust decentralized control laws for the ACES structure*, Contr. Sys. Mag., April (1991), pp. 62–70.

[100] D. ŽIGIĆ, L. T. WATSON, E. G. COLLINS, JR., AND D. S. BERNSTEIN, *Homotopy methods for solving the optimal projection equations for the H_2 reduced order model problem*, Internat. J. Control, 56 (1992), pp. 173–191.

[101] D. ŽIGIĆ, L. T. WATSON, E. G. COLLINS, JR., AND D. S. BERNSTEIN, *Homotopy approaches to the H_2 reduced order model problem*, J. Math. Systems, Estimation, Control, 3 (1993), pp. 173–205.

HOW TO PROPERLY RELAX DELAYED CONTROLS

JAVIER F. ROSENBLUETH
IIMAS-UNAM, Apartado Postal 20-726, México, DF, México

Abstract. Few attempts have been made to solve the question of how optimal control problems involving delays in the controls should be properly relaxed. Recently, several relaxation procedures have been proposed and, in this paper, we summarize the main features of these models, and explain in what sense they overcome certain difficulties encountered when previous techniques are used.

Key words: Optimal control problems, relaxation theory, delayed controls

1. Introduction

A relaxation procedure for optimal control problems is a device for extending the set of original controls in such a way that existence of minimizers for the extended (or relaxed) problem can be assured. In general, the set of original controls is such that, except for quite restricted classes of optimal control problems, existence theorems cannot be proved. For this reason, as Clarke points out in [1], there are many who deem a relaxed problem the only reasonable problem to consider in practice.

But existence of a minimizer is not the only desirable property of an extension. The problem of primary interest is the original one, and an important objective of relaxation is, once a solution of the relaxed problem is assured, to be able to approximate it with ordinary controls. When this can be achieved, the relaxation procedure is said to be *proper*.

For a wide range of optimal control problems, this theory has been successfully developed, by forming the relaxed controls a compactification of ordinary controls with respect to an appropriate topology (the weak star topology). This class of problems, however, does not include problems involving transformations of the control functions and, until recently, little attention has been paid to the important question of how such problems should be properly relaxed.

The purpose of this paper is to summarize the main results obtained so far in the search of a proper relaxation for problems with constant delays in the controls.

We begin by describing the standard relaxation procedure for delay free problems, and briefly discuss why properness of an extension is desirable. In Section 3 we explain a procedure for relaxing problems with nonseparable commensurate delays, based on a well-known technique for reducing systems of this nature into a delay free problem. Section 4 is devoted to a description of other models of relaxation which have been recently proposed, and we explain in detail the main features of each procedure.

S. Gomez and J-P. Hennart (eds.), Advances in Optimization and Numerical Analysis, 145–154.
© 1994 *Kluwer Academic Publishers.*

2. Standard relaxed controls

In this section we describe the standard relaxation procedure for delay free problems, and briefly discuss why properness of an extension is desirable even in situations where existence of a minimizing ordinary control is assured.

The basic concepts we shall deal with can be summarized by considering the following optimal control problem. Let $T = [0,1]$ and suppose we are given a point $\xi \in \mathbf{R}^n$, a compact set $\Omega \subset \mathbf{R}^m$, and functions g mapping $\mathbf{R}^n$ to $\mathbf{R}$ and f mapping $\mathbf{R} \times \mathbf{R}^n \times \mathbf{R}^m$ to $\mathbf{R}^n$. The problem we address is that of minimizing $g(x(1))$ subject to

$$\begin{aligned} &\dot{x}(t) = f(t, x(t), u(t)) \quad \text{a.e. in } T \\ &x(0) = \xi \\ &u(t) \in \Omega \quad \text{a.e. in } T \end{aligned}$$

where u is any measurable function mapping T to $\mathbf{R}^m$. Denote by $\mathcal{U}(T,\Omega)$ the set of measurable functions $u : T \to \mathbf{R}^m$ satisfying $u(t) \in \Omega$ a.e. in T. Elements of $\mathcal{U}(T,\Omega)$ are called *ordinary* (or *original*) *controls*. A pair (x,u) comprising an ordinary control u and an absolutely continuous function $x : T \to \mathbf{R}^n$ which satisfies the differential equation is called an *ordinary process* and, if $x(0) = \xi$, the ordinary process is called *admissible* (under the usual hypotheses imposed on the data, given an ordinary control u, there is a unique x such that (x,u) is an admissible ordinary process). The optimization problem, posed over admissible ordinary processes, is called the *original problem* ($\mathrm{P}_{\text{original}}$).

For problems like the above, in situations where existence of a minimizing admissible ordinary process is not assured, one is usually interested in finding admissible ordinary processes which come close to achieving the infimum cost. A methodology for finding such processes involves introduction of the notion of *relaxed process*.

A *relaxed control* is defined to be a measurable function μ mapping T to the space of Radon probability measures on Ω, where "measurable" is understood in the sense that $t \mapsto \int c(r)\mu(t)(dr)$ is measurable for all $c \in C(\Omega)$, the Banach space of continuous real valued functions on Ω with the *sup* norm. We denote by $\mathcal{M}(T,\Omega)$ the set of relaxed controls. It can be regarded as a subset of the topological dual space of $L^1(T, C(\Omega))$ acting on elements φ in the primal space according to

$$\varphi \mapsto \int \varphi(t, \mu(t))dt = \int dt \int \varphi(t,r)\mu(t)(dr).$$

We embed the set $\mathcal{U}(T,\Omega)$ of ordinary controls into $\mathcal{M}(T,\Omega)$ by identifying each $u \in \mathcal{U}(T,\Omega)$ with the function $t \mapsto \delta_{u(t)}$ where δ_a, the Dirac measure at a, denotes the unit measure concentrated at the point a. A *relaxed process* is defined to be a pair (x,μ) comprising a relaxed control μ and an absolutely continuous function x which satisfies the differential equation

$$\dot{x}(t) = \int f(t, x(t), r)\mu(t)(dr)$$

and it is *admissible* if $x(0) = \xi$. The problem posed over admissible relaxed processes is called the *relaxed problem* ($\mathrm{P}_{\text{relaxed}}$).

In [8] it is shown that, if we equip the set $\mathcal{M}(T,\Omega)$ with the relative weak star topology of $L^1(T,C(\Omega))^*$ and regard $\mathcal{U}(T,\Omega)$ as a subspace of $\mathcal{M}(T,\Omega)$, then $\mathcal{M}(T,\Omega)$ is a compact set which coincides with the closure of $\mathcal{U}(T,\Omega)$. This result implies the existence of a relaxed minimizer and the fact that it can be approximated with ordinary controls. We summarize these two properties by the statement "the set of relaxed controls provides a *proper extension* to the set of original controls".

For problems like the one we are addressing, observe that the set of controls u for which there exists x such that (x,u) is an admissible ordinary process, coincides with the whole set of ordinary controls. This fact clearly implies that the notion of proper relaxation fulfils the objective of finding admissible ordinary processes which come close to achieving the infimum cost. Moreover, properness is equivalent to the statement "the minimum cost for the relaxed problem coincides with the infimum cost for the original one", i.e.,

$$\inf\{\mathrm{P}_{\mathrm{original}}\} = \min\{\mathrm{P}_{\mathrm{relaxed}}\}.$$

For problems involving side constraints, these two notions of properness need not be equivalent. The effect of relaxation may very well reduce the infimum cost and the purpose in extending the set of ordinary controls may differ from the previous one. A simple example illustrates this fact.

Suppose we want to minimize $g(x(1),y(1),z(1))$ subject to

$$\begin{aligned}
&(\dot{x}(t),\dot{y}(t),\dot{z}(t)) = f(t,x(t),y(t),z(t),u(t)) \quad \text{a.e. in } T\\
&(x(0),y(0),z(0)) = (0,0,0)\\
&u(t)\in\Omega \quad \text{a.e. in } T
\end{aligned}$$

where $T=[0,1]$, $\Omega=[-1,1]$, and $g:\mathbf{R}^3\to\mathbf{R}$ and $f: T\times\mathbf{R}^3\times\mathbf{R}\to\mathbf{R}^3$ are given by

$$g(x,y,z)=z \qquad \text{and} \qquad f(t,x,y,z,u)=(u,x^2,x^2-u^2).$$

This is a particular case of the problem posed above so that all the previous theory can be applied. However, suppose we add one more restriction to the set of admissible processes, and we impose the condition $y(1)=0$.

Observe that, if (x,y,z,u) is any admissible ordinary process, then

$$y(t)=\int_0^t [x(s)]^2 ds \qquad (t\in T)$$

implying that

$$x(t)=\int_0^t u(s)ds = 0 \qquad (t\in T).$$

Therefore, $u(t)=0$ a.e. in T, and so also

$$z(t)=\int_0^t ([x(s)]^2-[u(s)]^2)ds = 0 \qquad (t\in T).$$

Hence, $(x,y,z,u)=(0,0,0,0)$, and this is the only admissible ordinary process, yielding the minimum $g(x(1),y(1),z(1))=0$.

Consider now the relaxed control

$$\mu(t) = \frac{1}{2}\delta_{-1} + \frac{1}{2}\delta_1 \qquad (t \in T).$$

As one readily verifies, the associated trajectory is given by $(0, 0, -t)$ and, moreover, this control minimizes the relaxed problem. Thus,

$$\min\{\mathrm{P}_{\text{original}}\} = 0 > -1 = \min\{\mathrm{P}_{\text{relaxed}}\}.$$

This behaviour is clearly a consequence of the fact that, for this problem, the set of controls u whose associated processes are admissible is strictly contained in the set of original controls. Observe, however, that the relaxed minimizer can be approximated with ordinary controls whose associated processes are "nearly" admissible in the sense that they satisfy the side constraint in the limit.

The notion of proper relaxation takes into account that the relation $y(1) = 0$ may represent restrictions measured with some error and one is inclined to consider points that, "nearly" satisfying this relation, may yield lower values of g than do minimizing ordinary solutions. The objective, then, of finding admissible ordinary processes which come close to achieving the infimum cost is replaced by that of finding ordinary processes, not necessarily admissible, which come close to achieving the minimum cost for the relaxed problem. Under the usual assumptions on the data of a wide range of optimal control problems, this is obtained via the standard relaxation procedure.

Now, a methodology for approximating a given relaxed control with ordinary ones is explicitly derived in [8]. In particular, for the above example, it leads to the construction of ordinary controls u_i that equal alternately 1 and -1 on succesive intervals of length $1/2i$. One can show that the corresponding trajectories satisfy

$$-1 \leq g(x_i(1), y_i(1), z_i(1)) \leq \frac{1}{(2i)^2} - 1 \qquad \text{and} \qquad 0 \leq y_i(1) \leq \frac{1}{(2i)^2}$$

so that one can lower the value of g to nearly -1 and, in doing so, the restriction $y(1) = 0$ is violated by an arbitrarily small amount. For a further discussion of these ideas, we refer to [4,8].

3. The reduced problem technique

The first attempts to solve the question of how optimal control problems involving time delays in the controls should be properly relaxed, were made by Warga in [7,8,9] treating the special cases when the delayed control functions are separable (additively coupled) or when the delays in the controls are constant and commensurate (the quotient of any two delays is rational). For the first case, the task of finding an appropriate notion of relaxation was achieved by a straightforward adaptation of standard delay free theory.

For the second case, involving constant commensurate delays not necessarily separable, one can easily find an example for which it is no longer sufficient to consider the standard relaxed version of the control function. The approach followed by Warga to introduce a new model of relaxation is based on a well-known technique to reduce a control system of this nature to a delay free problem.

To explain this procedure, let us consider a problem similar to that of Section 2. Suppose the data of the problem are as before except for a given integer p (the number of delays), a constant $0 \leq \theta \leq 1/p$, and f is now a function mapping $\mathbf{R} \times \mathbf{R}^n \times \mathbf{R}^{m(p+1)}$ to $\mathbf{R}^n$. Consider the problem, which we label (P), of minimizing $g(x(1))$ subject to

$$\begin{aligned}&\dot{x}(t) = f(t, x(t), u(t), u(t-\theta), \ldots, u(t-p\theta)) \quad \text{a.e. in } T\\&x(0) = \xi\\&u(t) \in \Omega \quad \text{a.e. in } \hat{T}\end{aligned}$$

where $\hat{T} = [-p\theta, 1]$. An *ordinary control* is a measurable function $u : \hat{T} \to \mathbf{R}^m$ satisfying $u(t) \in \Omega$ a.e. in $\hat{T}$. *Ordinary processes, admissible ordinary processes* and the *original problem* are defined by analogy with the earlier definitions.

The basic idea is to section ordinary controls and the corresponding trajectories into segments of length θ, and to stack these segments to form higher dimensional vector valued functions on the interval $[0, \theta]$. The resulting functions satisfy a delay free differential equation.

To be precise, let $k := \max\{i \in \mathbf{N} \mid i\theta < 1\}$, and extend f to $[0, (k+1)\theta] \times \mathbf{R}^n \times \mathbf{R}^{m(p+1)}$, if necessary, by setting $f(t, x, r) := 0$ for all $t \in (1, (k+1)\theta]$ and $(x, r) \in \mathbf{R}^n \times \mathbf{R}^{m(p+1)}$. Define the function $\hat{f} = (\hat{f}_0, \hat{f}_1, \cdots, \hat{f}_k)$ by setting, for all $i = 0, 1, \ldots, k$, $t \in [0, \theta]$, $\hat{x} = (\hat{x}_0, \hat{x}_1, \ldots, \hat{x}_k) \in \mathbf{R}^{n(k+1)}$ and $\hat{u} = (\hat{u}_{-p}, \hat{u}_{-p+1}, \ldots, \hat{u}_k) \in \mathbf{R}^{m(k+p+1)}$,

$$\hat{f}_i(t, \hat{x}, \hat{u}) := f(t + i\theta, \hat{x}_i, \hat{u}_i, \hat{u}_{i-1}, \ldots, \hat{u}_{i-p}).$$

Set $\hat{\Omega} := \Omega^{k+p+1}$ and, for all $(\hat{x}_0, \hat{x}_1, \ldots, \hat{x}_k) \in \mathbf{R}^{n(k+1)}$, let

$$\hat{g}(\hat{x}_0, \hat{x}_1, \ldots, \hat{x}_k) := g(\hat{x}_k)$$

and

$$c(\hat{x}_0, \hat{x}_1, \ldots, \hat{x}_k) := (\xi, \hat{x}_0, \hat{x}_1, \ldots, \hat{x}_{k-1}).$$

The reduced problem, which we label $(\hat{\mathrm{P}})$, is that of minimizing $\hat{g}(\hat{x}(\theta))$ subject to

$$\begin{aligned}&\dot{\hat{x}}(t) = \hat{f}(t, \hat{x}(t), \hat{u}(t)) \quad \text{a.e. in } [0, \theta]\\&\hat{x}(0) = c(\hat{x}(\theta))\\&\hat{u}(t) \in \hat{\Omega} \quad \text{a.e. in } [0, \theta].\end{aligned}$$

One can easily show that (P) and $(\hat{\mathrm{P}})$ are equivalent in the sense that there is a one to one mapping from admissible ordinary processes for (P) to those for $(\hat{\mathrm{P}})$, that is, ordinary processes for $(\hat{\mathrm{P}})$ which satisfy the mixed boundary conditions $\hat{x}(0) = c(\hat{x}(\theta))$. Furthermore, this mapping can be defined so that the value of the cost is preserved when we pass between the two problems. Explicitly, given an admissible process (x, u) for (P), extend x by setting $x(t) := x(1)$ for $t \in (1, (k+1)\theta]$. For all $i = 0, 1, \ldots, k$ and $t \in [0, \theta]$, let $\hat{x}_i(t) := x(t + i\theta)$ and set $\hat{x} = (\hat{x}_0, \hat{x}_1, \ldots, \hat{x}_k)$. Defining $\hat{u} = (\hat{u}_{-p}, \hat{u}_{-p+1}, \ldots, \hat{u}_k)$ in a similar way, one readily verifies that $(\hat{x}, \hat{u})$ is an admissible ordinary process for $(\hat{\mathrm{P}})$ and, clearly, this mapping satisfies the desired properties.

If we now relax $(\hat{\mathrm{P}})$, which has no delays, along the lines of Section 2, we obtain a proper extension for this problem since, as we know, the set $\mathcal{M}([0,\theta],\hat{\Omega})$ of relaxed controls for the reduced problem coincides with the weak star closure of the set $\mathcal{U}([0,\theta],\hat{\Omega})$ of ordinary controls.

But what are the connections with the original delayed problem? Our initial objective was, since this is a problem without side constraints, to find admissible ordinary processes for (P) which come close to achieving the infimum cost. Can we achieve this objective by knowing that existence of a relaxed minimizer for $(\hat{\mathrm{P}})$ is assured and there is a sequence of ordinary controls for $(\hat{\mathrm{P}})$ converging to the minimizer?

To answer these questions, observe first that the reduced problem is not free of side constraints since admissible processes must satisfy a relation involving mixed boundary conditions. Therefore, if a relaxed minimizer $\hat{\mu}$ for $(\hat{\mathrm{P}})$ is given, and we construct a sequence of ordinary controls converging to $\hat{\mu}$, there is no guarantee that the corresponding processes will be admissible. On the other hand, the one to one mapping between the two problems is defined only for admissible ordinary processes. Thus, we might not be able to use the inverse of this mapping, and the purpose of relaxing the original delayed problem will not be achieved.

What we clearly need is to ensure that, given a relaxed minimizer for $(\hat{\mathrm{P}})$, there exists a suitable ordinary process approximating it and satisfying the mixed boundary conditions. In other words, we need to prove precisely the statement concerning the infima costs, that is, that the minimum cost for $(\hat{\mathrm{P}}_{\mathrm{relaxed}})$ coincides with the infimum cost for $(\hat{\mathrm{P}}_{\mathrm{original}})$.

The example of Section 2 shows that, for problems with side constraints, properness of a relaxation procedure does not necessarily imply this relation. However, as we prove in [2], it does hold for these specific constraints, i.e.,

$$\min\{\hat{\mathrm{P}}_{\mathrm{relaxed}}\} = \inf\{\hat{\mathrm{P}}_{\mathrm{original}}\}.$$

In view of this result we can apply the one to one mapping between processes for (P) and $(\hat{\mathrm{P}})$, and obtain admissible ordinary processes for (P) which come close to achieving the infimum cost. Thus, we achieve our objective through this technique. However, several respects make it unsatisfactory.

Notice, first of all, that the dimension of the state and control spaces in the reduced problem can be very large ($(k+1)\times n$ and $(k+p+1)\times m$ respectively). Now, the fact that we have posed (P) on the time interval $T=[0,1]$ is merely a normalization procedure, and the above technique for eliminating delays works on an arbitrary compact interval. However, the dimension of the spaces involved increases rapidly with the length of the underlying time interval. Apart from this, in passing to the reduced problem, the connections with the original problem are somewhat obscured. In particular, though both problems are equivalent in the sense that there is a one to one mapping from admissible ordinary processes for the original problem to those for the reduced one, no assertions in this respect seemed clear for admissible relaxed processes. In other words, this technique exhibits a set of relaxed controls for the reduced problem, but not for the original one. Finally, this relaxation procedure was achieved only for problems with commensurate delays.

These features of the reduced problem technique led to the search of new procedures whose description will be the content of the following section.

4. New models of relaxation

In a recent paper (see [10]), Warga made the first attempts to solve some of the difficulties mentioned above, and proposed a new model (weak relaxation) applicable to problems with delays in the controls which may be nonseparable and noncommensurate. The basic ideas underlying this and other procedures which were later proposed, can be explained by considering the following problem which generalizes those of the two previous sections.

Suppose that, instead of the commensurate delays $0 < \theta < \cdots < p\theta \leq 1$, we are now given arbitrary constants $0 < \theta_1 < \cdots < \theta_p \leq 1$, and we want to minimize $g(x(1))$ subject to

$$\begin{aligned}&\dot{x}(t) = f(t, x(t), u(t), u(t-\theta_1), \ldots, u(t-\theta_p)) \quad \text{a.e. in } T\\ &x(0) = \xi\\ &u(t) \in \Omega \quad \text{a.e. in } \hat{T}\end{aligned}$$

where $\hat{T} = [-\theta_p, 1]$.

The approach initiated by Warga is based on the idea of transforming the original problem into one for which the delays are no longer present in the dynamics but in certain compatibility conditions imposed on the controls. Weakly relaxed controls are then defined to be standard relaxed controls satisfying some conditions generalizing those on the ordinary controls.

Explicitly, suppose (x, u) is an ordinary process. Let $\theta_0 := 0$ and define $u_i(t) := u(t-\theta_i)$ for $i = 0, 1, \ldots, p$ and $t \in T$. For all $i = 1, 2, \ldots, p$, let $\alpha_i := \theta_i - \theta_{i-1}$ and $T_i := [\alpha_i, 1]$. Then $(x, \tilde{u})$, with $\tilde{u} = (u_0, u_1, \ldots, u_p)$, satisfies

$$\begin{aligned}&\dot{x}(t) = f(t, x(t), \tilde{u}(t)) \quad \text{a.e. in } T\\ &x(0) = \xi\\ &\tilde{u}(t) \in \Omega^{p+1} \quad \text{a.e. in } T\end{aligned}$$

together with

$$u_i(t) = u_{i-1}(t-\alpha_i) \quad \text{a.e. in } T_i, \quad (i = 1, 2, \ldots, p).$$

In this way we separate the $p+1$ functions on T which are inserted as the last $p+1$ arguments of f in the dynamics, and impose conditions reflecting the fact that these functions are delayed versions of each other. We can use the standard theory for the new system and generalize the compatibility conditions.

A *weakly relaxed control* is defined to be an element μ of $\mathcal{M}(T, \Omega^{p+1})$ satisfying the compatibility conditions

$$\mathcal{P}_i\mu(t) = \mathcal{P}_{i-1}\mu(t-\alpha_i) \quad \text{a.e. in } T_i, \quad (i = 1, 2, \ldots, p)$$

where $\mathcal{P}_i : C(\Omega^{p+1})^* \to C(\Omega)^*$ and $\mathcal{P}_i\mu(t)$ denotes the projection onto the i-th coordinate of $\mu(t)$. An alternative, equivalent, statement of these conditions is:

$$\int_{T_i} dt \int \varphi(t, r_i)\mu(t)(dr) = \int_{T_i} dt \int \varphi(t, r_{i-1})\mu(t-\alpha_i)(dr)$$

for all $i = 1, \ldots, p$ and $\varphi \in L^1(T, C(\Omega))$, where $r = (r_0, r_1, \ldots, r_p)$.

The set of weakly relaxed controls, regarded as a subspace of the space of standard relaxed controls, is shown in [10] to be convex and compact, and existence theorems and first and higher order controllability conditions (generalizing Pontryagin's maximum principle) were established. However, several examples were found in [2,3], of systems involving two or more commensurate delays, for which solutions of this model could not be approximated with original controls.

A modified form of Warga's model (strong relaxation) was introduced in [2] and shown to be proper in the sense that the strongly relaxed problem has a solution and, for optimal control problems without endpoint state constraints, the minimum cost for the relaxed problem coincides with the infimum cost for the original one. This procedure coincides with Warga's model for systems involving a single delay in the controls. For problems with two or more delays, it is defined only for the commensurate case, and it solves some of the difficulties encountered when the reduced problem technique is used.

To explain this model, let us return to the commensurate case posed in Section 3, that is, we assume that $\theta_i = i\theta$ for some $0 < \theta \leq 1/p$.

A *strongly relaxed control* is defined to be an element μ of $\mathcal{M}(T, \Omega^{p+1})$ satisfying the compatibility conditions

$$\mathcal{P}_{1,2,\ldots,p}\mu(t) = \mathcal{P}_{0,1,\ldots,p-1}\mu(t-\theta) \quad \text{a.e. in } [\theta, 1].$$

As before, the last conditions are equivalent to:

$$\int_\theta^1 \int \varphi(t, r_1, r_2, \ldots, r_p)\mu(t)(dr) = \int_\theta^1 \int \varphi(t, r_0, r_1, \ldots, r_{p-1})\mu(t-\theta)(dr)$$

for all $\varphi \in L^1(T, C(\Omega^p))$, where $r = (r_0, r_1, \ldots, r_p)$.

Observe that this model shows explicitly which subset of the space of standard relaxed controls provides a proper extension of the delayed problem. Due to this fact, the dimension of the spaces involved in the original problem remains unaltered when the set of controls is strongly relaxed, thus solving one of the main disadvantages of the relaxation scheme via the reduced problem. A second feature is that the conditions defining membership of the set of strongly relaxed controls are easily verified in terms of projections onto the coordinates of the control functions so that, again, this set is characterized without the need of transforming the original problem. Apart from this, the proof given in [2] of its being proper includes an answer to the question of how admissible strongly relaxed controls may be associated to admissible relaxed controls for the reduced problem.

This proof can be briefly described as follows. Given a strongly relaxed control μ, one constructs a relaxed control $\hat{\mu}$ for the reduced problem using known conditions concerning existence of probability distributions whose marginal distributions satisfy certain relationships. One then constructs an ordinary control $\hat{u}$ for the reduced problem whose cost is arbitrarily close to that of $\hat{\mu}$. Using the one to one mapping between admissible ordinary processes for the original problem and those for the reduced one, one obtains an ordinary control whose cost is arbitrarily close to that of μ.

The question of finding a proper relaxation procedure was then solved for the commensurate case (for more general commensurate delayed problems than the one we are considering we refer to [5]).

For problems with possibly noncommensurate delays, another relaxation procedure, which we refer to as the $\mathcal{D}$-model, was introduced in [2] and shown to be proper for problems characterized by a linear cost function and dynamics which are separable in the state and control variables and affine in their dependence. The question of its being proper in the general case, was solved in the affirmative by Warga and Zhu in [11].

For the problem we are considering in this section, a *$\mathcal{D}$-relaxed control* is defined to be an element μ of $\mathcal{M}(T, \Omega^{p+1})$ such that, for all φ in $\mathcal{D}(\theta_1, \ldots, \theta_p)$,

$$\int_0^1 dt \int \varphi(t, r_0, r_1, \ldots, r_p)\mu(t)(dr) \leq 0$$

where $r = (r_0, \ldots, r_p)$, and

$$\mathcal{D}(\theta_1, \ldots, \theta_p) = \{\varphi \in L^1(T, C(\Omega^{p+1})) \mid \text{for every } u \in \mathcal{U}([-\theta_p, 1], \Omega), \int_0^1 \varphi(t, u(t), u(t-\theta_1), \ldots, u(t-\theta_p))dt \leq 0\}.$$

For problems with commensurate delays, this model and that of strong relaxation are equivalent, since the spaces of strongly and $\mathcal{D}$-relaxed controls coincide with the weak star closure of the space of original controls. For problems with noncommensurate delays, however, no characterization of the $\mathcal{D}$-model has been achieved. A question which remained open for some time was whether weak relaxation could still be proper for the noncommensurate case, but it was finally solved in [6] by providing a problem for which a weakly relaxed control cannot be approximated with ordinary noncommensurate delayed controls.

Since the $\mathcal{D}$-model is an abstract relaxation procedure, it is still of interest to know whether the conditions defining membership of the space of $\mathcal{D}$-relaxed controls can be replaced by constraints of a more verifiable nature, as is the case for problems with commensurate delays.

References

1. Clarke FH (1983) *Optimization and nonsmooth analysis*, Wiley Interscience, New York
2. Rosenblueth JF and Vinter RB (1991) *Relaxation procedures for time delay systems*, Journal of Mathematical Analysis and Applications, **162**: 542-563
3. Rosenblueth JF (1992) *Strongly and weakly relaxed controls for time delay systems*, SIAM Journal on Control and Optimization, **30**: 856-866
4. Rosenblueth JF (1992) *Proper relaxation of optimal control problems*, Journal of Optimization Theory and Applications, **74**: 509-526
5. Rosenblueth JF (1993) *Approximation of strongly relaxed minimizers with ordinary delayed controls*, submitted to Applied Mathematics and Optimization
6. Rosenblueth JF (1993) *Weak relaxation of delayed controls*, submitted to Transactions of the American Mathematical Society
7. Warga J (1968) *The reduction of certain control problems to an 'ordinary differential type'*, SIAM Review **10**: 219-222
8. Warga J (1972) *Optimal control of differential and functional equations*, Academic Press, New York

9. Warga J (1974) *Optimal controls with pseudodelays*, SIAM Journal on Control and Optimization, **12**: 286-299
10. Warga J (1986) *Nonadditively coupled delayed controls*, privately circulated
11. Warga J and Zhu QJ (1992) *A proper relaxation of shifted and delayed controls*, Journal of Mathematical Analysis and Applications, **169**: 546-561

ON OPERATOR EXTENSIONS: THE ALGEBRAIC THEORY APPROACH

ISMAEL HERRERA
Instituto de Geofísica, UNAM
Apartado postal 22-582, 14000 México, D.F., MEXICO

Abstract. The Localized Adjoint Method (LAM) is a new and promising methodology for discretizing partial differential equations, which is based on Herrera's Algebraic Theory of Boundary Value Problems. A large number of numerical applications have already been made. Herera's Algebraic Theory implies a kind of operator extensions of great generality, which can be applied to fully discontinuous trial and test functions, simultaneously. This is in contrast with standard theory of distributions, which can be applied to discontinuous trial functions, only if test functions satisfy a corresponding degree of regularity, or viceversa. This paper is devoted to make a brief presentation of such extensions.

1. Introduction

The Localized Adjoint Method (LAM) is a new and promising methodology for discretizing partial differential equations, which is based on Herrera's Algebraic Theory of Boundary Value Problems [1]–[5]. Applications have successively been made to ordinary differential equations, for which highly accurate algorithms were developed [4], [6]–[8], multidimensional steady state problems [9] and optimal spatial methods for advection-diffusion equations [10]–[17]. More recently, in a pair of articles [18, 19], generalizations of Characteristic Methods that we refer to as Eulerian-Lagrangian Localized Adjoint Method (ELLAM), were provided. Related work has been published separately [20]–[23] and some more specific applications have already been made [24]–[29].

For differential operators, Herrera's Algebraic Theory of Boundary Value Problems imply a kind of operator extensions of great generality, since using it, fully discontinuous trial and test functions can be applied simultaneously. Actually, the operator extensions implied by the Algebraic Theory (the "algebraic extensions"), yield extensions of distributional operators, because the distributional extensions coincide with the algebraic extensions, whenever the former are defined. However, the operator extensions implied by the Algebraic Theory are well defined, in cases for which the distributional definitions are not. This is the case, for example, when trial and test functions are fully discontinuous.

The definition of the algebraic extensions is based on an algebraic structure which systematically occurs in boundary value problems [2, 5]. In the present paper a comparison is made with the distributional approach [30, 31]. It must be mentioned that although in previous work, attention has been mainly devoted to analyze the implications of the theory for single differential equations, the manner of applying it to systems of equations has been explained in [22]. The interested reader may

S. Gomez and J-P. Hennart (eds.), Advances in Optimization and Numerical Analysis, 155–163.
© 1994 *Kluwer Academic Publishers.*

find more thorough expositions of the algebraic structure in [2, 5]. A more recent exposition presenting several aspects of the algebraic structure in a more complete manner, is given in [32] and a more systematic derivation of the operator extensions from such algebraic structure, will appear in [33]. A monograph, in which the discussion was restricted to symmetric operators, has already appeared in book form [1].

The operator extensions implied by the author's Algebraic Theory (the algebraic extensions), are introduced in Section 2. In Section 3, a sketch of the proof that the algebraic extension is indeed an extension of the distributional definition, is given. Section 4 is devoted to present simple illustrations of the results produced by the algebraic extensions.

2. Operator extensions

Consider a region Ω and for simplicity, assume the spaces of trial and test functions, defined in Ω, are the same linear space: D (i.e., $D = D_1 = D_2$). Assume further, that functions belonging to D may have jump discontinuities across some internal boundaries whose union will be denoted by Σ. For example, in applications of the theory to finite element methods, the set Σ would be the union of all the interelement boundaries.

To be specific, consider a linear differential operator $\mathcal{L}$ of order m and assume $\{\Omega_1, \ldots, \Omega_E\}$ is a partition of Ω. More precisely, $\{\Omega_1, \ldots, \Omega_E\}$ is a collection of disjoint open regions (the "elements") of Ω, such that Ω is contained in the closure of the union of $\{\Omega_1, \ldots, \Omega_E\}$. Then, one can define $D = H^m(\Omega_1) \oplus \cdots \oplus H^m(\Omega_E)$. In this case $\Sigma = \Omega - (\Omega_1 \cup \ldots \cup \Omega_E)$.

The definition of formal adjoint requires that a differential operator $\mathcal{L}$ and its formal adjoint $\mathcal{L}^*$, satisfy the condition that $w\mathcal{L}u - u\mathcal{L}^*w$ be a divergence; i.e.:

$$w\mathcal{L}u - u\mathcal{L}^*w = \nabla \cdot \{\underline{\mathcal{D}}(u, w)\} \tag{1}$$

for a suitable vector-valued bilinear function $\underline{\mathcal{D}}(u, w)$, which involves derivatives up to order $m-1$. Integration of (1) over Ω and application of generalized divergence theorem [34], yield:

$$\sum_i \int_{\Omega_i} \{w\mathcal{L}u - u\mathcal{L}^*w\}\, dx = \int_{\partial\Omega} \mathcal{R}_\partial(u, w)\, dx + \int_\Sigma \mathcal{R}_\Sigma(u, w)\, dx, \tag{2}$$

where

$$\mathcal{R}_\partial(u, w) = \underline{\mathcal{D}}(u, w) \cdot \underline{n} \quad \text{and} \quad \mathcal{R}_\Sigma(u, w) = -[\underline{\mathcal{D}}(u, w)] \cdot \underline{n}. \tag{3}$$

Here, as in what follows, the square brackets stand for the "jumps" across Σ of the function contained inside; i.e., limit on the positive side minus limit on the negative one. The positive side of Σ is chosen arbitrarily and then the unit normal vector $\underline{n}$, is taken pointing towards the positive side of Σ. The operators $\mathcal{L}$ and $\mathcal{L}^*$ are understood in a distributional sense, and since they are of order m, both $\int_{\Omega_i} w\mathcal{L}u\, dx$ and $\int_{\Omega_i} u\mathcal{L}^*w\, dx$ are well defined for every $i = 1, \ldots, E$. However, observe that $D \subset H^0(\Omega)$, but the relation $D \subset H^1(\Omega)$ does not hold, so that when $u \in D$ one can only grant that $\mathcal{L}u \in H^{-m}(\Omega)$ and $\mathcal{L}^*w \in H^{-m}(\Omega)$. Thus, $\int_\Omega w\mathcal{L}u\,dx$ and $\int_\Omega u\mathcal{L}^*w\,dx$ are not well defined for every $u \in D$ and $w \in D$. This Section is devoted to present extensions $\hat{\mathcal{L}}$ and $\hat{\mathcal{L}}^*$ (the algebraic extensions), of $\mathcal{L}$ and $\mathcal{L}^*$, respectively, for which $\int_\Omega w\hat{\mathcal{L}}u\,dx$ and $\int_\Omega u\hat{\mathcal{L}}^*w\,dx$ are well defined for every $u \in D$ and $w \in D$.

In the general theory of partial differential equations, Green's formulas are used extensively [31]. For the construction of such formulas, it is standard to introduce a decomposition of the bilinear function $\mathcal{R}_\partial$ (see, for example, Lions and Magenes [31], Vol. I, pp. 114–115). Indicating, as it is usual, transposes of bilinears forms by means of a star, the general form of such decomposition is:

$$\mathcal{R}_\partial(u,w) \equiv \underline{\mathcal{D}}(u,w)\cdot\underline{n} = \mathcal{B}(u,w) - \mathcal{C}^*(u,w) \tag{4}$$

where $\mathcal{B}(u,w)$ and $\mathcal{C}^*(u,w)$ are two bilinear functions, which involve derivatives up to order $m-1$. When considering initial-boundary value problems, the definitions of these bilinear forms depend on the type of boundary and initial conditions to be prescribed. A basic property required of $\mathcal{B}(u,w)$ is that for any u which satisfies the prescribed boundary and initial conditions, $\mathcal{B}(u,w)$ is a well-defined linear function of w, independent of the particular choice of u. This linear function will be denoted by g_∂ (thus, its value for any given function w, will be $g_\partial(w)$) and the boundary conditions can be specified by requiring that $\mathcal{B}(u,w) = g_\partial(w)$, for every $w \in D$ (or more briefly: $\mathcal{B}(u,\cdot) = g_\partial$). For example, for Dirichlet problem of Laplace Equation, $\mathcal{B}(u,w)$ can be taken to be $u\partial w/\partial n$, on $\partial\Omega$ [19]. Thus, if u_∂ is the prescribed value of u on $\partial\Omega$, one has $\mathcal{B}(u,w) = u_\partial \partial w/\partial n$, for any function u which satisfies the boundary conditions. Thus, $g_\partial(w) = u_\partial\partial w/\partial n$, in this case.

The linear function $\mathcal{C}^*(u,\cdot)$, on the other hand, can not be evaluated in terms of the prescribed boundary values, but it also depends exclusively, on certain boundary values of u (the "complementary boundary values"). Generally, such boundary values can only be evaluated after the initial-boundary value problem has been solved. Taking again the example of Dirichlet problem for Lapalce Equation, $\mathcal{C}^*(u,w) = w\partial u/\partial n$ and the complementary boundary values, correspond to the normal derivative on $\partial\Omega$ [19].

In a similar fashion, convenient formulations of boundary value problems with prescribed jumps, requires constructing Green's formulas in discontinuous fields. This can be done by means of a general decomposition of the bilinear function $\mathcal{R}_\Sigma(u,w)$ that has been introduced by the author [22] (see also [19]) and whose definition is point-wise on Σ. The general theory includes the treatment of differential operators with discontinuous coefficients [4]. However, for simplicity in this article only continuous coefficients will be considered. In this case, such decomposition is easy to obtain and it stems from the algebraic identity:

$$[\underline{\mathcal{D}}(u,w)] = \underline{\mathcal{D}}([u],\dot{w}) + \underline{\mathcal{D}}(\dot{u},[w]) \tag{5}$$

where

$$[u] = u_+ - u_-, \qquad \dot{u} = (u_+ + u_-)/2 \tag{6}$$

The desired decomposition is obtained combining the second of Equs. (3) and (5):

$$\mathcal{R}_\Sigma(u,w) - \mathcal{J}(u,w) - \mathcal{K}^*(u,w) \tag{7}$$

where

$$\mathcal{J}(u,w) = -\underline{\mathcal{D}}([u],w)\cdot\underline{n} \tag{8a}$$

$$\mathcal{K}^*(u,w) = \mathcal{K}(w,u) = \underline{\mathcal{D}}(\dot{u},[w])\cdot\underline{n} \tag{8b}$$

Observe that the expressions for $\mathcal{J}(u,w)$ and $\mathcal{K}^*(u,w)$, involve jumps and averages across Σ, of u, w and their derivatives up to order $m-1$.

An important property of the bilinear functional $\mathcal{J}(u,w)$ is that when the jumps of u and its derivatives up to order $m-1$, are specified, it defines a unique linear function of w, which is independent of the particular choice of the function u, as long as it satisfies the prescribed jump conditions. When considering initial-boundary value problems with prescribed jumps, the linear function defined by the prescribed jumps in this manner, is denoted by j_Σ (thus, its value for any given function w, will be $j_\Sigma(w)$) and the jump conditions at any point of Σ, can be specified by means of the equation: $\mathcal{J}(u,\cdot) = j_\Sigma$ [19]. In problems with prescribed jumps, the linear function $\mathcal{K}^*(u,\cdot)$, plays a role similar to the complementary boundary values $\mathcal{C}^*(u,\cdot)$. It can only be evaluated after the initial-boundary value problem has been solved and certain information about the average of the solution and its normal derivatives on Σ, is known (see Equ. (8b)). Such information, is called the "generalized averages" [2, 4, 19].

Introducing the notation

$$\langle Pu,w\rangle = \sum_i \int_{\Omega_i} w\mathcal{L}u\,dx; \qquad \langle Q^*u,w\rangle = \sum_i \int_{\Omega_i} u\mathcal{L}^*w\,dx \tag{9a}$$

$$\langle Bu,w\rangle = \int_{\partial\Omega} \mathcal{B}(u,w)\,dx; \qquad \langle C^*u,w\rangle = \int_{\partial\Omega} \mathcal{C}(w,u)\,dx \tag{9b}$$

$$\langle Ju,w\rangle = \int_{\Sigma} \mathcal{J}(u,w)\,dx \quad \text{and} \quad \langle K^*u,w\rangle = \int_{\Sigma} \mathcal{K}(w,u)\,dx \tag{9c}$$

equation (2), can be written as:

$$\langle Pu,w\rangle - \langle Q^*u,w\rangle = \langle Bu,w\rangle - \langle C^*u,w\rangle + \langle Ju,w\rangle - \langle K^*u,w\rangle \tag{10}$$

This is an identity between bilinear forms and as such, can be written more briefly, after rearranging, as:

$$P - B - J = Q^* - C^* - K^* \tag{11}$$

This is *Green-Herrera formula for operators in discontinuous fields* [2, 5, 19].

It can be shown [33] that the pair of operators $\{J, -K^*\}$ constitutes a *weak decomposition* of $(P-B)-(Q-C)^*$, that B and J are *boundary operators* for P, which are *fully disjoint* and that (11) is indeed a Green's formula, in the *weak sense*. On the other hand, when $\mathcal{J}$ and $\mathcal{K}^*$ are defined by (8), then the pair of bilinear functionals $\{\mathcal{J}, -\mathcal{K}^*\}$, constitutes a strong decomposition, point-wise, of the bilinear functional $\mathcal{R}_\Sigma$ which is defined point-wise, also [33].

The algebraic extension $\hat{\mathcal{L}}$ of the distributional operator $\mathcal{L}$, is defined to be the bilinear functional $P-J$. More precisely, $\hat{\mathcal{L}}$ is defined by:

$$\int_\Omega w\hat{\mathcal{L}}u\,dx \equiv \langle (P-J)u,w\rangle \tag{12}$$

which holds whenever $u \in D$ and $w \in$D. Similarly, the operator extension corresponding to $\hat{\mathcal{L}}^*$ is defined to be the bilinear functional $Q^* - K^*$; i.e.:

$$\int_\Omega u\hat{\mathcal{L}}^* w\, dx \equiv \langle (Q-K)^* u, w\rangle \tag{13}$$

which also holds when both u and w belong to D. Thus, using these operator extensions, Green-Herrera formula (11) can be written as:

$$\int_\Omega w\hat{\mathcal{L}} u\, dx - \int_\Omega u\hat{\mathcal{L}}^* w\, dx \equiv \langle (B-C^*)u, w\rangle \tag{14}$$

for elements $u \in D$ and $w \in D$.

3. Comparison between $\hat{\mathcal{L}}$ and $\mathcal{L}$

Since the definitions for $\int_\Omega w\mathcal{L}u\, dx$ and $\int_\Omega u\mathcal{L}^* w\, dx$, which are standard in the theory of distributions, can not be applied to all possible pairs $\{u, w\}$, such that $u \in D$ and $w \in D$, the algebraic extensions $\hat{\mathcal{L}}$ and $\hat{\mathcal{L}}^*$ were introduced in the last Section, for which both $\int_\Omega w\hat{\mathcal{L}}u\, dx$ and $\int_\Omega u\hat{\mathcal{L}}^* w\, dx$ are well defined, whenever $u \in D$ and $w \in D$. It can be shown that the operators $\hat{\mathcal{L}}$ and $\hat{\mathcal{L}}^*$, defined by Equs. (12) and (13), are indeed extensions of the distributional operators $\mathcal{L}$ and $\mathcal{L}^*$, respectively, and the main purpose of this Section is to briefly explain a proof of this result. To achieve this goal, it is only necessary to prove that for every $u \in D$ and $w \in D$, the following two implications hold:

$$\int_\Omega w\mathcal{L}u\, dx \text{ is defined } \Rightarrow \int_\Omega w\mathcal{L}u\, dx = \int_\Omega w\hat{\mathcal{L}}u\, dx \tag{15a}$$

and

$$\int_\Omega u\mathcal{L}^* w\, dx \text{ is defined } \Rightarrow \int_\Omega u\mathcal{L}^* w\, dx = \int_\Omega u\hat{\mathcal{L}}^* w\, dx \tag{15b}$$

We only sketch a proof of implications (15) for the case when the order of the operator is 1 (i.e., $\mathcal{L} \equiv A(\underline{x})\partial/\partial x_i + B(\underline{x})$, where i may be $1, \ldots, N$, while the coefficients $A(\underline{x})$ and $B(\underline{x})$ are given functions of $\underline{x}$), since the result for the case when $\mathcal{L}$ is of arbitrary order, can be derived from this case, by induction on the order of the operator (see [33] for details). For this choice of $\mathcal{L}$, one has $\mathcal{L}^* w \equiv -\partial(Aw)/\partial x_i + Bw$ and $\underline{\mathcal{D}}(u, w)\cdot \underline{n} = Auwn_i$, so that

$$\mathcal{J}(u, w) = A[u]\dot{w}n_i \quad \text{and} \quad \mathcal{K}^*(u, w) = -A\dot{u}[w]n_i \tag{16}$$

by virtue of Equ. (8a). Actually, only the implication (15a) will be shown, since the proof of (15b) is similar. When $u \in D \subset H^0(\Omega)$ and $w \in H^1(\Omega)$ or when $u \in H^1(\Omega)$ and $w \in D \subset H^0(\Omega)$, $\int_\Omega w\mathcal{L}u\, dx$ is defined. Consider first the case when $u \in H^1(\Omega)$ and $w \in D \subset H^0(\Omega)$. In this case

$$\int_\Omega w\mathcal{L}u\, dx = \sum_i \int_{\Omega_i} w\mathcal{L}u\, dx \tag{17}$$

since both w and $\mathcal{L}u$ belong to $H^0(\Omega) = L^2(\Omega)$. In addition, when $u \in H^1(\Omega)$, u is continuous and $\mathcal{J}(u,w) \equiv 0$ on Σ, by virtue of Equ. (16). Thus, $Ju = 0$. This proves that

$$\int_\Omega w\hat{\mathcal{L}}u\,dx = \langle (P-J)u, w\rangle = \langle Pu, w\rangle = \sum_i \int_{\Omega_i} w\mathcal{L}u\,dx \tag{18}$$

Comparing Equs. (17) and (18), the desired equality follows.

If $u \in D \subset H^0(\Omega)$ and $w \in H^1(\Omega)$, a standard Green's formula used in the theory of distributions (see p. 115 of Lions and Magenes [31]), yields:

$$\int_\Omega w\mathcal{L}u\,dx = \int_\Omega u\mathcal{L}^*w\,dx + \langle (B-C^*)u, w\rangle = \sum_i \int_{\Omega_i} u\mathcal{L}^*w\,dx + \langle (B-C^*)u, w\rangle \tag{19}$$

The last equality holds because u and $\mathcal{L}^*w$ belong to $H^0(\Omega) = L^2(\Omega)$. On the other hand, using Green-Herrera formula (11), it is seen that

$$\int_\Omega w\hat{\mathcal{L}}u\,dx = \langle (P-J)u, w\rangle = \langle (Q^*-K^*)u, w\rangle + \langle (B-C^*)u, w\rangle \tag{20}$$

However, w is continuous, because $w \in H^1(\Omega)$. Thus, $\mathcal{K}(w,\cdot) \equiv 0$ on Σ, by virtue of Equ. (16), and $Kw = 0$. Hence, $\langle K^*u, w\rangle = \langle Kw, u\rangle = 0$. Using this fact, Equ. (20) reduces to

$$\int_\Omega w\hat{\mathcal{L}}u\,dx = \langle Q^*u, w\rangle + \langle (B-C^*)u, w\rangle = \sum_i \int_{\Omega_i} u\mathcal{L}^*w\,dx + \langle (B-C^*)u, w\rangle \tag{21}$$

Comparing this equation with (19), the desired result follows.

4. Examples

As a first illustration, let us consider the operators $\mathcal{L}$ and $\hat{\mathcal{L}}$, in the case when the distributional operator $\mathcal{L} \equiv d/dx$, the region Ω is the interval $(-1,1)$ of the real line and the partition of Ω is made of two subintervals: $\Omega_1 = (-1,0)$ and $\Omega_2 = (0,1)$. Then $\mathcal{L}^* \equiv -d/dx$, while $\mathcal{D}(u,w) \equiv uw$. Let the function u be defined by: $u = 0$ for $-1 < x < 0$ and $u = 1$ for $0 \le x < 1$. Thus, u is essentially, a Heaviside step function. The test function w will be taken having different degrees of smoothness.

Case A. $w \in H^1(\Omega)$, so that w is continuous.

i) In this case, application of a Green's formul operators (see [31], p.115) yields:

$$\int_{-1}^{1} w\mathcal{L}u\,dx = \int_{-1}^{1} u\mathcal{L}^*w\,dx + (uw)|_{-1}^{1}$$

and evaluating, it is obtained

$$\int_{-1}^{1} w\mathcal{L}u\,dx = -\int_0^1 \frac{dw}{dx}\,dx + w(1) = -w|_0^1 + w(1) = w(0). \tag{22a}$$

This result is standard. In essence, it establishes that du/dx is a Dirac's Delta function when u is a Heaviside step function.

ii) Using the fact that $\mathcal{D}(u,w) \equiv uw$ and applying Equ. (8a), it is seen that

$$\int_{-1}^{1} w\hat{\mathcal{L}}u\,dx = \sum_i \int_{\Omega_i} w\mathcal{L}u\,dx + (\dot{w}[u])_{x=0} = w(0) \tag{22b}$$

since $[u]_{x=0} = 1$, while $\dot{w}(0) = w(0)$, because w is continuous.

Case B. w has a jump discontinuiy at $x = 0$, so that $w \in D$ but $w \notin H^1(\Omega)$.

i) $\int_{-1}^{1} w\mathcal{L}u\,dx$ is not defined.

ii) $\int_{-1}^{1} w\hat{\mathcal{L}}u\,dx$ is well defined and it is still given by (22b), except that $\dot{w}(0) \neq w(0)$, so that

$$\int_{-1}^{1} w\hat{\mathcal{L}}u\,dx = \dot{w}(0) \tag{23}$$

It is recalled that $\dot{w}(0) = (w(0^+) + w(0^-))/2$.

As a second illustration, replace d/dx by d^2/dx^2, in the previous example. Then $\mathcal{L}^* \equiv \mathcal{L}$, while $\mathcal{D}(u,w) \equiv w\dfrac{du}{dx} - u\dfrac{dw}{dx}$ and proceeding as before:

Case A. $w \in H^2(\Omega)$, so that w is continuous, with continuous first order derivative.

i) In this case, as before, application of a Green's formula yields:

$$\int_{-1}^{1} w\mathcal{L}u\,dx = \int_{-1}^{1} u\mathcal{L}^*w\,dx + (wu' - uw')|_{-1}^{1}$$

and evaluating, it is obtained

$$\int_{-1}^{1} w\mathcal{L}u\,dx = \int_{0}^{1} w''\,dx - w'(1) = -w'(0) \tag{24a}$$

This is a standard result. In essence, it establishes that u'' is the derivative of Dirac's Delta function, when u is a Heaviside step function.

ii) Using the fact that $\mathcal{D}(u,w) \equiv wu' - uw'$ and applying Equ. (8a), it is seen that

$$\int_{-1}^{1} w\hat{\mathcal{L}}u\,dx = \sum_i \int_{\Omega_i} w\mathcal{L}u\,dx + (\dot{w}[u'] - \dot{w}'[u])_{x=0} = -w'(0) \tag{24b}$$

where the fact that $\dot{w}'(0) = w'(0)$, because w' is continuous, has been used.

Case B. w' has a jump discontinuity at $x = 0$, so that $w \in D$ but $w \notin H^2(\Omega)$.

i) $\int_{-1}^{1} w\mathcal{L}u\,dx$ is not defined.

ii) $\int_{-1}^{1} w\hat{\mathcal{L}}u\,dx$ is well defined and it is still given by (24b), except that $\dot{w}'(0) \neq w'(0)$, so that

$$\int_{-1}^{1} w\hat{\mathcal{L}}u\,dx = -\dot{w}'(0) \tag{25}$$

References

1. Herrera, I., "Boundary Methods: An Algebraic Theory", Pitman Advanced Publishing Program, London, 1984.
2. Herrera, I., "Unified Formulation of Numerical Methods. I Green's Formulas for Operators in Discontinuous Fields", Numerical Methods for Partial Differential Equations, Vol. **1**, pp. 25-44, 1985.
3. Herrera, I., "Unified Approach to Numerical Methods, Part 2. Finite Elements, Boundary Methods, and its coupling", Numerical Methods for Partial Differential Equations, **3**, pp. 159-186, 1985.
4. Herrera, I, Chargoy, L., Alduncin, G., "Unified Approach to Numerical Methods. III. Finite Differences and Ordinary Differential Equations", Numerical Methods for Partial Differential Equations, **1**, pp. 241-258, 1985.
5. Herrera, I., "Some unifying concepts in applied mathematics". In: "The Merging of Disciplines: New Directions in Pure, Applied, and Computational Mathematics". Edited by R.E. Ewing, K.I. Gross and C.F. Martin. Springer Verlag, New York, pp. 79-88, 1986 (Invited paper).
6. Celia, M.A., and Herrera, I., "Solution of General Ordinary Differential Equations Using The Algebraic Theory Approach", Numerical Methods for Partial Differential Equations, **3**(1) pp. 117-129, 1987.
7. Herrera, I., "The Algebraic Theory Approach for Ordinary Differential Equations: Highly Accurate Finite Differences", Numerical Methods for Partial Differential Equations, **3**(3), pp. 199-218, 1987.
8. Herrera, I. and Chargoy, L., "An Overview of the Treatment of Ordinary Differential Equations by Finite Differences", Pergamon Press, Oxford, Vol. **8**, pp. 17-19, 1987.
9. Celia, M.A., Herrera, I., and Bouloutas, E.T., "Adjoint Petrov-Galerkin Methods for Multi-Dimensional Flow Problems", In Finite Element Analysis in Fluids, T.J. Chung and Karr R., Eds., UAH Press, Huntsville Alabama. pp. 953-958, 1989. (Invited Lecture).
10. Herrera, I., "New Method for Diffusive Transport", Groundwater Flow and Quality Modelling, by D. Reidel Publishing Co. pp. 165-172, 1988.
11. Herrera, I., "New Approach to Advection-Dominated Flows and Comparison with other Methods", Computational Mechanics' 88, Springer Verlag, Heidelberg, Vol 2, 1988.
12. Celia, M.A., Herrera, I., Bouloutas, E.T., and Kindred, J.S., "A New Numerical Approach for the Advective-Diffusive Transport Equation", Numerical Methods for Partial Differential Equations, **5** pp. 203-226, 1989.
13. Celia, M.A., Kindred, J.S., and Herrera, I., "Contaminant Transport and Biodegradation: 1. A Numerical Model for Reactive Transport in Porous Media", Water Resources Research, **25**(6) pp. 1141-1148, 1989.
14. Herrera, I., G. Hernández. "Advances on the Numerical Simulation of Steep Fronts". Numerical Methods for Transport and Hydrologic Processes, Vol. 2, M.A. Celia, L.A. Ferrand and G. Pinder Eds. of the Series Developments in Water Science Computational Mechanics Publications, Elsevier, Amsterdam Vol. 36 pp. 139-145, 1988 (Invited paper).
15. Herrera, I., Celia, M.A., Martínez, J.D., "Localized Adjoint Method as a New Approach to Advection Dominated Flows". In Recent Advances in Ground-Water Hydrology, J.E. Moore, A.A. Zaporozec, S.C. Csallany and T.C. Varney, Eds. American Institute of Hydrology, pp. 321-327, 1989. (Invited paper).
16. Herrera, I., "Localized Adjoint Methods: Application to advection dominated flows". Groundwater Management: Quantity and Quality. IAHS Publ. No 188, pp. 349-357, 1989.89.
17. Herrera, G. and I. Herrera, "An Eulerian-Lagrangian Method of Cells, Based on Localized Adjoint Method", Numerical Methods for Partial Differential Equations (in press), 1993.
18. Celia, M.A., Russell, T.F., Herrera, I., and Ewing R.E., "An Eulerian-Langrangian Localized Adjoint Method for the Advection-Diffusion Equation", Advances in Water Resources, Vol. **13**(4), pp. 187-206, 1990.
19. Herrera, I., Ewing R.E., Celia, M.A. and Russell, T.F., "An Eulerian-Langrangian Localized Adjoint Method: The Theoretical Frame-work", Numerical Methods for Partial Differential Equations (in press), 1993.
20. Russell, T.F., "Eulerian-Lagrangian Localized Adjoint Methods for Advection-Dominated Problems", Numerical Analysis 1989, G.A. Watson and D.F. Griffiths, eds., Pitman Research

Notes in Mathematics Series, vol. 228, Longman Scientific and Technical, Harlow, U.K., 1990, 206–228.

21. Herrera, I., "Localized Adjoint Methods in Water Resources Problems", in Computational Methods in Surface Hydrology, G. Gambolati, A. Rinaldo and C.A. Brebbia, Eds., Springer-Verlag, 433–440, 1990 (Invited paper).
22. Herrera, I., "Localized Adjoint Methods: A New Discretization Methodology", Chapter 6 of the book: "Computational Methods in Geosciences", W.E. Fitzgibbon and M.F. Wheeler, eds., SIAM, 1992, 66–77 (Invited).
23. Russell, T.F and R.V. Trujillo., "Eulerian-Lagrangian Localized Adjoint Methods with Variable Coefficients in Multiple Dimensions", Computational Methods in Surface Hydrology, Eds. G. Gambolati et al., Computational Mechanics Publications, Springer Verlag, pp. 357–363, 1990.
24. Ewing, R.E.,"Operator Splitting and Eulerian-Lagrangian Localized Adjoint Methods for Multiphase Flow", J. Whiteman, ed., MAFELAP 1990, Academic Press, San Diego, 1991, pp. 215–232.
25. Herrera, I., R.E. Ewing.,"Localized Adjoint Methods: Applications to Multiphase Flow Problems". Proceedings Fifth Wyoming Enhanced Oil Recovery Symposium, Mayo 10-11, 1989, Enhanced Oil Recovery Institute, University of Wyoming, pp. 155–173, 1990.
26. Ewing, R.E. and Celia. M.A., "Multiphase Flow Simulation in Groundwater Hydrology and Petroleum Engineering". Computational Methods in Subsurface Hydrology, Eds, G. Gambolati et al., Computational Mechanics Publications, Springer Verlag, pp. 195–202, 1990.
27. Zisman, S., "Simulation of contaminant transport in groundwater systems using Eulerian-Langrangian localized adjoint methods", MS Thesis, Dept. Civil Eng., MIT, 1989.
28. Celia, M.A and Zisman S.,"Eulerian-Lagrangian Localized Adjoint Method for Reactive Transport in Groundwater", Computational Methods in Subsurface Hydrology, Eds, G. Gambolati et al., Computational Mechanics Publications, Springer Verlag, pp. 383–390. 1990.
29. Neuman, S.P., "Adjoint Petrov-Galerkin Method with Optimum Weight and Interpolation Functions Defined on Multi-dimensional Nested Grids", Computational Methods in Surface Hydrology, Eds G. Gambolati et al., Computational Mechanics Publications, Springer Verlag, pp. 347–356, 1990.
30. Schwartz, L., "Theorie des Distributions I, II". Paris: Hermann 1950–1951.
31. Lions, J.L. and E. Magenes, "Non-Homogeneous Boundary Value Problems and Applications, I", Springer-Verlag, New York, 1972.
32. Herrera, I., "The Algebraic Theory of Boundary Value Problems: Basic Concepts and Results", Comunicaciones Técnicas, Instituto de Geofísica, UNAM, 1992.
33. Herrera, I., "Algebraic Extensions of Operators", Applicable Analysis, 1993 (Submitted).
34. Allen, M.B., Herrera, I., Pinder, G.F., "Numerical Modeling in Science and Engineering", A Wiley-Interscience Publication, John Wiley and Sons, 1988.

GLOBAL SPACE-TIME FINITE ELEMENT METHODS FOR TIME-DEPENDENT CONVECTION DIFFUSION PROBLEMS

O. AXELSSON
O. Axelsson, Faculty of Mathematics and Informatics, University of Nijmegen, Toernooiveld, 6525 ED Njimegen The Netherlands

and

J. MAUBACH
Department of Mathematics and Statistics, University of Pittsburgh, PA. 15260, U.S.A

Abstract. Even if one adjusts the mesh points in space to the behaviour of the solution, the classical method of lines to solve time-dependent parabolic equations can be costly to apply when the solution exhibits boundary or interior layers, because very small time-steps may have to be taken when one wants to approximate the solution accurately.

On the other hand, for a global time-space finite element method, where the finite elements are – adaptively– adusted to the behaviour of the solution in both space and time, it suffices to use an of order magnitude mesh points fewer. In practice, such a method uses finite elements on big time-slabs. The stability and discretization error estimate of such methods applied to convection dominated diffusion problems is presented in detail, and illustrated with some numerical results.

Key words: Global continuous time-space finite elements, convection diffusion

1. Introduction

By way of illustration, consider first the following time-dependent parabolic, partial differential equation problem

$$\begin{aligned} u_t(x,t) &= Lu(x,t) + f(x,t) \quad \text{in } (0,1)\times(0,\infty) \\ u(x,0) &= u_o(x) \quad \text{for } 0 < x < 1 \\ u(0,t) &= l(t), u(1,t) = r(t) \quad \text{for } 0 < t < \infty\,, \end{aligned} \tag{1.1}$$

where

$$Lu(x,t) = \epsilon u_{xx}(x,t) + b(x,t)u_x(x,t)\,.$$

Here ϵ is assumed to be a small, positive parameter. For simplicity it is assumed that $b_x > 0$ uniformly, for all $t > 0$ and all $0 < x < 1$. It can readily be seen that this guarantees that the problem is parabolic, even when $\epsilon \longrightarrow 0^+$. For small epsilon, the solution usually has layers, i.e, it has sharp gradients near the boundary $x = 0, t > 0$ and in addition possibly in the interior of the domain of definition. Equation (1.1) can be solved numerically using either the method of lines, or a global time-space integration method.

S. Gomez and J-P. Hennart (eds.), Advances in Optimization and Numerical Analysis, 165–184.
© 1994 *Kluwer Academic Publishers.*

The method of lines first employs a semidiscretization of the time or the space variable derivatives. This can be obtained using a finite difference approximation method, or with a finite element variational method. Suppose that the time derivative is discretized first, using the familiar Crank-Nicolson method. If τ is the time-step, this method applied to (1.1) leads to

$$u(x,t+\tau) = u(x,t) + \frac{1}{2}\tau\left[Lu(x,t) + f(x,t) + Lu(x,t+\tau) + f(x,t+\tau)\right]$$

for $t = 0, \tau, 2\tau, \ldots$, assuming that $u(x,0)$ satisfies the initial solution and boundary conditions. Next, in order to get a fully discretized problem, the space operator is discretized. Using a variational formulation to this end, the above equation reduces to the system

$$(B_h + A_h)u(x,t+\tau) = R_h u(x,t) \tag{1.2}$$

where h is the – space – finite element mesh parameter, B_h is the finite element discretization of $u(t+\tau, x)$, and A_h is the finite element discretization of $-\frac{1}{2}\tau Lu(x,t+\tau)$, and where R_h is the finite element discretization accounting for the $f(x,t)$ and remaining terms involving $u(x,t)$. System (1.2) must be solved for each time step. The mesh-points on the line $t = k\tau$, $0 < x < 1$, $k \geq 1$ can be chosen adaptively to be concentrated near points where the solution has layers. This can significantly decrease the order of the system (1.2) to be solved, as compared with the case where one uses a uniform distribution of the space-mesh points for all lines $k \geq 1$. However, to approximate the solution accurately, one must choose small time-steps with the consequence that (1.2) must be solved many times, in order to obtain a solution approximation for $u(x,t)$, for large t. If one uses an explicit Euler type time-integration method, instead of the implicit Crank-Nicolson method presented above, then in addition one must guarantee that the standard criterion $\tau \leq O(h_m^2)$ is satisfied between the time-step τ and h_m, being the minimum element width over all elements in the non-uniform finite element mesh used on the line $t = k\tau$.

Now consider the use of a global time-space finite element method in order to solve the partial differential equation (1.1). Such methods exploit a big time-step, typically of fixed length T. The cross product of time and space domain $Q_k = (0,1)\times(kT,(k+1)T]$ for $k \geq 0$ is called a time-slab, and the global approach integrates (1.1) over such a time-slab in time and space simultaneously, using a variational formulation involving both the time and the space variables. For the k-th time-slab Q_k, this leads to the relation

$$\int_{kT}^{(k+1)T}\int_0^1 \left[u_t^h(x,t) - Lu^h(x,t)\right] v(x,t)\,dx\,dt = \int_{kT}^{(k+1)T}\int_0^1 f(x,t)v(x,t)\,dx\,dt\,,$$

where all test functions $v \in \mathrm{H}^0((0,1)\times(kT,(k+1)T])$ are functions in time and space. They are assumed to be square integrable, and to have zero trace at the boundary of the time-slab $(0,1)\times(kT,(k+1)T]$. Here, u^h stands for the corresponding variational solution. It will be demonstrated in section 2 that there is no need to impose any boundary conditions on the line $t = (k+1)T$, $0 < x < 1$. When one uses a finite element subspace of the above mentioned Sobolev space for the test functions, to obtain an approximate solution U^h, the above equation reduces to the system of equation

$$C_h U^h = F_h , \tag{1.3}$$

where C_h is the discrete equivalent of

$$\int_{kT}^{(k+1)T} \int_0^1 \left[u_t^h(x,t) - Lu^h(x,t)\right] v(x,t) dx dt ,$$

and where F_h is the discrete right hand side. Note that C_h is non-symmetric, even when $b = 0$. In addition, its dimension is much higher than the before mentioned matrices A_h and B_h, used in the method of lines. However, this is partially compensated for by the fact that there will be many more systems of the latter type to be solved. In addition, it will be shown that the total number of mesh-points which have to be used by the global time-space integration method will be of an order of magnitude smaller than those for the method of lines. Furthermore, there exist efficient preconditioned iterative solvers which only require an arithmetic computational complexity, about proportional to the number of mesh-points used. The iterative solution methods will not be discussed further in the present paper.

To integrate in time-space on time-slabs $Q_k = (0,1) \times (kT, (k+1)T]$ for $k > 1$, one can use a continuous integration method, where the finite element solution found on Q_{k-1} at the line $t = kT$ provides the initial condition for the solution to be approximated on Q_k. The boundary conditions on Q_k are provided for by the functions $l(t)$ and $r(t)$, for $t \in (kT, (k+1)T]$. Alternatively, one can impose the initial condition at the line $t = kT$ weakly, permitting the use of fewer and different mesh-points on time $t = kT$ than used in the computation on the previous time-slabs. This results in a – slightly – discontinuous solution across the line $t = kT$. Both methods, without, and with imposing the initial value weakly, can be continued this way for every $k \geq 1$, over an arbitrary number of time-slabs. The continuous time-slabbing method requires the use of the same, or more mesh-points on the line $t = kT$ for the integration of the new time-slab, as used for the previous time-slab. However, the continuous method is easier to implement and the above restriction is of minor importance. Therefore, this paper will focus on the analysis of the continuous time-space integration method.

The remainder of the paper is organized as follows. First, in section 2, the time-slabbing method is introduced, and its stability is shown, independent on the number of time-slabs taken into account. In section 3, the discretization error for this method is derived, using both the standard and the streamline upwind Galerkin finite element method. Then, section 4 discusses the application of local adaptive grid refinement in order to control the interpolation errors. Further, the method is illustrated by means of some numerical tests. Finally, section 5 provides some conclusions, and is followed by a list of references.

2. Global time-space integration methods

This section will focus on the application of the global time-space finite element on the, in general, higher dimensional problem

$$\begin{aligned} u_t(\mathbf{x},t) &= Lu(\mathbf{x},t) + f(\mathbf{x},t) \quad \text{in } \Omega \times (0,\infty) \\ u(\mathbf{x},0) &= u_o(\mathbf{x}) \quad \text{for } \mathbf{x} \in \Omega \\ u(\mathbf{x},t) &= u_c(t) \quad \text{for } \mathbf{x} \in \partial\Omega \text{ and } 0 < t < \infty\,, \end{aligned} \tag{2.1}$$

where

$$Lu(\mathbf{x},t) = \epsilon \nabla \cdot \underline{\nabla} u(\mathbf{x},t) + b(\mathbf{x},t)\underline{\nabla} u(\mathbf{x},t) - c(\mathbf{x},t)u(\mathbf{x},t)$$

in $\Omega \times (0,\infty)$. The space-domain Ω is a bounded polygonal – or polyhedral – subset of $\mathbf{R}^n$, for $n = 1$, 2,or 3. It is assumed that $\frac{1}{2}\nabla \cdot b + c \geq 0$ in $\Omega \times (0,\infty)$, and that $|b|$ and c are bounded uniformly. In the above equation,

$$\underline{\nabla} u = [\frac{\partial}{\partial x_1} u, \ldots, \frac{\partial}{\partial x_n} u]^t\,.$$

Note that equation (2.1) requires the solution to satisfy Dirichlet boundary conditions at $\partial\Omega \times (0,\infty)$. This assumption is not necessary for the theory to be presented, it is only made in order to simplify notations. Using a discontinuous global time-space integration, the Galerkin finite element variational formulation of equation (2.1) takes the form

$$\begin{aligned} &\int_{kT}^{(k+1)T} \int_\Omega \left[u_t^h(\mathbf{x},t) - Lu^h(\mathbf{x},t)\right] v(\mathbf{x},t)\, dx\, dt = \\ &\int_{kT}^{(k+1)T} \int_\Omega [f(\mathbf{x},t)]\, v(\mathbf{x},t)\, dx\, dt - \int_\Omega \left[u^h(\mathbf{x},kT) - u_I(\mathbf{x},kT)\right] v(\mathbf{x},kT)\, dx\,, \end{aligned} \tag{2.2}$$

for all $v \in V$, where

$$V = \{v \in \mathrm{H}^1(\Omega \times (kT,(k+1)T]) \colon v(\mathbf{x},t) = 0 \text{ for } (\mathbf{x},t) \in \partial\Omega \times (kT,(k+1)T]\}\,,$$

and $u_I(\mathbf{x},kT)$, is the initial solution $u(\mathbf{x},0)$ if $k = 0$, and equal to the approximate solution found on the previous time-slab Q_{k-1} on the line $t = kT$ for $k > 0$. This method results in a discontinuous approximate solution across the interfaces $t = kT$, $k = 1, 2, \ldots$, but it allows for adjustments of the computational grid at those interfaces between subsequent time-slabs. It can be readily seen, that this method, formulated globally on $(0, mT)$, is a result of the summation of the variational equations (2.2) for $k = 0, 1, \ldots, m-1$.

In the continuous global time-space method, the last term in (2.2) is dropped, leading to the variational formulation

$$\int_{kT}^{(k+1)T} \int_\Omega \left[u_t^h(\mathbf{x},t) - Lu^h(\mathbf{x},t)\right] v\, dx\, dt = \int_{kT}^{(k+1)T} \int_\Omega [f(\mathbf{x},t)]\, v\, dx\, dt \tag{2.3}$$

for all $v \in V$, where in this case

$$\begin{aligned} V = \{v \in \mathrm{H}^1(\Omega \times (kT,(k+1)T]) \colon & v(\mathbf{x},t) = 0 \text{ for } t = kT \wedge \\ & v(\mathbf{x},t) = 0 \text{ for } (\mathbf{x},t) \in \partial\Omega \times (kT,(k+1)T]\}\,. \end{aligned}$$

Here, the discrete solution u^h at time kT is equal to the discrete solution on the previous time-slab at the same time by definition, requiring the test functions v to have value zero at the interface $t = kT$. In addition, the continuous global time-stepping method has the following properties:

- The variational formulation has no global summation property.
- Using a grid with gridlines parallel or orthogonal to a layer moving in space and time, one can obtain a continuous approximate solution of a higher order of accuracy for all times t, than when using elements not oriented along the layer.
- In some cases, using a regular grid geometry, it can be seen that the continuous global finite element method is equivalent to the application of certain time-stepping methods of the implicit Runge-Kutta type.
- Finally, note that the discontinuous and continuous global time-space methods give identical approximations on the first time-slab.

Note that there are no boundary conditions on the interface at time $(k+1)T$ between subsequent time-slabs, and that the variational formulations given in (2.2) and (2.3) do not involve an integral over these interfaces. The reason for this is that the second order derivatives do not involve the time-variable. Hence,

$$\begin{aligned}
-\int_{kT}^{(k+1)T}\int_{\Omega} \nabla\cdot\underline{\nabla} uv\, dx\, dt = &\int_{kT}^{(k+1)T}\int_{\Omega} \underline{\nabla} u\underline{\nabla} v\, dx\, dt - \\
&\int_{\partial\Omega}\int_{kT}^{(k+1)T} v\underline{\nabla} u\cdot\mathbf{n}\, dx\, dt - \\
&\int_{\Omega} [v(\mathbf{x},t)\underline{\nabla} u(\mathbf{x},t)\cdot\mathbf{n}(\mathbf{x},t)]_{|t=(k+1)T}\, dx - \\
&\int_{\Omega} [v(\mathbf{x},t)\underline{\nabla} u(\mathbf{x},t)\cdot\mathbf{n}(\mathbf{x},t)]_{|t=kT}\, dx = \\
&\int_{kT}^{(k+1)T}\int_{\Omega} \underline{\nabla} u\underline{\nabla} v\, dx\, dt\,,
\end{aligned} \tag{2.4}$$

because the outward pointing unit normal $\mathbf{n}$ of the time-slab Q_k has no component in the space-variable directions at the lower and upper interface at $t = kT$, respectively $t = (k+1)T$, and because $v(\mathbf{x}, t) = 0$ at $\partial\Omega$. In addition to the above,

$$\begin{aligned}
\int_{kT}^{(k+1)T}\int_{\Omega} u_t v - b\underline{\nabla} uv\, dx\, dt = &\int_{kT}^{(k+1)T}\int_{\Omega} [-uv_t + b\underline{\nabla} vu + \nabla\cdot buv]\, dx\, dt \\
&+ \int_{\Omega} u(\mathbf{x},(k+1)T)v(\mathbf{x},(k+1)T)\, dx \\
&- \int_{\Omega} u(\mathbf{x},kT)v(\mathbf{x},kT)\, dx \\
= &\int_{kT}^{(k+1)T}\int_{\Omega} [-uv_t + b\underline{\nabla} vu + \nabla\cdot buv]\, dx\, dt \\
&+ \int_{\Omega} u(\mathbf{x},(k+1)T)v(\mathbf{x},(k+1)T)\, dx
\end{aligned}$$

because $v = 0$ at $\partial\Omega$, and at $t = kT$. In particular

$$\int_{kT}^{(k+1)T}\int_{\Omega}[u_t u - b\underline{\nabla} uu]\,dx\,dt = \frac{1}{2}\int_{kT}^{(k+1)T}\int_{\Omega}\nabla\cdot bu^2\,dx\,dt + \frac{1}{2}\int_{\Omega}u^2(\mathbf{x},(k+1)T)\,dx\,. \tag{2.5}$$

Now consider the stability of the global time-space integration method. Below, it will be shown that it is stable for the general evolution equation of the form

$$\frac{\partial u}{\partial t} = F(u,t) \quad \text{for } t > 0\,, \tag{2.6}$$

where $u(\mathbf{x},0) \in V$, V a Sobolev space in the space-variables, satisfies a prescribed initial condition. The functional F is assumed to be a – nonlinear – monotone operator

$$<F(u,t) - F(v,t), u - v> \leq 0 \quad \forall u,v \in V\,, \tag{2.7}$$

at least in the conservative sense. The duality paring $<,>$ is defined on the space-variables. In the case that the functional F is linear in u, $F(u,t) = F(t)u$, and equation (2.7) reduces to

$$<F(t)v, v> \leq 0 \quad \forall v \in V^{\circ}\,, \tag{2.8}$$

where $V^{\circ} = \{v \in V : v = 0 \text{ at } \partial\Omega\}$. Applying $F(u,t) = F(t)u$ and (2.4) and (2.5) to (2.1), one finds that

$$<F(t)v, v> = -\int_{\Omega}\left[\epsilon|\underline{\nabla} v|^2 + (\frac{1}{2}\nabla\cdot b + c)v^2\right]\,dx\,, \tag{2.9}$$

where the assumption $\frac{1}{2}\nabla\cdot b + c \geq 0$ shows that (2.7) holds. Strenghtening this assumption to

$$\frac{1}{2}\nabla\cdot b + c \geq c_0 > 0 \tag{2.10}$$

will cause the stronger monotonicity condition

$$<F(t)v, v> \leq -\rho_0\|v\|^2 \quad \forall v \in V^{\circ}\,, \tag{2.11}$$

to be satisfied. Here ρ_0 is a positive number and

$$\|v\|^2 = <v, v> = \int_{\Omega}v^2(\mathbf{x},t)\,dx \quad \forall v \in V\,.$$

Now, multiplying (2.6) by v, and integrating it over Ω, leads to

$$\frac{1}{2}\frac{\partial}{\partial t}\|v\|^2 \leq -\rho_0\|v\|^2 \quad \forall v \in V^{\circ}\,,$$

implying that

$$\frac{\partial}{\partial t}\left[\exp^{2\rho_1 t}\|v\|^2\right] \leq 0\,, \tag{2.12}$$

where

$$\begin{cases} \rho_1 = 0 & \text{in the conservative case, and} \\ \rho_1 = \rho_0 & \text{in the strongly monotone case.} \end{cases}$$

Replacing v with $u - v$ for $u, v \in V$, and integrating (2.12) with respect to time, this finally leads to

$$\|(u-v)(\mathbf{x},(k+1)T)\| \leq \exp^{-\rho_1 T} \|(u-v)(\mathbf{x},kT)\| \,,$$

or, since this is valid for all $k \geq 0$,

$$\|(u-v)(\mathbf{x},kT)\| \leq \begin{cases} k & \|u(\mathbf{x},0)-v(\mathbf{x},0)\| \quad \text{if } \rho_1 = 0 \,, \\ \exp^{-\rho_0 kT} & \|u(\mathbf{x},0)-v(\mathbf{x},0)\| \quad \text{if } \rho_1 = \rho_0 \,. \end{cases} \tag{2.13}$$

This shows the stability of the global time-space method with respect to pertubations of the initial values. Note that in the conservative case, there is at most a linear increase of the errors with respect to the number of time-slabs in the solution at time $t = kT$. In practice, the number of time-slabs is small, as one may integrate over large time-intervals using large T. The conservative case holds in particular when $\epsilon = 0$ and $\frac{1}{2}\nabla \cdot b + c = 0$ in equation (2.1), i.e., where one has to solve a pure hyperbolic problem.

Consider now the stability with respect to errors in the solution caused by discretization and other errors, such as rounding errors, integration errors, errors due to stopping of an iterative solver before reaching the machine precision accuracy. These errors may be made at every time-slab. Now, consider time-slab $Q_{k-1} = (0,1) \times ((k-1)T, kT]$, and denote the exact solution on Ω at time $t = kT$ of (2.1) with $u(kT)$, and the approximate solution in Ω at the same time with $u^h(kT)$. The approximate solution may contain various errors of the type referred to above, and is computed using the previous approximate solution on Q_{k-2} at $t = kT$ as an initial value. Contrary to this, define $\hat{u}(kT)$ on Q_{k-1} to be the exact solution in Ω at time $t = kT$ of (2.1), satisfying the same initial value as $u^h(kT)$. Finally, define on time-slab Q_{k-1} the local error $r(kT) = \hat{u}(kT) - u^h(kT)$. The local error $r(kT)$ contains all the local errors committed at time-slab Q_{k-1}. The functions $u(kT)$, $u^h(kT)$, and $\hat{u}(kT)$ satisfy

$$\begin{aligned} \|u(kT) - u^h(kT)\| \leq & \|u(kT) - \hat{u}(kT)\| + \|\hat{u}(kT) - u^h(kT)\| \\ = & \|u(kT) - \hat{u}(kT)\| + \|r(kT)\| \,, \end{aligned}$$

and (2.13) shows that

$$\|u(kT) - u^h(kT)\| \leq \exp^{-\rho_1 T} \|u((k-1)T) - \hat{u}((k-1)T)\| + \|r(kT)\| \,.$$

Hence,

$$\|u(kT) - u^h(kT)\| \leq \exp^{-\rho_1 kT} \|u(0) - \hat{u}(0)\| + \sum_{s=1}^{k} \exp^{-\rho_1 (s-1)T} \|r(sT)\| \,. \tag{2.14}$$

The right hand side of this inequality can be estimated above by

$$\begin{cases} \|u(0)-\hat{u}(0)\| + & k \max_{1\le s\le k} \|r(sT)\| \quad \text{if } \rho_1 = 0\,, \\ \exp^{-\rho_0 kT}\|u(0)-\hat{u}(0)\| + & \dfrac{1}{1-\exp^{-\rho_0 T}} \max_{1\le s\le k} \|r(sT)\| \quad \text{if } \rho_1 = \rho_0\,. \end{cases}$$

This shows the stability of the continuous global time-space method with respect to various kinds of errors, when this method is applied in order to solve (2.1). Furthermore, one can control the errors globally, for instance making them arbitrarily small, by making the local errors sufficiently small. The major component of the local errors is the discretization error which will be analysed in the next section.

3. Discretization error estimates

Singularly perturbed convection-diffusion problems have layers at those places where the solution has steep gradients and special care must be taken when one wants to solve such problems numerically. For instance, although the solution is bounded – even in the maximum norm – derivatives of the solution are in general not bounded uniformly in the perturbation parameter ϵ. This property of the solution is illustrated by the lemma below. It applies to the partial differential equation

$$\epsilon\Delta u(\mathbf{x}) + b\underline{\nabla}\, u(\mathbf{x}) - c(\mathbf{x})u(\mathbf{x}) = f(\mathbf{x}) \quad \text{in } \Omega\,,$$

where the solution is assumed to satisfy some standard homogeneous boundary conditions on the boundary $\partial\Omega$ of the domain. The provided estimates are sharp, i.e., there exists problems for which the orders of ϵ in the upper bounds are sharp.

Lemma (3.1) *Assume that* $\frac{1}{2}\nabla\cdot b(\mathbf{x}) + c(\mathbf{x}) \ge c_0 > 0$ *in* Ω.

a. Then for all $2 \le p \le \infty$ *there exists a scalar* C *such that*

$$\|u\|_{L^p(\Omega)} \le C\|f\|_{L^p(\Omega)}\,.$$

b. If Ω *is either a convex polygon, or if* $\partial\Omega$ *is smooth, then there exists a scalar* C *such that*

$$\epsilon^{3/2}\|\Delta u\| + \epsilon^{1/2}\|u\|_1 + \|u\| \le C\|f\|\,,$$

where $\|\,.\,\|_1$ *stands for the Sobolev norm of order 1 on* Ω.

c. If $\partial\Omega$ *is smooth, then there exists a scalar* C *such that*

$$\|u\|_2 \le \epsilon^{-3/2}\|f\|\,,$$

where $\|\,.\,\|_2$ *stands for the Sobolev norm of order 2 on* Ω.

Proof. The inequalities above are derived in [5].

The last inequality in lemma (3.1) is an elliptic regularity type inequality, as it bounds the second order Sobolev norm of the solution by the L^2 norm of the right hand side source f. In general, this inequality shows that elliptic regularity can not hold uniformly in ϵ.

In order to derive discretization error estimates for the global time-space finite element method, note first that one can apply estimates for stationary convection-diffusion problems to each time-slab. This is made possible by treating the time-variable as an auxilliary space-variable. Discretization error estimates for convection diffusion problems can be found in [1], [2], [4], and [11]. However, since there is no second order derivative in the time-variable, one must modify these estimates somewhat. In order to demonstrate this, consider the partial differential equation

$$L_\epsilon u(\mathbf{x},t) := -\epsilon\Delta u(\mathbf{x},t) - \hat{\mathbf{b}}(\mathbf{x},t)\hat{\underline{\nabla}} u(\mathbf{x},t) + c(\mathbf{x},t)u(\mathbf{x},t) = f(\mathbf{x},t) \tag{3.1}$$

in $\Omega \times (0,\infty)$, where the solution u has to satisfy the same boundary conditions as in equation (2.1), and where $\hat{\mathbf{b}}(\mathbf{x},t)$ and $c(\mathbf{x},t)$ are assumed to satisfy

$$\frac{1}{2}\nabla\cdot b(\mathbf{x},t) + c(\mathbf{x},t) \geq c_0 > 0 \quad \text{in } \Omega \times (0,\infty)$$

uniformly in time and space. In the above equation (3.1), the vector $\hat{\mathbf{b}}$ is defined by $\hat{\mathbf{b}} = [-1, b(\mathbf{x},t)]^t$, and $\hat{\underline{\nabla}} u(\mathbf{x},t) = [u_t(\mathbf{x},t), \underline{\nabla} u(\mathbf{x},t)]^t$.

Now consider the standard Galerkin finite element method. Integrating (2.1) over the time-slab Q_k in time and space simultaneuously, leads to a global time-space variational formulation: Find $u \in V$, such that

$$a_\epsilon(u,v) := \int_{Q_k} [L_\epsilon u]\, v \, dx\, dt = \int_{Q_k} f v \, dx\, dt\,,$$

or

$$a_\epsilon(u,v) = \int_{Q_k} \left[\epsilon\underline{\nabla} u\underline{\nabla} v + u\hat{\mathbf{b}}\underline{\nabla} v + (c + \nabla\cdot b)uv\right] dx\,dt + \int_\Omega [uv](\mathbf{x},kT)\,dx\,, \tag{3.2}$$

for all $v \in V^\circ$. Note that $a_\epsilon(v,v)$ is positive, because

$$\begin{aligned} a_\epsilon(v,v) &= \int_{Q_k} \left[\epsilon|\underline{\nabla} v|^2 + (\tfrac{1}{2}\nabla\cdot b + c)v^2\right] dx\,dt + \frac{1}{2}\int_\Omega v^2(\mathbf{x},kT)\,dx \\ &\geq \int_{Q_k} \left[\epsilon|\underline{\nabla} v|^2 + c_0 v^2\right] dx\,dt + \frac{1}{2}\int_\Omega v^2(\mathbf{x},kT)\,dx \\ &=: |||v|||_\epsilon^2 \quad \forall v \in V^\circ\,. \end{aligned} \tag{3.3}$$

For the coming error estimate derivations, it is assumed that the domain Q_k is covered with a regular finite element mesh, globally in time and space. Regular means that the ratio between the radius of the circumscribed and inscribed circles for any element e_l is bounded below by a a fixed scalar. On this mesh, a finite element basis V_h is constructed in the usual manner. Then, the standard Galerkin finite element formulation takes the form: Find $u^h \in V_h$, such that

$$a_\epsilon(u^h, v) = \int_{Q_k} f v\, dx\,dt \quad \forall v \in V_h^\circ\,.$$

In order to find an estimate of the discretization error $u - u^h$, let $u_I \in V_h$ be the interpolant of u, set $\theta = u^h - u_I$, and let $\eta = u - u_I$. Since

$$u^h - u = \theta - \eta$$

and

$$a_\epsilon(u^h - u, v) = 0 \quad \forall v \in V_h^\circ ,$$

substituting $v = \theta$ in this latter equation leads to

$$a_\epsilon(\theta, \theta) = a_\epsilon(\eta, \theta) . \tag{3.4}$$

In order to complete the error estimate, assume that C is a generic positive constant, independent of ϵ and the solutions u and u^h. Using elementary inequalities, relation (3.4) leads to

$$\begin{aligned} |a_\epsilon(\eta, \theta)| \leq & \epsilon\|\nabla\eta\|\|\nabla\theta\| + C\sum_l \int_{e_l} |\eta|\left|\hat{\nabla}\theta\right| dx\, dt + \\ & C\|\eta\|\,\|\theta\| + \int_\Omega [\eta\theta](\mathbf{x}, kT)\, dx , \end{aligned} \tag{3.5}$$

where e_l is an element in time-space globally, and where now

$$\|v\| = \left\{\int_{Q_k} v^2(\mathbf{x}, t)\, dx\, dt\right\}^{1/2} .$$

Now, consider the inverse estimate, valid for any function in V_h,

$$\int_{e_l} \left|\hat{\nabla}\theta\right|^2 dx\, dt \leq C h_l^{-2} \int_{e_l} |\theta|^2\, dx\, dt ,$$

where h_l is the diameter of the element. This inverse estimate shows that

$$\begin{aligned} \sum_l \int_{e_l} |\eta|\left|\hat{\nabla}\theta\right| dx\, dt & \leq C\sum_l \left[\int_{e_l} |\eta|^2\, dx\, dt \cdot \int_{e_l} \left|\hat{\nabla}\theta\right|^2 dx\, dt\right]^{1/2} \\ & \leq C\sum_l h_l^{-1} \left[\int_{e_l} |\eta|^2\, dx\, dt\right]^{1/2} \cdot \left[\int_{e_l} |\theta|^2\, dx\, dt\right]^{1/2} \\ & \leq C\|\eta\|'_{h^{-1}}\|\theta\| \end{aligned} \tag{3.6}$$

where

$$\|u\|'_h = \sum_l h_l \left[\int_{e_l} |u|^2\, dx\, dt\right]^{1/2} \quad \forall u \in V . \tag{3.7}$$

Then, using the standard arithmetic-geometric inequality, and (3.4), (3.5), and (3.6), one can show that

$$\{a_\epsilon(\theta, \theta)\}^{1/2} \leq C\left[\epsilon^{1/2}\|\nabla\eta\| + \|\eta\|'_{h^{-1}} + |||\eta|||_\epsilon\right] ,$$

where $|||\,.\,|||_\epsilon$ is defined in (3.3). Since

$$|||u-u^h|||_\epsilon \leq |||\theta|||_\epsilon + |||\eta|||_\epsilon\,,$$

the above, in combination with the substitution of $u = \theta$ in (3.3), lead to the following theorem.

Theorem (3.1) *s Let u be the solution to (3.1), let u^h be the related standard Galerkin finite element solution, and let $\frac{1}{2}\nabla\cdot b(\mathbf{x},t) + c(\mathbf{x},t) \geq c_0 > 0$ in Q_k. Then, the local discretization error at time-slab Q_k satisfies*

$$|||u-u^h|||_\epsilon \leq C\left[\epsilon^{1/2}\|\underline{\nabla}\eta\| + \|\eta\|'_{h^{-1}} + |||\eta|||_\epsilon\right]\,,$$

where $\|\,.\,\|'$, and $|||\,.\,|||_\epsilon$ are as defined in (3.7) and (3.3) respectively.

The theorem shows that if $\epsilon = O(1)$, and if $h_l = h \downarrow 0$, then the discretization error is of optimal order, i.e., the same order as the interpolation erro. However, if $\epsilon \leq \min_l\{h_l^2\}$, then one can not prove a better estimate than

$$\|u-u^h\| \leq C\left[\max_l\{h_l\}\|\underline{\nabla}\eta\| + \|\eta\|'_{h^{-1}} + |||\eta|||_\epsilon\right]\,,$$

i.e., an estimate of one order less than for η. In addition, in this latter case the interpolation error grows with some power of ϵ, because the derivatives of u grow this way, as can be concluded from lemma (3.1). As is well known, this causes in general the solution to oscillate around the layer. However, since $\epsilon > 0$ the gradients are bounded and one can therefore 'resolve the layer' using a sufficiently fine mesh where the solution has steep gradients. Then these oscillations do not occur and the interpolation errors can be bounded in the same way as for a smooth solution, typically leading to

$$\|\eta\|_s \leq Ch^{r-s+1}\|u\|_{r+1} \qquad \text{for } s = 0, 1\,,$$

where $\|\,.\,\|_i$ are Sobolev norms of order i on Q_k, and where r is the degree of polynomial basis functions used for the finite element approximation. Theorem (3.1) provides the corresponding orders of the discretization error. Hence, if one resolves the layers with a sufficiently fine mesh about the layer, the discretization error estimate

$$\|u-u^h\| \leq C\max_l\{h_l\}\|u\|_{r+1}\,,$$

holds for all $\epsilon > 0$.

Now, in order to solve (3.1) numerically, consider the streamline upwind finite element method, which was originally presented in [7]. This method uses the finite element variational formulation: Find $u^h \in V_h$, such that

$$b_\epsilon(u^h,v) := \int_{Q_k}\left[L_\epsilon u^h\right](v - \delta\hat{\mathbf{b}}\hat{\underline{\nabla}}v)\,dx\,dt = \int_{Q_k} f(v-\delta\hat{\mathbf{b}}\hat{\underline{\nabla}}v)\,dx\,dt \qquad (3.8)$$

for all $v \in V_h^\circ$. The parameter δ is a method function, which is non-negative and which may be chosen to be variable, $\delta = \delta_l$ for each finite element e_l. If $\delta = 0$, then the streamline upwind method reduces to the standard Galerkin finite element method. Rewriting (3.8) leads to

$$\begin{aligned} b_\epsilon(u^h, v) = & \sum_l \int_{e_l} \left[\epsilon \underline{\nabla} u^h \underline{\nabla} v + \delta_l (\hat{\mathbf{b}} \cdot \hat{\underline{\nabla}} u^h) \cdot (\hat{\mathbf{b}} \cdot \hat{\underline{\nabla}} v)\right] \, dx\,dt + \\ & \sum_l \int_{e_l} \left[-v \hat{\mathbf{b}} \cdot \hat{\underline{\nabla}} u^h + \epsilon \delta_l \nabla \cdot \underline{\nabla} u^h \hat{\mathbf{b}} \cdot \hat{\underline{\nabla}} v - \delta_l c u^h \hat{\mathbf{b}} \cdot \hat{\underline{\nabla}} v\right] \, dx\,dt \,. \end{aligned} \tag{3.9}$$

Now, assume that $\delta_l \epsilon \leq c_1 h_l^2$, where c_1 is a sufficiently small constant, which does not depend on ϵ and the solution. Using equation (3.9) and an inverse inequality to bound

$$\int_{e_l} |\nabla \cdot \underline{\nabla} v|^2 \, dx\,dt \leq C h_l^{-2} \int_{e_l} |\underline{\nabla} v|^2 \, dx\,dt$$

for all $v \in V_h$, one can show that there exists a positive scalar ρ_0, such that

$$\begin{aligned} b_\epsilon(v, v) \geq \rho_0 & \left[\sum_l \int_{e_l} \left[\epsilon |\underline{\nabla} v|^2 + \delta_l (\hat{\mathbf{b}} \cdot \hat{\underline{\nabla}} v)^2 + c v^2\right] \, dx\,dt + \frac{1}{2} \int_\Omega v^2(\mathbf{x}, kT) \, dx\right] \\ =: & \, \rho_0 |||v|||_{\epsilon,\delta}^2 \end{aligned} \tag{3.10}$$

for all $v \in V_h^\circ$ (see, for instance, [4], [8] and [11]). Compared to the norm $\|\,.\,\|_\epsilon$ in (3.3), the norm $|||\,.\,|||_{\epsilon,\delta}$ in the coercivety estimate (3.10) contains an additional term

$$\sum_l \int_{e_l} \delta_l (\hat{\mathbf{b}} \cdot \hat{\underline{\nabla}} v)^2 \, dx\,dt \,.$$

One now finds, analogously to (3.4), that

$$b_\epsilon(\theta, \theta) = b_\epsilon(\eta, \theta) \,, \tag{3.11}$$

whence (3.9) leads to

$$|b_\epsilon(\eta, \theta)| \leq C \left[\int_{Q_k} \left[\epsilon |\underline{\nabla} \eta|^2 + (\overline{\delta}^{1/2} \hat{\mathbf{b}} \cdot \underline{\nabla} \eta)^2 + \eta^2\right] \, dx\,dt + \left(\|\eta\|'_{\delta^{-1/2}}\right)^2 + \int_\Omega \eta^2(kT) \, dx\right] ,$$

where $\overline{\delta} = \max_l\{\delta_l\}$, and where analogous to definition (3.7),

$$\|u\|'_{\delta^{-1/2}} = \sum_l \delta_l^{-1/2} \left[\int_{e_l} |\eta|^2 \, dx\,dt\right]^{1/2} . \tag{3.12}$$

Summarizing the above, the following theorem can be stated.

Theorem (3.2) *o Let u be the solution to (3.1), let u^h be the related streamline upwind Galerkin finite element solution, assume that $\delta_l \epsilon \leq c_1 h_l^2$ for c_1 sufficiently* *small, and assume that* $\frac{1}{2}\nabla\cdot[b(\mathbf{x},t)+\min_l \delta_l c(\mathbf{x},t)b(\mathbf{x},t)]+c(\mathbf{x},t)\geq c_0>0$ *in* Q_k. *Then the local discretization error at* Q_k *satisfies*

$$|||u|||_{\epsilon,\delta}\leq C\left[\epsilon^{1/2}\|\underline{\nabla}\eta\|+\|\eta\|'_{\delta^{-1/2}}+\||\underline{\nabla}\eta|\|'_{\delta^{1/2}}+\|\eta\|_\epsilon\right],$$

where $|||\,.\,|||_{\epsilon,\delta}$ *is as defined by (3.10), and* $\|\,.\,\|'_{\delta^{1/2}}$ *is defined by (3.12).*

If $\epsilon=O(1)$ and $\delta_l=O(h_l^2)$, theorem (3.2) shows that the interpolation error estimate is of optimal order, compared to the interpolation errors, as it is in the case of the standard Galerkin method. However, the above estimate holds for any positive value of ϵ, and even when $\epsilon=0$. If one chooses

$$\delta_l=\min_l\{\frac{c_1}{\epsilon}h_l^2,c_1h_l\},$$

then theorem (3.2) provides an estimate which is never worse than half an order less than the optimal one, and half an order better than the estimate in theorem (3.1) for the standard Galerkin finite element method. In addition, it can be seen that there occur no, or minor, oscillations in the numerical solution along the layers, even when one does not resolve the layers. Finally, note that Theorem (3.2) indicates that one should use variable δ_l, satisfying $\delta_l\epsilon\leq c_1h_l^2$ in the case of adaptive mesh refinement, involving finite elements e_l of different sizes h_l.

4. Applications

In this section, the properties of the streamline upwind global time-space finite element method are examined by means an example of a convection diffusion equation with a known solution. After the presentation of this example, the continuous global time-space finite element solution method is described globally, followed by some paragraphs focussing on details, such as the iterative solution mehod used, the initial solution used for this method, and the stopping criterion used. In addition, the type of grid refinement, used to compute the finite element solution at each time-slab Q_k, is discussed. Finally, some numerical tests and tables are provided.

All numerical global time-space finite element numerical tests in this section focus on the following convection diffusion equation

$$\begin{aligned}
&u_t(x,t)-\epsilon u_{xx}(x,t)+\frac{2}{3}u_x(x,t)+2u(x,t)=f(x,t) && \text{in } \Omega\times(0,\infty)\\
&u(x,0)=\arctan(60x)+\pi/2 \quad \text{in } \Omega\\
&u(1,t)=\arctan(60-40t)+\pi/2 \quad \text{at } (0,\infty)\\
&u(0,t)=\arctan(-40t)+\pi/2 \quad \text{at } (0,\infty)
\end{aligned}\tag{4.1}$$

where $\Omega=(0,1)$, and $f(x,t)$ is such that the solution to (4.1) is given by the function $\hat{u}(x,t)=\arctan(60x-40t)+\pi/2$. This function is close to a stepfunction at $x=\frac{2}{3}t$ and is representative for the hyperbolic case, where $\epsilon=0$ and where no damping of initial stepfunctions occurs. The numerical tests will make use of large time intervals $(kT, (k+1)T)$, where $T = \frac{1}{4}$. The solution $\hat{u}$ has a layer moving in time and space along the line $6x - 4t = 0$, and all its equi-contour lines are parallel to this line. It will be approximated with a piecewise linear finite element approximation u_h on subsequent time-slabs Q_0, Q_1, etc. Before going into detail about the continuous global time-space finite element solution procedure, first a global overview is presented in the next paragraph.

Consider the computation of the finite element approximation of the solution on a time-slab Q_k. This approximation will be computed using local grid refinement, starting from an initial grid. On the first time-slab Q_0, the initial grid $\mathcal{Q}^0$ is imposed. In this example, it is the 4 by 1 grid as shown in figure 1. On all other time-slabs, the initial grid is such that it will be compatible with the previous grid, i.e., a vertex of the last grid covering the previous time-slab Q_k at time $t = (k+1)T$ will also be a vertex of the initial grid $\mathcal{Q}^0$ on time-slab Q_{k+1}. After the construction of an initial solution for each computation grid $\mathcal{Q}^i$, an iterative solution method is used to compute the finite element approximation on this grid. Thereafter, the new computational grid $\mathcal{Q}^{i+1}$ is created using either uniform refinement, or adaptive refinement. This procedure is repeated, until the finite element approximation on time-slab Q_k is found to be accurate enough. In all tests below, thirteen grids $\mathcal{Q}^0, \ldots, \mathcal{Q}^{12}$ are used. The finite element approximation obtained on the last computational grid on Q_k will be called the finite element approximation on time-slab Q_k. After computing the finite element approximation on time-slab Q_k, its trace on time $t = (k+1)T$ will be used as an initial condition for the solution to be approximated at the next time-slab Q_{k+1}. In order to ensure a continuous finite element solution across different time-slabs, the initial grid $\mathcal{Q}^0$ on Q_{k+1} shares all Q_k vertices situated at the time $t = (k+1)T$. This can be seen in figures 2 – 4. Proceeding in this way, a piecewise linear finite element solution is computed for three time-slabs. The finite element solution is continuous across the different time-slabs.

For all tests, the iterative solution method used to solve the matrix vector equations, is the Conjugate Gradient Square CGS method by Sonneveld in [12]. It has been chosen since it is reasonably insensitive to the magnitude of the skew-symmetric part of the matrix under consideration. This is of importance, as the global time-space integration causes the linear system to be strongly skew-symmetric. In the case, where one uses the standard Galerkin finite element method, this part may dominate the symmetric part of the matrix. As is usual for iterative solution methods, CGS has to be provided with an initial solution. The construction of the initial solution for grids $\mathcal{Q}^i$, $i > 0$, will be different from the construction on the first grid $\mathcal{Q}^0$, for all time-slabs. On the initial grid $\mathcal{Q}^0$, the initial solution is set to zero at all vertices not situated at the Dirichlet or the initial value boundary. Further, it is set to the value of the solution $\hat{u}$ at the vertices situated at the Dirichlet boundary, and it is set to the value of u^h at the vertices at the line $t = kT$. Note, that most of the vertices situated at the line $t = (k+1)T$ are not part of the Dirichlet and initial value boundary, as was shown in (2.4).Further, it should be observed that, due to the above procedure, the Dirichlet boundary conditions are actually linearly interpolated. For all grids $\mathcal{Q}^i$, $i = 1, 2, \ldots$, the initial solution on grid $\mathcal{Q}^i$ is constructed using linear interpolation applied to the previous finite element approximation on grid Q^{i-1}. At the Dirichlet boundary, the new initial approximation is set to the value of the solution $\hat{u}$, as in the previous case, but in the remaining vertices, including those on the line $t = kT$, its value is determined via linear interpolation from the finite element approximation on grid Q^{i-1}.

For the sake of convenience, the iterative solution method uses the stopping criterion

$$\|A\mathbf{x} - \mathbf{b}\|_2 < 10^{-10}$$

for all grids Q^i on all time-slabs Q_k. As initially the Dirichlet boundary conditions are poorly taken into account, one could better use a stopping criterion reflecting the discretization error, like

$$\|A\mathbf{x} - \mathbf{b}\|_2 < \max\{ch, 10^{-10}\}.$$

Here c stands for a to be chosen scalar, and h stands for the average mesh-size parameter.

Table (4.1) Degrees of freedom.

Slab	Q^0	Q^2	Q^4	Q^6	Q^8	Q^{10}	Q^{12}
$Q_1^{(1)}$	10	27	85	297	1105	4257	16705
$Q_2^{(1)}$	643	647	680	843	1554	4513	16705
$Q_3^{(1)}$	643	647	680	843	1554	4513	16705
$Q_1^{(2)}$	10	27	85	101	128	322	989
$Q_2^{(2)}$	67	78	129	138	170	369	1050
$Q_3^{(2)}$	68	78	128	139	166	367	1059
$Q_4^{(2)}$	71	79	128	136	168	364	1055
$Q_5^{(2)}$	69	78	129	139	172	371	1050
$Q_6^{(2)}$	67	78	128	140	154	314	999

The exploited local refinement is the n-dimensional local bisection algorithm introduced in [9], applied for $n = 2$. Here, the triangles in the initial coarse grid Q^0 are of level 0. When a triangle is refined, its descendants are said to be of one level higher. Suppose that T is a triangle of level k', having vertices ordered $\mathbf{x}_0$, $\mathbf{x}_1$ and $\mathbf{x}_2$. Then, the local bisection of T takes place as follows. First, compute $k = n - k' \bmod n$, n being the dimension. Here $n = 2$, whence $k = 2 - k' \bmod 2$. Then the new vertex $\mathbf{z} = \frac{1}{2}\{\mathbf{x}_0 + \mathbf{x}_k\}$ is created. If $k = 0$, the newly created triangles will have vertices ordered $\mathbf{x}_0, \mathbf{x}_1, \mathbf{z}$ and $\mathbf{x}_1, x_2, \mathbf{z}$. Alternatively, if $k = 1$, then the new triangles will have vertices ordered $\mathbf{x}_0, \mathbf{z}, \mathbf{x}_2$ and $\mathbf{x}_1, \mathbf{z}, \mathbf{x}_2$. In the case of two dimensions, this local bisection method (applied in for instance figure 2) leads to grids identical to those obtained by the application of the newest vertex method of Mitchell in [10], but applications of the latter method are restricted to grids of triangles in two dimensions.

The numerical tests performed involve either uniform or adaptive refinement. In the case of uniform refinement, Q^{i+1} is constructed from Q^i, by bisecting every triangle in the latter grid. In the case of adaptive refinement, every triangle in Q^i on which the approximate finite element solution attains a gradient greater than 2.0, will be refined. The adaptive refinement is combined with de-refinement, deleting all triangles for which the solution does not meet the gradient criterion. Independent of the type of refinement, uniform refinement is always applied for the creation of the computational grids $Q^1, \ldots, Q^5$, even in the case of adaptive refinement.

Table (4.2) Number of CGS iterations.

Slab	Q^0	Q^2	Q^4	Q^6	Q^8	Q^{10}	Q^{12}
$Q_1^{(1)}$	1	4	9	14	20	26	36
$Q_2^{(1)}$	4	5	9	14	21	25	37
$Q_3^{(1)}$	4	5	9	15	18	26	33
$Q_1^{(2)}$	1	4	9	9	15	19	32
$Q_2^{(2)}$	4	4	9	10	15	19	33
$Q_3^{(2)}$	4	4	9	10	13	19	31
$Q_4^{(2)}$	4	4	9	9	14	20	34
$Q_5^{(2)}$	4	4	8	11	14	19	33
$Q_6^{(2)}$	4	4	8	19	13	18	28

The tests involving the equation (4.1) will only deal with the hyperbolic case, where $\epsilon = 0$. For results regarding the case of small positive ϵ, the reader is referred to results in [6]. In the hyperbolic case, the streamline upwind finite element basis functions will be used in order to compute the finite element approximation. The streamline upwind parameter δ_l is taken to be equal to $\sqrt{|J_l|}$, where $|J_l|$ stands for the determinant of the affine transformation related to element e_l (see for instance [3]). This determinant is equal to twice the area of the element e_l. Now, consider some tables providing information on the number of degrees of freedom involved, the number of CGS iterations counted, and the calculated vertex-wise maximum error estimates.

Table (4.1) provides the number of degrees of freedom for every time-slab Q_k, $k = 1, 2, 3$, for every computational grid Q^i, $i = 0, 2, \ldots, 12$. The time-slabs have a super-index $^{(1)}$, if all the grids Q^i are created using uniform refinement, and have super-index $^{(2)}$, if adaptive refinement is used. As two subsequent bisection step reduce the descendants element size with a factor $\frac{1}{2}$, only even numbered grids are listed. Note that, in the case of uniform refinement, the number of degrees of freedom at the second and every time-slab thereafter will be identical, due to the uniform nature of the refinement. The number of degrees of freedom in the case of adaptive refinement is considerably less than that in the case where the uniform refinement is applied. For grid Q^{12}, the increase in degrees of freedom from time-slab Q_1 to Q_2, and similar decrease from Q_5 to Q_6 reflects the fact that the layers is departing from the left hand side boundary $x = 0$, respectively is arriving at the right hand side boundary $x = 1$. The adaptive refinement is functioning fine, as Q^i contains about the same number of degrees of freedom for every time-slab Q_k, $k = 1, \ldots, 6$, reflecting the fact that the layer is a straight line in all time-slabs.

Table (4.3) Vertex-wise maximum error.

Slab	Q^0	Q^2	Q^4	Q^6	Q^8	Q^{10}	Q^{12}
$Q_1^{(1)}$	0.640	0.885	0.603	0.362	0.181	0.068	0.017
$Q_2^{(1)}$	0.391	0.586	0.547	0.303	0.171	0.063	0.016
$Q_3^{(1)}$	0.625	0.375	0.493	0.351	0.167	0.062	0.016
$Q_1^{(2)}$	0.640	0.885	0.603	0.393	0.186	0.057	0.034
$Q_2^{(2)}$	0.348	0.634	0.558	0.332	0.166	0.057	0.034
$Q_3^{(2)}$	0.642	0.336	0.504	0.423	0.158	0.060	0.034
$Q_4^{(2)}$	0.462	0.793	0.551	0.337	0.168	0.061	0.041
$Q_5^{(2)}$	0.350	0.627	0.561	0.330	0.165	0.063	0.041
$Q_6^{(2)}$	0.583	0.258	0.497	0.317	0.138	0.064	0.038

Table (4.2) shows the number of CGS iterations necessary to meet the above formulated stopping criterion. In the uniform case, there are small deviations in the number of iterations of the CGS iterative solution method for, for instance for grid Q^{10} on different time-slabs. As the discretized differential equation and the computational grid is the same for every grid Q^{10}, this must be caused by the differences in the initial solution or the small difference in the degrees of freedom involved. It is interesting to note the small difference in the number of CGS iterations between the uniformly and adaptively refined grids. This indicates that it is the elements with the largest condition numbers which influence the number of iterations most. Even though the number of degrees of freedom between uniformly refined grids Q^{12} at time-slabs Q_k differs more than a factor 10 from that of adaptively refined grids Q^{12}, the number of iterations differ less than 10 percent.

Finally, table (4.3) shows the vertex-wise infinity-norm error $\|u^h - \hat{u}\|_\infty$ of the difference between the computed finite element approximation u^h on grid Q^i on time-slabs Q_k, $k = 1, 2, 3$. Initially, this error increases due to the fact that the Dirichlet boundary conditions, including their steep gradients, are gradually better represented by their piecewise linear approximation. This initial change of the linear approximation of the Dirichlet boundary conditions will also cause the error to decay slower during the first six to eight computational grids on every time-slab. From grid Q^{10} to grid Q^{12}, the decrease of the residuals in the case of uniform refinement is of the order $h^{1.98}$, using that $\log(0.16/0.63)/\log(0.5) \approx 1.977$. In the case of adaptive refinement the corresponding decrease is approximately of the order $h^{0.75}$. This is less than in the case of uniform refinement, but likely to be due to the fact that there are not yet enough grid points situated along the layer. Finally, note that even though a linear increase in the error is predicted below equation (2.14) – though in a different norm – this is hardly the case for the errors measured in the maximum norm above. Using uniform refinement, there is no visible increase in the error during the first three time-slabs. Using adaptive refinement, there is a small increase in the error on time-slabs Q_4 and Q_5, which most likely accounts for the possible linear increase of the global error mentioned below (2.14). The small decrease in error at Q^{12} at Q_6 is likely due to the fact that at the time-slab, the layer will reach the left hand side of the domain, which is a Dirichlet boundary. The fact that the error does not increase from Q_1 to Q_2 to Q_3 in this case could be explained by assuming that the largest vertex-wise error was measured at a vertex inside these time-slabs, not lying on the line $t = kT$, for $k = 1, 2$.

5. Conclusions

It has been demonstrated that the continuous global time-space finite element method is very well applicable for the solution of convection diffusion problems. The global error is bounded above by the maximum of the finite element errors over all time-slabs, and can therefore easily be controlled. In addition to its capability of effectively dealing with singularly perturbed parabolic problems, the continuous global time-space method has been shown to be effectively dealing with hyperbolic equations. In this case, the global error was proven to be at most linearly growing with the amount of time-slabs. Numerical experiments for the hyperbolic case were provided, and supported the theory presented for that case. The adaptive bisection refinement used for the location of the layer tracks the position of the layer accurately. Finally, both the presented theory and the local bisection refinement are applicable to grids of n-simplices in n dimensions, enabling the application of the continuous global time-space finite element method to problems in two and more space-variables.

References

1. Axelsson O.: Finite element methods for convection-diffusion problems, in Numerical Treatment of Differential Equations, (Strehmel K. ed.) Leipzig: Teubner 1988 (Teubner-Texte zur Mathematik; Bd. 104), 171-182 [Proceedings of the Fourth Seminar "Numdiff-4", Halle, 1987]
2. Axelsson O.: Stability and error estimates of Galerkin finite element approximations for convection-diffusion equations, IMA Journal of Numerical Analysis 1 (1981), 329-345
3. Axelsson O. and Barker V.A.: Finite Element Solution of Boundary Value Problems, Academic Press, Orlando, Florida, 1984
4. Axelsson O., Eijkhout V., Polman B. and Vassilevski P.: Iterative solution of singular pertubation 2nd order boundary value problems by use of incomplete block-factorization methods, BIT 29(1989), 867-889
5. Axelsson O. and Layton W.: Defect correction methods for convection-diffusion problems, Mathematical Modelling and Numerical Analysis, 423-455
6. Axelsson O. and Maubach J.: A time-space finite element discretization technique for the calculation of the electromagnetic field in ferromagnetic materials, Journal for Numerical Methods in Engineering 29(1989), 2085-2111
7. Hughes T.J. and Brooks A.: A multi-dimensional upwind scheme with no crosswind diffusion, in AMD 34(1979), Finite element methods for convection dominated flows (Hughes T.J. ed.), ASME, New York
8. Maubach J.: Iterative Methods for Non-Linear Partial Differential Equations, C.W.I., Amsterdam, 1991, ISBN 90-9004007-2
9. Maubach J.: Parallel local bisection refinement for n-dimensional simplicial grids, Submitted to the Fifth ISMM International Conference on Parallel and Distributed Computing and Systems, October 1-3, 1992, Pittsburgh, Pennsylvania, U.S.A.

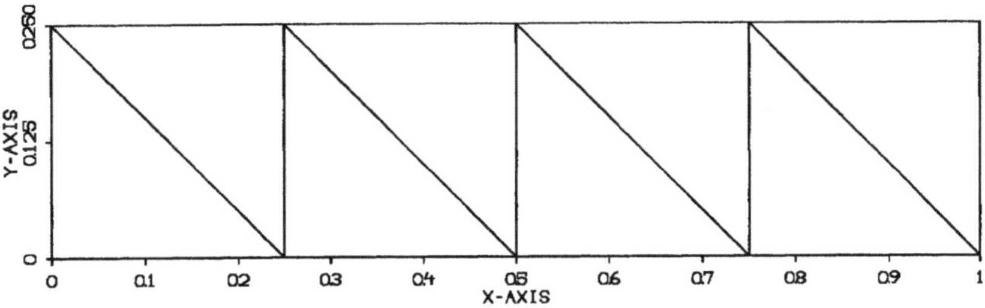

Fig. 1. Initial coarse 4 by 1 grid.

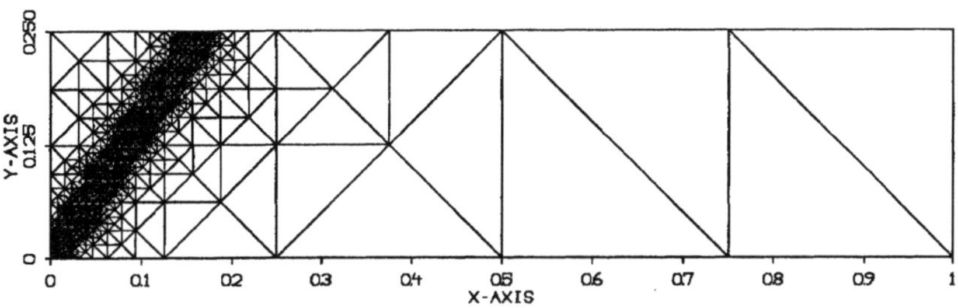

Fig. 2. Grid $\mathcal{Q}^{12}$ on Q_1, using adaptive refinement.

10. Mitchell W.F.: Optimal multilevel iterative methods for adaptive grids, SIAM Journal on Scientific and Statistical Computing 13(1992), 146-167
11. Nävert U.: A finite element method for convection diffusion problems, Ph.D. thesis, Department of Computer Sciences Chalmers University of Technology, Göteborg, Sweden, 1982
12. Sonneveld P.: CGS, a fast Lanczos-type solver for non-symmetric linear systems, SIAM Journal on Scientific and Statistical Computing 10(1989), 36-52

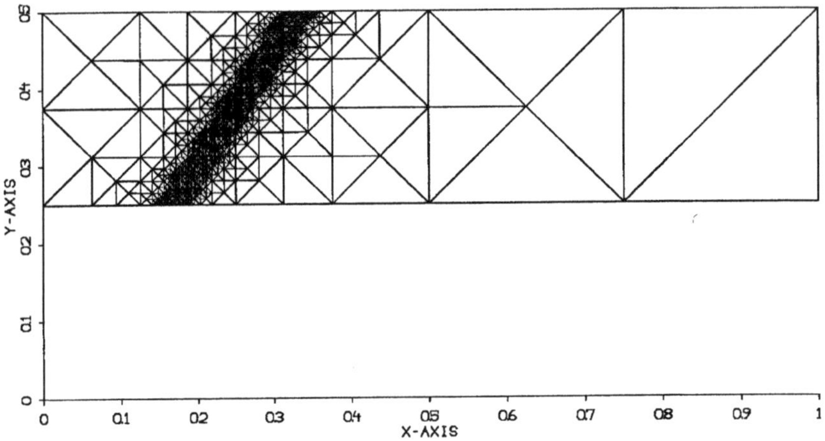

Fig. 3. Grid $\mathcal{Q}^{12}$ on Q_2, using adaptive refinement.

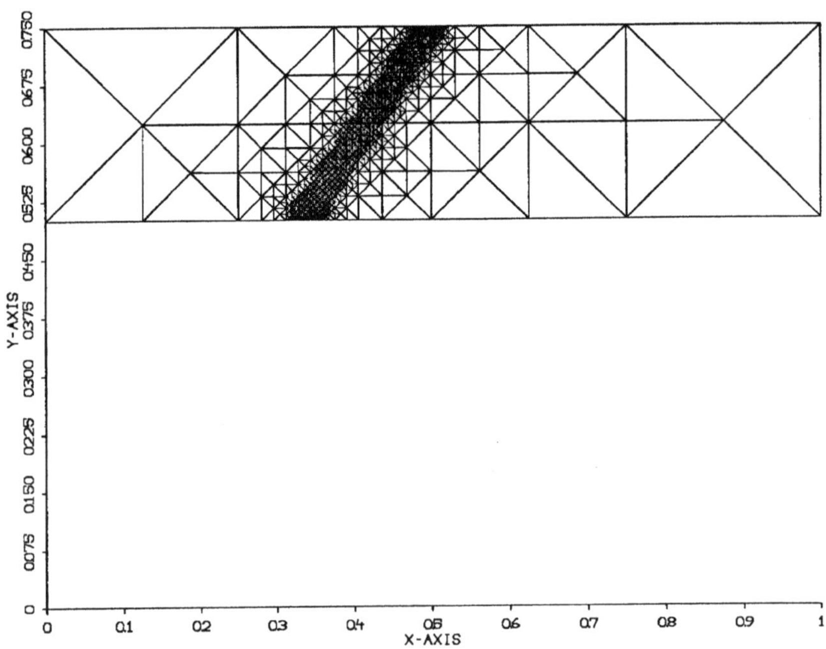

Fig. 4. Grid $\mathcal{Q}^{12}$ on Q_3, using adaptive refinement.

EULERIAN-LAGRANGIAN LOCALIZED ADJOINT METHODS FOR VARIABLE-COEFFICIENT ADVECTIVE-DIFFUSIVE-REACTIVE EQUATIONS IN GROUNDWATER CONTAMINANT TRANSPORT

RICHARD E. EWING and HONG WANG
Institute for Scientific Computation, Texas A&M University, College Station, TX 77843-3404, U.S.A.

Abstract. In this paper, we develop Eulerian-Lagrangian localized adjoint methods (ELLAM) to solve variable-coefficient advective-diffusive-reactive transport equations governing contaminant transport in groundwater flowing through a porous medium. The derived numerical schemes can treat various combinations of boundary conditions and conserve mass. Moreover, our numerical schemes provide an alternative approach to reduce the dependence of the numerical solutions on the accurate tracking of the characteristics, which is essential in many Eulerian-Lagrangian methods (ELM). Numerical results are presented to demonstrate the strength of our schemes.

1. Introduction

The growing incidence of the contamination of our groundwater from a variety of sources makes a proper description and understanding of contaminant transport in porous media very important. It is rapidly becoming clear that the clean-up of contamination cannot be effective without a thorough knowledge of the mechanism for transport, chemical and biological reactions, and remediation. However, the mathematical models are usually advective-diffusive-reactive transport equations, which need to be solved numerically in general. This kind of problems, which also arise in numerical simulation of oil reservoir and a multitude of other applications, often present serious numerical difficulties. Standard finite difference or finite element methods often suffer from severe nonphysical oscillations. While upstream weighting methods can eliminate these oscillations, they tend to significantly smear the sharp fronts and suffer from grid-orientation problems. The sharp fronts of the solutions of the physical problems need to be resolved very accurately in applications [10, 17], so these methods can not be used. Due to the Lagrangian nature of the problems, Eulerian-Lagrangian methods (ELM) can stabilize and symmetrize the problems, and have been successfully applied in these applications [5, 7, 8, 9, 20]. However, the principal drawbacks of ELM are their failure to conserve mass and their difficulty to treat boundary conditions. ELM usually impose a periodic assumption on the problems considered.

Eulerian-Lagrangian localized adjoint methods (ELLAM) are formulated to maintain mass conservation and to treat general boundary conditions [3, 4, 6, 11, 12, 13, 18, 19, 21, 22]. Thus, ELLAM overcome the two principal shortcomings of ELM while maintaining their numerical advantages. In this paper, we develop two ELLAM schemes to solve variable-coefficient advective-diffusive-reactive transport

S. Gomez and J-P. Hennart (eds.), Advances in Optimization and Numerical Analysis, 185–205.
© 1994 *Kluwer Academic Publishers.*

equations. We focus on how to treat variable-coefficients and how to treat reaction terms. For variable-coefficient problems, the numerical solutions with ELM often strongly rely on the accurate tracking of characteristics that may consume a large portion of CPU time. Our ELLAM schemes can significantly reduce the temporal error in the numerical solutions and provide an alternative approach to reduce the dependence of the numerical solutions on the accuracy of the tracking algorithm. Moreover, our ELLAM schemes can naturally be combined with domain decomposition and local refinement techniques to solve problems with interfaces, to solve the problems in parallel [22].

Now we briefly describe the ELLAM ideas here. ELLAM are motivated by localized adjoint methods (LAM) [2, 14, 15]. Let

$$\mathcal{L}u = f, \qquad x \in \Omega \quad \text{or} \quad (x,t) \in \Omega, \tag{1.1}$$

denote a partial differential equation in space or space-time. Integrating against a test function w, we obtain the weak form

$$\int_\Omega \mathcal{L}u \; w \; d\Omega = \int_\Omega f w \; d\Omega \, . \tag{1.2}$$

If we choose test functions w to satisfy the formal adjoint equation $\mathcal{L}^* w = 0$, except at certain nodes or edges denoted by l_i on Ω, then integrating by parts (the divergence theorem in higher dimensions) yields

$$\sum_i \int_{l_i} u \; \mathcal{L}^* w \; = \int_\Omega f w \; d\Omega \, . \tag{1.3}$$

Various test functions can be used to focus upon different types of sought information [14, 15]. Different choices of test functions lead to different classes of approximations. Motivated by the ideas of LAM and ELM, in ELLAM we choose the space-time test functions to be oriented along the characteristics. By careful analyses, we can derive ELLAM schemes that overcome the drawbacks of ELM while maintaining their numerical advantages. Although we have successfully applied ELLAM to solve multidimensional problems [23], in this paper we focus on the following one-dimensional problems to demonstrate the ideas and to develop our ELLAM schemes.

Consider the following one-dimensional linear variable-coefficient advective-diffusive-reactive transport equation in a conservative form:

$$\mathcal{L}u \equiv \left(\Phi \; u\right)_t + \left(V \; u - D \; u_x\right)_x + K \; u = f, \qquad x \in (a,b), \quad t \in (0,T], \tag{1.4}$$

subject to one of the following boundary conditions at the inflow boundary $x = a$:

$$\begin{aligned} u(a,t) &= g_1(t), & t \in [0,T], \\ -D(a,t)u_x(a,t) &= g_2(t), & t \in [0,T], \\ V(a,t)u(a,t) - D(a,t)u_x(a,t) &= g_3(t), & t \in [0,T], \end{aligned} \tag{1.5}$$

and subject to one of the following boundary conditions at the outflow boundary $x = b$:

$$\begin{aligned} u(b,t) &= h_1(t), & t \in [0,T], \\ -D(b,t)u_x(b,t) &= h_2(t), & t \in [0,T], \\ V(b,t)u(b,t) - D(b,t)u_x(b,t) &= h_3(t), & t \in [0,T], \end{aligned} \tag{1.6}$$

and the initial condition

$$u(x,0) = u_0(x), \qquad x \in [a,b], \tag{1.7}$$

where $u_t = \frac{\partial u}{\partial t}$ and $u_x = \frac{\partial u}{\partial x}$. The nomenclature is such that $\Phi(x,t)$ is a retardation coefficient that has positive lower and upper bounds, $V(x,t)$ is a fluid velocity field that we assume to be positive, $D(x,t)$ is a diffusion coefficient that has positive lower and upper bounds too, $K(x,t)$ is a first-order reaction coefficient, $f(x,t)$ is a given forcing function, and $u(x,t)$ is a measure of concentration of a dissolved substance. We write the accumulation term $(\Phi u)_t$ in a conservative form, because in applications we often end up with solving nonlinear analogy of (1.4) that leads to (1.4) after some linearization techniques are used.

The remainder of this paper is organized as follows. In Section 2, we derive an ELLAM formulation satisfied by the exact solution of problem (1.4). In Sections 3 and 4, we develop two ELLAM schemes. In Section 5, we present some numerical experiments to demonstrate the strength of our numerical schemes and indicate some further directions.

2. Variational Formulation

Let I and N be two positive integers. We define the partition of space and time as follows:

$$\begin{aligned} \Delta x &= \frac{b-a}{I}, \; x_i = a + i\Delta x, & i = 0,1,\ldots,I; \\ \Delta t &= \frac{T}{N}, \quad t^n = n\Delta t, & n = 0,1,\ldots,N. \end{aligned} \tag{2.1}$$

Let w be any test function. We can write out the space-time variational formulation for problem (1.4) as follows:

$$\begin{aligned} &\int_0^T \int_a^b \left(\Phi\, u\, w\right)_t dxdt + \int_0^T \int_a^b D\, u_x\, w_x dxdt + \int_0^T \int_a^b K\, u\, w dxdt \\ &\quad - \int_0^T \int_a^b u\left(\Phi\, w_t + V\, w_x\right) dxdt + \int_0^T \int_a^b \left(\left(Vu - Du_x\right)w\right)_x dxdt \\ &\quad = \int_0^T \int_a^b f\, w dxdt. \end{aligned} \tag{2.2}$$

In numerical schemes, we consider space-time test functions w that vanish outside $[a,b] \times (t^n, t^{n+1}]$ and are discontinuous at each time level t^n, whose exact form will be given below as part of the ELLAM development. With these test functions, we can rewrite (2.2) as

$$\begin{aligned}
&\int_a^b \Phi(x,t^{n+1})u(x,t^{n+1})w(x,t^{n+1})dx \\
&\quad + \int_{t^n}^{t^{n+1}} \int_a^b Du_x w_x dxdt + \int_{t^n}^{t^{n+1}} \int_a^b Kuwdxdt \\
&\quad + \int_{t^n}^{t^{n+1}} \Big(Vu - Du_x\Big) w \,\Big|_a^b \, dt - \int_{t^n}^{t^{n+1}} \int_a^b u\Big(\Phi w_t + V w_x\Big) dxdt \\
&\quad = \int_a^b \Phi(x,t^n)u(x,t^n)w(x,t_+^n)dx + \int_{t^n}^{t^{n+1}} \int_a^b f\ wdxdt,
\end{aligned} \tag{2.3}$$

where $w(x,t_+^n) = \lim_{t\to t_+^n} w(x,t)$ and $V^\Phi = V/\Phi$.

If we choose the test functions to be constant along the characteristics, the last term on the left-hand side of (2.3) vanishes. However, in the variable-coefficient case we cannot track the characteristics exactly as in the constant-coefficient case. Thus, generally $\Phi\ w_t + V\ w_x \neq 0$. Nevertheless, the residual should be small as long as we can approximate the characteristics in some crude ways. Moreover, the term $\int_{t^n}^{t^{n+1}} \int_a^b u(\Phi w_t + V w_x) dxdt$ in (2.3) vanishes if $w \equiv 1$, so this term does not affect mass conservation [19, 21, 24]. Second, this term accounts for the advection missed by errors in tracking characteristics along which w is constant; its accurate approximation serves to advect mass to the correct location and reduce the time truncation error dominance that is common in the numerical methods for problem (1.4). Based on these observations, we define our test functions and develop our numerical schemes.

First, we discuss the approximations of the characteristics. At the time t^{n+1}, we define an approximate characteristic $X(\theta; x, t^{n+1})$, $\theta \in [t^n, t^{n+1}]$, which emanates backward from (x, t^{n+1}), by the following:

$$X(\theta; x, t^{n+1}) - x \;=\; V^\Phi(x, t^{n+1})(\theta - t^{n+1}), \qquad \theta \in [t^n, t^{n+1}]. \tag{2.4}$$

For a given point x of time t^{n+1}, when it is clear from the context, we shall also write $x(\theta)$ in place of $X(\theta; x, t^{n+1})$.

Similarly, at the outflow boundary $\{x = b,\ t^n \le t \le t^{n+1}\}$, the approximate characteristic $X(\theta; x_I, t)$, $\theta \in [t^n, t]$, emanating backward from (x_I, t) is defined by the following:

$$X(\theta; x_I, t) - x_I = V^\Phi(x_I, t)(\theta - t), \qquad \theta \in [t^n, t]. \tag{2.5}$$

Also, let $x^* = X(t^n; x, t^{n+1})$ or $x_I^*(t) = X(t^n; x_I, t)$ be the foot of the approximate characteristic defined by (2.4) or (2.5); and let $x_i^* = X(t^n; x_i, t^{n+1})$, $i = 0, 1, \ldots, I$, or $x_i^* = X(t^n; x_I, t_i)$, $i = I+1, \ldots, I+IC$, where t_i is given in (2.9) below.

In order that all the approximate characteristics (2.4) emanating backward from the points at t^{n+1} not intersect each other, we need to impose the following restriction on the time step Δt:

$$\max_{a\le x\le b} \left|V_x^\Phi(x, t^{n+1})\right| \Delta t < 1. \tag{2.6}$$

In order that all the approximate characteristics (2.5) extending backward from the outflow boundary $[t^n, t^{n+1}]$ not intersect each other, recalling that $V(x,t)$ is strictly positive, we need to impose the following additional restriction on the time step Δt:

$$\max_{t^n \leq t \leq t^{n+1}} \left| V_t^{\Phi}(x_I, t) \right| \Delta t < \min_{t^n \leq t \leq t^{n+1}} V^{\Phi}(x_I, t). \tag{2.7}$$

Conditions (2.6)–(2.7) and the elementary implicit function's theorem conclude that for any $(y, \theta) \in [a, b] \times [t^n, t^{n+1}]$, there must be a unique approximate characteristic backward from (x, t^{n+1}) (or (x_I, t)) that passes through y at the time θ. That is, $y = X(\theta; x, t^{n+1})$ (or $y = X(\theta; x_I, t)$). Conditions (2.6) and (2.7) impose a strict restriction on the time step Δt where the velocity field V has a rapid change, as in the case of almost all ELM. A local time refinement technique can be used to ease this restriction. In this case, the local time step Δt_c instead of the global time step Δt should satisfy conditions (2.6) and (2.7). However, a numerical interface is usually introduced between a coarse mesh region and a fine mesh region. We have successfully combined ELLAM with domain decomposition and local refinement techniques to solve this problem. For detail, please see [22].

In order to discuss inflow boundary conditions effectively, we define the notation

$$\Delta t(x) = \begin{cases} t^{n+1} - t^n, & \text{if } x \geq \tilde{x}_0, \\ t^{n+1} - t^*(x), & \text{if } x < \tilde{x}_0, \end{cases} \tag{2.8}$$

where $\tilde{x}$ is the point at time t^{n+1} such that the approximate characteristic extending backward from $(\tilde{x}, t^{n+1})$ meets (x, t^n) (*i.e.* $x = X(t^n; \tilde{x}, t^{n+1}) = \tilde{x}(t^n)$) and $t^*(x)$ refers to the time when the approximate characteristic $x(\theta)$ traced backward from (x, t^{n+1}) meets the inflow boundary, *i.e.* $x_0 = X(t^*(x); x, t^{n+1}) = x(t^*(x))$; $t_i^* = t^*(x_i)$, $i = 0, 1, \ldots, IC_1$, where $Cu_1 = (\tilde{x}_0 - x_0)/\Delta x$ and $IC_1 = [Cu_1]$ is the integer part of Cu_1. For later convenience, we also define a notation $t^*_{IC_1+1} = t^n$. However, $t^*_{IC_1+1}$ may not be $t^*(x_{IC_1+1})$, in general.

In order to discuss outflow boundary conditions effectively, we define the notation

$$Cu_2 = \frac{V_m^{\Phi} \Delta t}{\Delta x}, \tag{2.9}$$

where

$$V_m^{\Phi} = \max_{t \in [t^n, t^{n+1}]} V^{\Phi}(x_I, t).$$

$$t_i = t^{n+1} - \frac{(i - I)\Delta x}{V_m^{\Phi}}, \qquad (i = I, I+1, \ldots, I+IC),$$

and IC is given by the following:

$$IC = \begin{cases} IC_2, & \text{if } Cu_2 > IC_2 = [Cu_2], \\ IC_2 - 1, & \text{if } Cu_2 = IC_2 = [Cu_2]. \end{cases}$$

For later convenience, in addition to t_i, $i = I, I+1, \ldots, I+IC$ introduced in (2.9), we also define a notation $t_{I+IC+1} = t^n$.

From (2.9), we partition the outflow boundary $[t^n, t^{n+1}]$ based on Cu_2. For $Cu_2 = IC_2$, we partition $[t^n, t^{n+1}]$ into IC_2 subintervals with length $\Delta t/Cu_2$. For $Cu_2 > IC_2$, we partition $[t^n, t^{n+1}]$ into $IC_2 + 1$ subintervals. The first IC_2 subintervals are of length $\Delta t/Cu_2$. The last one, which may be up to the same size of the others, is of length $(Cu_2 - IC_2)\Delta t/Cu_2$.

Given the partitions above, we define the test functions w to be the standard hat functions at the new time level (or at the outflow boundary) and to be constant along the approximate characteristics from t^{n+1} (or the outflow boundary) to t^n (or the inflow boundary if the approximate characteristic meets it), and to be discontinuous at t^n. With the test functions in hand, we derive a formulation satisfied by the exact solution of problem (1.4). Following the discussions under (2.7), we can rewrite the first term on the right-hand of (2.3) as

$$
\begin{aligned}
&\int_a^b \Phi(y,t^n)u(y,t^n)w(y,t^n_+)dy \\
&\quad = \int_{x_0}^{x_I^*} \Phi(x^*,t^n)u(x^*,t^n)w(x^*,t^n_+)dx^* + s\int_{x_I^*}^{x_I} \Phi(x^*,t^n)u(x^*,t^n)w(x^*,t^n_+)dx^* \\
&\quad = \int_{\tilde{x}_0}^{b} \Psi_1(t^n;x,t^{n+1})\Phi(x^*,t^n)u(x^*,t^n)w(x,t^{n+1})dx \\
&\qquad + \int_{t^n}^{t^{n+1}} \Psi_2(t^n;x_I,t)\Phi(x_I^*(t),t^n)u(x_I^*(t),t^n)w(x_I,t)dt,
\end{aligned}
\tag{2.10}
$$

where we have used the fact that the test functions w are constant along the approximate characteristics (2.4) and (2.5). $\Psi_1(s;x,t^{n+1})$ and $\Psi_2(s;x_I,t)$ are the Jacobians of the transformations (2.4) and (2.5), which are given by

$$
\begin{aligned}
\Psi_1(s;x,t^{n+1}) &= 1 - V_x^\Phi(x,t^{n+1})(t^{n+1}-s), \qquad s\in[t^*(x),t^{n+1}], \quad x\in[a,b],\\
\Psi_2(s;x_I,t) &= V^\Phi(x_I,t) + V_t^\Phi(x_I,t)(t-s), \qquad s\in[t^n,t], \quad t\in[t^n,t^{n+1}].
\end{aligned}
\tag{2.11}
$$

We now consider the last term on the right-hand side of (2.3). To avoid confusion, we write $\int_{t^n}^{t^{n+1}}\int_a^b f(x,t)w(x,t)dxdt = \int_{t^n}^{t^{n+1}}\int_a^b f(y,t)w(y,t)dydt$. If we evaluate the temporal integral by the backward Euler quadrature at t^{n+1}, we obtain

$$
\begin{aligned}
&\int_{t^n}^{t^{n+1}}\int_a^b f(y,s)w(y,s)dyds \\
&\quad = \int_{t^n}^{t^{n+1}}\int_a^{x_I^*(s)} f(X(s;x,t^{n+1}),s)w(X(s;x,t^{n+1}),s)dX(s;x,t^{n+1})ds \\
&\qquad + \int_{t^n}^{t^{n+1}}\int_{x_I^*(s)}^{b} f(X(s;x_I,t),s)w(X(s;x_I,t),s)dX(s;x_I,t)ds \\
&\quad = \int_{t^n}^{t^{n+1}}\int_a^{x_I^*(s)} f(x,t^{n+1})w(x,t^{n+1})dX(s;x,t^{n+1})ds
\end{aligned}
$$

$$
\begin{aligned}
&+\int_{t^n}^{t^{n+1}}\int_{x_I^*(s)}^{b} f(x_I,t)w(x_I,t)dX(s;x_I,t)ds\\
&-\int_{t^n}^{t^{n+1}}\int_{a}^{x_I^*(s)} \left\{f(x,t^{n+1})-f(X(s;x,t^{n+1}),s)\right\}w(x,t^{n+1})dX(s;x,t^{n+1})ds\\
&-\int_{t^n}^{t^{n+1}}\int_{x_I^*(s)}^{b} \left\{f(x_I,t)-f(X(s;x_I,t),s)\right\}w(x_I,t)dX(s;x_I,t)ds\\
=&\int_a^b\int_{t^*(x)}^{t^{n+1}}\Psi_1(s;x,t^{n+1})f(x,t^{n+1})w(x,t^{n+1})dsdx\\
&+\int_{t^n}^{t^{n+1}}\int_{t^n}^{t}\Psi_2(s;x_I,t)f(x_I,t)w(x_I,t)dsdt\\
&-\int_a^b\int_{t^*(x)}^{t^{n+1}}\left\{\int_s^{t^{n+1}} f_\theta(x(\theta),\theta)d\theta\right\}w(x,t^{n+1})\Psi_1(s;x,t^{n+1})dsdx\\
&-\int_{t^n}^{t^{n+1}}\int_{t^n}^{t}\left\{\int_s^{t} f_\theta(X(\theta;x_I,t),\theta)d\theta\right\}w(x_I,t)\Psi_2(s;x_I,t)dsdt\\
=&\int_a^b\Psi_3(x,t^{n+1})f(x,t^{n+1})w(x,t^{n+1})dx\\
&+\int_{t^n}^{t^{n+1}}\Psi_4(x_I,t)f(x_I,t)w(x_I,t)dt+R_f(w).
\end{aligned}
\tag{2.12}
$$

In the derivation above, we have used the transformations (2.4)–(2.5) and the discussion under (2.7). $R_f(w)$, $\Psi_3(x,t^{n+1})$ and $\Psi_4(x_I,t)$ are given below:

$$
\begin{aligned}
R_f(w) = &-\int_a^b\int_{t^*(x)}^{t^{n+1}}\left\{\int_s^{t^{n+1}} f_\theta(x(\theta),\theta)d\theta\right\}w(x,t^{n+1})\Psi_1(s;x,t^{n+1})dsdx\\
&-\int_{t^n}^{t^{n+1}}\int_{x_I^*(s)}^{b}\left\{\int_s^{t} f_\theta(X(\theta;x_I,t),\theta)d\theta\right\}w(x_I,t)\Psi_2(s;x_I,t)dsdt,
\end{aligned}
\tag{2.13}
$$

$$
\begin{aligned}
\Psi_3(x,t^{n+1}) &= \Delta t(x)-\frac{V_x^\Phi(x,t^{n+1})\left[\Delta t(x)\right]^2}{2}, && x\in[a,b],\\
\Psi_4(x_I,t) &= V^\Phi(x_I,t)(t-t^n)+\frac{V_t^\Phi(x_I,t)(t-t^n)^2}{2}, && t\in[t^n,t^{n+1}].
\end{aligned}
\tag{2.14}
$$

Similarly, we can rewrite the reaction term in (2.3) as follows:

$$
\begin{aligned}
\int_{t^n}^{t^{n+1}}\int_a^b Kuwdydt = &\int_a^b\Psi_3(x,t^{n+1})K(x,t^{n+1})u(x,t^{n+1})w(x,t^{n+1})dx\\
&+\int_{t^n}^{t^{n+1}}\Psi_4(x_I,t)K(x_I,t)u(x_I,t)w(x_I,t)dt+R_K(w),
\end{aligned}
\tag{2.15}
$$

$$
\begin{aligned}
R_K(w) &= -\int_{t^n}^{t^{n+1}} \int_a^{x_I^*(s)} \left\{ \int_s^{t^{n+1}} \Big[K(x(\theta),\theta) u(x(\theta),\theta) \Big]_\theta d\theta \right\} \\
&\qquad \cdot\ w(x,t^{n+1}) \Psi_1(s;x,t^{n+1}) ds dx \\
&= -\int_{t^n}^{t^{n+1}} \int_{x_I^*(s)}^{b} \left\{ \int_s^{t} \Big[K(X(\theta;x_I,t),\theta) u(X(\theta;x_I,t),\theta) \Big]_\theta d\theta \right\} \\
&\qquad \cdot\ w(x_I,t) \Psi_2(s;x_I,t) ds dt.
\end{aligned} \tag{2.16}
$$

The diffusion term can be rewritten as

$$
\begin{aligned}
\int_{t^n}^{t^{n+1}} \int_a^b D u_x w_x dy dt &= \int_a^b \Delta t(x) D(x,t^{n+1}) u_x(x,t^{n+1}) w_x(x,t^{n+1}) dx \\
&\quad - \int_{t^n}^{t^{n+1}} (t-t^n) D(x_I,t) u_x(x_I,t) w_t(x_I,t) dt + R_D(w),
\end{aligned} \tag{2.17}
$$

where $R_D(w)$ is the truncation error term given below:

$$
\begin{aligned}
R_D(w) &= -\int_a^b \int_{t^*(x)}^{t^{n+1}} \left\{ \int_s^{t^{n+1}} \Big[D(x(\theta),\theta) u_x(x(\theta),\theta) \Big]_\theta d\theta \right\} w_x(x,t^{n+1}) ds dx \\
&\quad + \int_{t^n}^{t^{n+1}} \int_{t^n}^{t} \left\{ \int_s^t \Big[D(X(\theta;x_I,t),\theta) u_x(X(\theta;x_I,t),\theta) \Big]_\theta d\theta \right\} w_t(x_I,t) ds dt.
\end{aligned} \tag{2.18}
$$

Putting (2.10), (2.12), (2.15) and (2.17) into (2.3), we obtain the following ELLAM formulation:

$$
\begin{aligned}
&\int_a^b \Phi(x,t^{n+1}) u(x,t^{n+1}) w(x,t^{n+1}) dx \\
&\quad + \int_a^b \Delta t(x) D(x,t^{n+1}) u_x(x,t^{n+1}) w_x(x,t^{n+1}) dx \\
&\quad + \int_a^b \Psi_3(x,t^{n+1}) K(x,t^{n+1}) u(x,t^{n+1}) w(x,t^{n+1}) dx \\
&\quad + \int_{t^n}^{t^{n+1}} \Psi_4(x_I,t) K(x_I,t) u(x_I,t) w(x_I,t) dt \\
&\quad - \int_{t^n}^{t^{n+1}} (t-t^n) D(x_I,t) u_x(x_I,t) w_t(x_I,t) dt \\
&\quad + \int_{t^n}^{t^{n+1}} (Vu - Du_x)(x_I,t) w(x_I,t) dt \\
&\quad - \int_{t^n}^{t^{n+1}} (Vu - Du_x)(x_0,t) w(x_0,t) dt
\end{aligned}
$$

$$\begin{aligned}
&= \int_{\tilde{x}_0}^{b} \Psi_1(t^n; x, t^{n+1})\Phi(x^*, t^n)u(x^*, t^n)w(x, t^{n+1})dx \\
&+ \int_{t^n}^{t^{n+1}} \Psi_2(t^n; x_I, t)\Phi(x_I^*(t), t^n)u(x_I^*(t), t^n)w(x_I, t)dt \\
&+ \int_a^b \Psi_3(x, t^{n+1})f(x, t^{n+1})w(x, t^{n+1})dx \\
&+ \int_{t^n}^{t^{n+1}} \Psi_4(x_I, t)f(x_I, t)dt + R(w),
\end{aligned} \tag{2.19}$$

where $R(w)$ is defined by

$$R(w) = \int_{t^n}^{t^{n+1}} \int_a^b \Phi u\Big(w_t + V^{\Phi} w_x\Big)dxdt - R_D(w) - R_K(w) + R_f(w). \tag{2.20}$$

By defining the test functions w to be constant along the approximate characteristics, we rewrite the variational formulation (2.3) as (2.19). Except for the $R(w)$ term on the right-hand side of (2.19) which is considered to be an error term, all other terms in (2.19) have been expressed as integrals at the time levels t^n and t^{n+1} or at inflow and outflow boundaries. Moreover, (2.19) is an identity. Thus, (2.19) provides a basis for the development of our numerical schemes in the next two sections.

3. ELLAM Scheme 1

In the last section, we derived an ELLAM formulation (2.19) satisfied by the exact solutions of problem (1.4). Based on this formulation, we develop two numerical schemes for problem (1.4) in this paper. We derive Scheme 1 in this section and Scheme 2 in the next section. First, we define the trial functions.

3.1. Trial Function

At the time t^{n+1}, we use a piecewise-linear trial function:

$$U(x, t^{n+1}) = \sum_{i=0}^{I} U(x_i, t^{n+1})w_i(x, t^{n+1}), \qquad x \in [a, b], \quad n = 0, 1, \cdots, N-1. \tag{3.1}$$

At the outflow boundary $[t^n, t^{n+1}]$, we need to consider two cases. For the outflow Neumann or flux boundary conditions, the trial function is unknown. We use a piecewise-linear trial function at the outflow boundary $[t^n, t^{n+1}]$:

$$U(x_I, t) = \sum_{i=I}^{I+IC+1} U(x_I, t_i)w_i(x_I, t), \qquad t \in [t^n, t^{n+1}], \quad n = 0, 1, \ldots, N-1. \tag{3.2}$$

For the outflow Dirichlet boundary condition, the trial function is known at the outflow boundary while its normal derivative is unknown. Thus, we use a piecewise-constant trial function for the normal derivative of the exact solution at the outflow boundary $(t^n, t^{n+1}]$:

$$U_x(x_I, t) = \sum_{i=I+1}^{I+IC+1} U_x(x_I, t_{i-1/2})\chi_i(t), \qquad t \in (t^n, t^{n+1}]. \tag{3.3}$$

where $\chi_i(t) = \chi_{(t_i,t_{i-1})}(t)$ is the standard characteristic function over the interval (t_i, t_{i-1}), which is 1 on (t_i, t_{i-1}) and 0 outside. $U_x(x_I, t_{i-1/2})$ denotes the value of $U_x(x_I, t)$ at $t_{i-1/2} = \frac{t_{i-1}+t_i}{2}$.

3.2. Scheme 1 on the Interior Nodes

We obtain Scheme 1 by dropping the error term $R(w)$ in (2.19) and replacing the exact solution u by the trial function U. At a typical interior node, the scheme is defined by

$$\begin{aligned}
&\int_{x_{i-1}}^{x_{i+1}} \Phi(x,t^{n+1})U(x,t^{n+1})w_i(x,t^{n+1})dx \\
&\quad + \int_{x_{i-1}}^{x_{i+1}} \Delta t D(x,t^{n+1})U_x(x,t^{n+1})w_{ix}(x,t^{n+1})dx \\
&\quad + \int_{x_{i-1}}^{x_{i+1}} \Psi_3(x,t^{n+1})K(x,t^{n+1})U(x,t^{n+1})w_i(x,t^{n+1})dx \\
&= \int_{x_{i-1}}^{x_{i+1}} \Psi_1(t^n;x,t^{n+1})\Phi(x^*,t^n)U(x^*,t^n)w_i(x,t^{n+1})dx \\
&\quad + \int_{x_{i-1}}^{x_{i+1}} \Psi_3(x,t^{n+1})f(x,t^{n+1})w_i(x,t^{n+1})dx.
\end{aligned} \tag{3.4}$$

Equation (3.4) has a symmetric and positive definite coefficient matrix and has more accurate spatial approximations than the modified method of characteristics (MMOC), due to the introduction of the Jacobian's $\Psi_1, \ldots, \Psi_4$. Moreover, Scheme 1 directly applies to problem (1.4) in a conservative form, while MMOC only applies to (1.4) in a nonconservative form [8, 19, 21].

When one or more of the approximate characteristics $x_{i-1}(\theta)$, $x_i(\theta)$, and $x_{i+1}(\theta)$ intersects the spatial boundary, Equation (3.4) needs to be modified. In the next two subsections, we present Scheme 1 related to the inflow or outflow boundaries.

3.3. Scheme 1 Related to the Inflow Boundary

In this part, we derive Scheme 1 for the inflow Dirichlet, Neumann and flux boundary conditions. The treatment of boundary conditions directly affects the accuracy, stability and mass conservation property of the numerical schemes. If we simply discretize the formulation (2.19), we may obtain schemes that are not of optimal convergence order, or are strongly time-dominant, or may not conserve mass, or may even tend to exhibit some oscillations. We have conducted detailed analyses for the different treatments of the boundary conditions in the context of ELLAM schemes for constant convection-diffusion equations [24]. We do not repeat the analysis here. Instead, we focus on the new difficulties arising from the treatments for variable-coefficient problem (1.4). We first present Scheme 1 for inflow flux boundary conditions. If we drop the last term on the right-hand side of (2.19) and replace the exact solution u by the trial function U, we obtain the scheme at a typical node related to the inflow boundary as follows:

$$\begin{aligned}\int_{x_{i-1}}^{x_{i+1}} &\Phi(x,t^{n+1})U(x,t^{n+1})w_i(x,t^{n+1})dx \\ &+ \int_{x_{i-1}}^{x_{i+1}} \Delta t(x)D(x,t^{n+1})U_x(x,t^{n+1})w_{ix}(x,t^{n+1})dx \\ &+ \int_{x_{i-1}}^{x_{i+1}} \Psi_3(x,t^{n+1})K(x,t^{n+1})U(x,t^{n+1})w_i(x,t^{n+1})dx \\ &= \int_{[x_{i-1},x_{i+1}]\cap\{x\geq\tilde{x}_0\}} \Psi_1(t^n;x,t^{n+1})\Phi(x^*,t^n)U(x^*,t^n)w_i(x,t^{n+1})dx \\ &+ \int_{x_{i-1}}^{x_{i+1}} \Psi_3(x,t^{n+1})f(x,t^{n+1})w_i(x,t^{n+1})dx + \int_{t^*_{i+1}}^{t^*_{i-1}} g_3(s)w_i(x_0,s)ds.\end{aligned} \tag{3.5}$$

We next derive Scheme 1 for inflow Neumann boundary conditions. If we repeat the derivation above, the resulting scheme contains a negative term involving $U(x_0,s)$ on its left-hand side. A constant or linear interpolation along the time direction in this term may introduce strong temporal dispersion and violate the regular structure of the coefficient matrix [24]. To overcome these difficulties, we should lump the negative term to the new time level, by moving the numerical solution from the inflow boundary to the new time level along the approximate characteristics. If we treat the negative term this way, we obtain the scheme at a typical node as follows:

$$\begin{aligned}\int_{[x_{i-1},x_{i+1}]\cap\{x\geq\tilde{x}_0\}} &\Phi(x,t^{n+1})U(x,t^{n+1})w_i(x,t^{n+1})dx \\ &+ \int_{x_{i-1}}^{x_{i+1}} \Delta t(x)D(x,t^{n+1})U_x(x,t^{n+1})w_{ix}(x,t^{n+1})dx \\ &+ \int_{x_{i-1}}^{x_{i+1}} \Psi_3(x,t^{n+1})K(x,t^{n+1})U(x,t^{n+1})w_i(x,t^{n+1})dx \\ &= \int_{[x_{i-1},x_{i+1}]\cap\{x\geq\tilde{x}_0\}} \Psi_1(t^n;x,t^{n+1})\Phi(x^*,t^n)U(x^*,t^n)w_i(x,t^{n+1})dx \\ &+ \int_{x_{i-1}}^{x_{i+1}} \Psi_3(x,t^{n+1})f(x,t^{n+1})w_i(x,t^{n+1})dx + \int_{t^*_{i+1}}^{t^*_{i-1}} g_2(s)w_i(x_0,s)ds.\end{aligned} \tag{3.6}$$

For inflow Dirichlet boundary conditions, (2.19) contains an unknown inflow boundary diffusive flux. If we simply discretize the unknown diffusive flux along the time direction, the derived algebraic system contains the inflow boundary dif fusive flux as unknowns, which are coupled with other interior unknowns involving only function values. This introduces strong temporal truncation errors and the numerical solutions deteriorate, characterized by some oscillations near the inflow boundary. To overcome this difficulty, Russell *et al.* [18, 19] proposed to derive an ELLAM scheme differently from that for inflow flux and Neumann boundary conditions. That is, we approximate the term $-\int_0^T \int_a^b (Du_x)_x w\,dx\,dt$ by the backward-Euler time integration along the approximate characteristics before integrating by parts in space. The resulting scheme has the following form:

$$\begin{aligned}
&\int_{x_{i-1}}^{x_{i+1}} \Phi(x,t^{n+1})U(x,t^{n+1})w_i(x,t^{n+1})dx \\
&\quad + \int_{x_{i-1}}^{x_{i+1}} \Psi_3(x,t^{n+1})D(x,t^{n+1})U_x(x,t^{n+1})w_{ix}(x,t^{n+1})dx \\
&\quad + \int_{x_{i-1}}^{x_{i+1}} \hat{V}^{\Phi}(x,t^{n+1})D(x,t^{n+1})U_x(x,t^{n+1})w_i(x,t^{n+1})dx \\
&\quad + \int_{x_{i-1}}^{x_{i+1}} \Psi_3(x,t^{n+1})K(x,t^{n+1})U(x,t^{n+1})w_i(x,t^{n+1})dx \\
&= \int_{[x_{i-1},x_{i+1}]\cap\{x\geq\tilde{x}_0\}} \Psi_1(t^n;x,t^{n+1})\Phi(x^*,t^n)U(x^*,t^n)w_i(x,t^{n+1})dx \\
&\quad + \int_{x_{i-1}}^{x_{i+1}} \Psi_3(x,t^{n+1})f(x,t^{n+1})w_i(x,t^{n+1})dx \\
&\quad + \int_{t^*_{i+1}}^{t^*_{i-1}} V(x_0,s)g_1(s)w_i(x_0,s)ds,
\end{aligned} \tag{3.7}$$

where

$$\hat{V}^{\Phi}(x,t^{n+1}) = \begin{cases} [\Psi_3(x,t^{n+1})]_x = \dfrac{1}{V^{\Phi}(x,t^{n+1})} - \dfrac{2V_x^{\Phi}(x,t^{n+1})\Delta t(x)}{V^{\Phi}(x,t^{n+1})} \\ \quad + \dfrac{\left[V_x^{\Phi}(x,t^{n+1})\Delta t(x)\right]^2}{V^{\Phi}(x,t^{n+1})} - \dfrac{V_{xx}^{\Phi}(x,t^{n+1})[\Delta t(x)]^2}{2}, & \text{if } x\in[x_0,\tilde{x}_0], \\ [\Psi_3(x,t^{n+1})]_x = \Delta t\left[1 - \dfrac{V_{xx}^{\Phi}(x,t^{n+1})\Delta t}{2}\right], & \text{if } x\in[\tilde{x}_0,x_I]. \end{cases} \tag{3.8}$$

$\hat{V}^{\Phi}(x,t^{n+1})$ in (3.8) contains V_{xx}^{Φ}. In applications the velocity V is usually given as a numerical solution of the associated pressure equation, which is a continuous piecewise polynomial and does not have the required regularity. The second term on the left-hand side of (3.7) is different from its counterparts in (3.4). Thus, Equation (3.7) may lose some mass. To avoid these drawbacks, we derive the scheme in a different way. We approximate $U_x(x_0,t^*(x))$ by $U_x(x,t^{n+1})$, where $t^*(x)$ is given in

(2.8). The error involves only the first order derivative of the exact solution along the approximate characteristics, which is usually much smaller than any temporal derivatives of the exact solution for problem (1.4). The resulting scheme is as follows:

$$\begin{aligned}
&\int_{x_{i-1}}^{x_{i+1}} \Phi(x,t^{n+1})U(x,t^{n+1})w_i(x,t^{n+1})dx \\
&\quad + \int_{x_{i-1}}^{x_{i+1}} \Delta t(x)D(x,t^{n+1})U_x(x,t^{n+1})w_{ix}(x,t^{n+1})dx \\
&\quad + \int_{[x_{i-1},x_{i+1}]\cap\{x<\tilde{x}_0\}} \Psi_5(x,t^{n+1})D(x_0,t^*(x))U_x(x,t^{n+1})w_i(x,t^{n+1})dx \\
&= \int_{[x_{i-1},x_{i+1}]\cap\{x\geq\tilde{x}_0\}} \Psi_1(t^n;x,t^{n+1})\Phi(x^*,t^n)U(x^*,t^n)w_i(x,t^{n+1})dx \\
&\quad + \int_{x_{i-1}}^{x_{i+1}} \Psi_3(x,t^{n+1})f(x,t^{n+1})w_i(x,t^{n+1})dx \\
&\quad + \int_{t^*_{i+1}}^{t^*_{i-1}} V(x_0,s)g_1(s)w_i(x_0,s)ds,
\end{aligned} \tag{3.9}$$

where $\Psi_5(x,t^{n+1})$ is given by

$$\Psi_5(x,t^{n+1}) = \frac{V^\Phi(x,t^{n+1}) - V_x^\Phi(x,t^{n+1})(x-x_0)}{\left[V^\Phi(x,t^{n+1})\right]^2}, \qquad x \in [x_0,\tilde{x}_0]. \tag{3.10}$$

Note that the integral $\int_{x_0}^{\tilde{x}_0} \Psi_5(x,t^{n+1})D(x_0,t^*(x))U_x(x,t^{n+1})w_i(x,t^{n+1})dx$ is equal to the integral $\int_{t^n}^{t^{n+1}} D(x_0,t^*(x))U_x(x_0,t^*(x))w_i(x_0,t^*(x))dt^*(x)$, with $U_x(x_0,t^*(x))$ replaced by $U_x(x,t^{n+1})$. Thus, Equation (3.9) conserves mass exactly as we approximate $U_x(x_0,t^*(x))$ by $U_x(x,t^{n+1})$.

To conserve mass, all test functions should sum to one exactly [3]. Russell [18, 19] chose the test function to be 1 on $[x_0,x_1]$ for the inflow Dirichlet boundary condition. Our theoretical analysis and numerical experiments show that his scheme has only an asymptotic convergence rate of $O((\Delta x)^{3/2}+\Delta t)$ instead of $O((\Delta x)^2+\Delta t)$. To recover second-order convergence rate in space, we introduce an extra equation at the node $i=0$ that solves for the unknown $U_x(x_0,t^{n+1})$ at the inflow boundary. This equation is introduced to maintain optimal-order convergence rate and mass conservation. It is completely decoupled from all other equations [24].

3.4. Scheme 1 Related to the Outflow Boundary

In this part, we derive Scheme 1 for outflow Dirichlet, Neumann or flux boundary conditions. The derivation is similar to the case for the inflow boundary. At a typical node on the outflow boundary, Scheme 1 for the outflow flux boundary condition is as follows:

$$
\begin{aligned}
&-\int_{t_{i+1}}^{t_{i-1}}(t-t^n)V(x_I,t)U(x_I,t)\hat{w}_{it}(x_I,t)dt \\
&\qquad+\int_{t^n}^{t^{n+1}}\Psi_4(x_I,t)K(x_I,t)U(x_I,t)\hat{w}_i(x_I,t)dt \\
&=\int_{t_{i+1}}^{t_{i-1}}\Psi_2(t^n;x_I,t)\Phi(x_I^*(t),t^n)U(x_I^*(t),t^n)\hat{w}_i(x_I,t)dt \\
&\qquad+\int_{t_{i+1}}^{t_{i-1}}\Psi_4(x_I,t)f(x_I,t)\hat{w}_i(x_I,t)dt-\int_{t_{i+1}}^{t_{i-1}}h_3(t)\hat{w}_i(x_I,t)dt \\
&\qquad-\int_{t_{i+1}}^{t_{i-1}}(t-t^n)h_3(t)\hat{w}_{it}(x_I,t)dt,
\end{aligned}
\tag{3.11}
$$

where $\hat{w}_i = w_i$, for $i = I+1,\ldots,I+IC$ and $\hat{w}_{I+IC} = w_{I+IC} + w_{I+IC+1}$.

Scheme 1 for the outflow Neumann boundary condition has the following form:

$$
\begin{aligned}
&\int_{t_{i+1}}^{t_{i-1}}V(x_I,t)U(x_I,t)\hat{w}_i(x_I,t)dt+\int_{t^n}^{t^{n+1}}\Psi_4(x_I,t)K(x_I,t)U(x_I,t)\hat{w}_i(x_I,t)dt \\
&=\int_{t_{i+1}}^{t_{i-1}}\Psi_2(t^n;x_I,t)\Phi(x_I^*(t),t^n)U(x_I^*(t),t^n)\hat{w}_i(x_I,t)dt \\
&\qquad+\int_{t_{i+1}}^{t_{i-1}}\Psi_4(x_I,t)f(x_I,t)\hat{w}_i(x_I,t)dt-\int_{t_{i+1}}^{t_{i-1}}h_2(t)\hat{w}_i(x_I,t)dt \\
&\qquad-\int_{t_{i+1}}^{t_{i-1}}(t-t^n)h_2(t)\hat{w}_{it}(x_I,t)dt.
\end{aligned}
\tag{3.12}
$$

We need to be careful for the outflow Dirichlet boundary condition. In this case, we need to approximate the normal derivative of the exact solution at the outflow boundary. Thus, we use a piecewise-constant trial function $U_x(x_I,t)$ given in (3.3). Scheme 1 is as follows:

$$
\begin{aligned}
&-\int_{t_{i+1}}^{t_{i-1}}D(x_I,t)U_x(x_I,t)\hat{w}_i(x_I,t)dt-\int_{t_{i+1}}^{t_{i-1}}(t-t^n)D(x_I,t)U_x(x_I,t)\hat{w}_{it}(x_I,t)dt \\
&=\int_{t_{i+1}}^{t_{i-1}}\Psi_2(t^n;x_I,t)\Phi(x_I^*(t),t^n)U(x_I^*(t),t^n)\hat{w}_i(x_I,t)dt \\
&\qquad+\int_{t_{i+1}}^{t_{i-1}}\Psi_4(x_I,t)f(x_I,t)\hat{w}_i(x_I,t)dt-\int_{t_{i+1}}^{t_{i-1}}V(x_I,t)h_1(t)\hat{w}_i(x_I,t))dt \\
&\qquad-\int_{t^n}^{t^{n+1}}\Psi_4(x_I,t)K(x_I,t)h_1(t)\hat{w}_i(x_I,t)dt.
\end{aligned}
\tag{3.13}
$$

Equation (3.13) solves for U_x and is completely decoupled from Equation (3.4), which is for the interior nodes. Note that Equation (3.13) can be solved successively for

$i = I+1, \ldots, I+IC$. However, the equations at $i = I+IC$ may have a degenerate coefficient as Cu_2 tends to an integer, since the last element $[t^n, t_{I+IC}]$ has only length $(Cu_2 - IC_2)\Delta t/Cu_2$. To avoid this problem, we combine the last two elements $[t^n, t_{I+IC}]$ and $[t_{I+IC}, t_{I+IC-1}]$ together to form a large element $[t^n, t_{I+IC-1}]$ with length $[1 + (Cu_2 - IC_2)]\Delta t/Cu_2$ in this case.

Scheme 1 developed in this section can treat various boundary conditions and maintains mass conservation. The scheme has a symmetric and positive definite coefficient matrix on the interior nodes. For an inflow flux boundary condition and outflow Dirichlet or Neumann boundary conditions, the resulting scheme has a global symmetric and positive definite coefficient matrix. For all other types of boundary conditions, the coefficient matrix is not symmetric and may be indefinite. The nonsymmetry and indefiniteness of the coefficient matrix are not due to our ELLAM scheme, they come from the intrinsic physics of problem (1.4). Problem (1.4) is strongly nonsymmetric due to the dominant advection term. While the use of Lagrangian coordinates symmetrizes the operator in (1.4), the domain becomes deformed in the Lagrangian coordinates if a boundary effect is taken into account. This is the fundamental reason why we obtain a nonsymmetric matrix. Nevertheless, the nonsymmetry and indefiniteness of the coefficient matrix raise the questions about the solvability, stability and convergence properties of our ELLAM scheme. We have conducted theoretical analyses on all these problems and prove that our scheme is solvable, stable and has optimal-order convergence rate. Due to the space limitation, we will present the detailed analyses elsewhere. In the last section of this paper, we will present some numerical experiments to demonstrate the strength of our schemes. In the next section, we develop our ELLAM scheme 2 that can significantly reduce the temporal error and provides an alternative approach to reduce the dependence of the numerical solution on the accurate tracking of the characteristics.

4. ELLAM Scheme 2

When we derived Scheme 1, we dropped the term $R(w)$ in (2.19). The last three terms on the right-hand side of (2.20) represent truncation-error terms, dropping it introduces negligible error. The first term on the right-hand side of (2.20) accounts for the advection missed by the errors in tracking characteristics. When the velocity field $V(x,t)$ varies slowly, this term is very small since the approximation to the characteristics is very accurate. When $V(x,t)$ varies rapidly, this term may introduce large temporal error that dominates the numerical solution. The conventional approach to reduce the tracking error is to approximate the characteristics more accurately. To do so, one can either use a higher-order single-step quadrature (such as Runge-Kutta's rule) or use a multi-step quadrature to approximate the characteristics. A higher-order single-step quadrature often needs certain spatial and characteristic derivatives of the velocity field. In applications the velocity is usually given as numerical solutions of an associated pressure equation and does not have the required regularity. While a multiple-step quadrature is difficult to implement when the Lagrangian coordinates are used. In this section we present an alternative approach to reduce the effects of the tracking errors on the numerical solutions. Note that the first term on the right-hand side of (2.20) represents the temporal error due to the approximation of the characteristics, so we should approximate this term instead of dropping it completely. We now discuss this approach in detail. By the definition of our test functions, they satisfy the following approximate adjoint equations:

$$\begin{aligned} w_s(x(s),s) + V^{\Phi}(x,t^{n+1})w_x(x(s),s) &= 0, \ s \in [t^*(x),t^{n+1}], \quad x \in [a,b], \\ w_s(X(s;x_I,t),s) + V^{\Phi}(x_I,t)w_x(X(s;x_I,t),s) &= 0, \ s \in [t^n,t], \quad t \in [t^n,t^{n+1}]. \end{aligned} \tag{4.1}$$

Thus, we can rewrite the term $\int_{t^n}^{t^{n+1}} \int_a^b \Phi u(w_t + V^{\Phi} w_x) dx dt$ as follows:

$$\begin{aligned}
&\int_{t^n}^{t^{n+1}} \int_a^b \Phi \ u \left(w_s + V^{\Phi} w_x\right) dx ds \\
&= \int_{t^n}^{t^{n+1}} \int_{x_0}^{x_I(s)} \Phi \ u \left(w_s + V^{\Phi} w_x\right) dx ds \\
&\quad + \int_{t^n}^{t^{n+1}} \int_{x_I(s)}^{x_I} \Phi \ u \left(w_s + V^{\Phi} w_x\right) dx ds \\
&= -\int_a^b \int_{t^*(x)}^{t^{n+1}} \left[V^{\Phi}(x,t^{n+1}) - V^{\Phi}(x(s),s)\right] \\
&\qquad \cdot \Phi(x(s),s) u(x(s),s) w_x(x,t^{n+1}) ds dx \\
&\quad - \int_{t^n}^{t^{n+1}} \int_{x_I(s)}^{x_I} \left[V^{\Phi}(x_I,t) - V^{\Phi}(X(s;x_I,t),s)\right] \Phi(X(s;x_I,t),s) \\
&\qquad \cdot u(X(s;x_I,t),s) w_x(X(s;x_I,t),s) dX ds \\
&= \int_{x_0}^{\tilde{x}_0} \Psi_6(x,t^{n+1}) \Phi(x_0,t^*(x)) u(x_0,t^*(x)) w_x(x,t^{n+1}) dx \\
&\quad + \int_{\tilde{x}_0}^{x_I} \Psi_6(x,t^{n+1}) \Phi(x^*,t^n) u(x^*,t^n) w_x(x,t^{n+1}) dx \\
&\quad + \int_{t^n}^{t^{n+1}} \Psi_7(t) \Phi(x_I^*(t),t^n) u(x_I^*(t),t^n) w_t(x_I,t) dt \\
&\quad - \int_a^b \int_{t^*(x)}^{t^{n+1}} \frac{(s-t^*(x))(t^{n+1}-s)}{2} \Theta_1(s;x,t^{n+1}) w_x(x,t^{n+1}) ds dx \\
&\quad + \int_{t^n}^{t^{n+1}} \int_{t^n}^{t} \frac{(s-t^n)(t-s)}{2} \Theta_2(s;x_I,t) w_t(x_I,t) ds dt.
\end{aligned} \tag{4.2}$$

Here we have used the adjoint equations (4.1) at the second equal sign and the trapezoidal rule at the last equal sign when we derived (4.2). $\Psi_6(x,t^{n+1})$, $\Psi_7(t)$,

$\Theta_1(s; x, t^{n+1})$ and $\Theta_2(s; x_I, t)$ are defined as follows:

$$\Psi_6(x, t^{n+1}) = \begin{cases} \dfrac{\Delta t(x) \left[V^{\Phi}(x, t^{n+1}) - V^{\Phi}(x_0, t^*(x))\right]}{2}, & \text{if } x \in [x_0, \tilde{x}_0], \\ \dfrac{\Delta t \left[V^{\Phi}(x, t^{n+1}) - V^{\Phi}(x^*, t^n)\right]}{2}, & \text{if } x \in [\tilde{x}_0, x_I], \end{cases} \tag{4.3}$$

$$\Psi_7(t) = \frac{(t - t^n) \left[V^{\Phi}(x_I, t) - V^{\Phi}(x_I^*(t), t^n)\right]}{2}, \qquad t \in [t^n, t^{n+1}].$$

$$\begin{aligned} \Theta_1\left(s; x, t^{n+1}\right) &= \frac{\partial^2}{\partial s^2}\left\{\left[\int_s^{t^{n+1}} V_\theta^{\Phi}(x(\theta), \theta) d\theta\right] \Phi(x(s), s) u(x(s), s)\right\}, \\ & \quad s \in [t^*(x), t^{n+1}], \quad x \in [a, b], \\ \Theta_2\left(s; x_I, t\right) &= \frac{\partial^2}{\partial s^2}\left\{\left[\int_s^{t} V_\theta^{\Phi}(X(\theta; x_I, t), \theta) d\theta\right] \right. \\ & \quad \left. \cdot\, \Phi(X(\theta; x_I, t), \theta) u(X(\theta; x_I, t), \theta)\right\}, \qquad s \in [t^n, t], \quad t \in [t^n, t^{n+1}]. \end{aligned} \tag{4.4}$$

When deriving (4.2), we have also used the fact that for any function $p(s) \in C^2[\alpha, \beta]$ with $p(\alpha) = p(\beta) = 0$, we have

$$\int_\alpha^\beta p(s) ds = \int_\alpha^\beta \frac{(s - \alpha)(s - \beta)}{2} p''(s) ds. \tag{4.5}$$

If we replace the exact solution u by the trial function U in (4.2), drop the last two terms on the right-hand side of the last equal sign in (4.2) and then add the remaining three terms to the right-hand side of Scheme 1, we obtain our Scheme 2. Even though the term $\int_{t^n}^{t^{n+1}} \int_a^b \Phi u(w_t + V^{\Phi} w_x) dx dt$ is a nonsymmetric term, Scheme 2 only adds a correction term on the right-hand side of Scheme 1. Thus, Scheme 2 maintains the same data structure as Scheme 1. At the same time, Scheme 2 significantly reduces the time truncation error in the numerical solution with Scheme 1. We will present some numerical results in the next section to show this. For simplicity, we omit the detailed formulae of Scheme 2 for all the boundary conditions here.

5. Computational Results and Related Issues

In this section we run some numerical experiments to observe the performance of our ELLAM schemes developed in this paper. We also discuss some related issues and related works. Our theoretical analyses presented in the subsequent paper prove the following estimates for any combinations of inflow and outflow boundary conditions

$$\max_{n=0,\ldots,N} \left\| U(x, t^n) - u(x, t^n) \right\|_{L^2(a,b)} \leq C\left[(\Delta x)^\alpha + (\Delta t)^\beta\right]. \tag{5.1}$$

where $\alpha = 2$ and $\beta = 1$.

In this section we present a representative example to verify the estimate and to observe the performance of our schemes. We perform two kinds of computations. One is to test the convergence rate α of the error with respect to the space, where we pick small Δt and observe the convergence order α with respect to Δx. The other is to test the convergence rate β of the error with respect to the time, where we pick small Δx and observe the convergence order β with respect to Δt. We use regression techniques to compute the convergence factors α and β. Due to the space limitation, we cannot present our numerical results for all the combinations of boundary conditions into the tables, we only put those for one combination of boundary conditions into the tables. The numerical results are representative.

In [13, 24], we have presented the numerical results of our ELLAM schemes for constant-coefficient advective-diffusive-reactive transport equations for all combinations of boundary conditions. We theoretically analysed and numerically compared the effects of different combinations of boundary conditions on the numerical solutions. We still have similar observations for our ELLAM schemes for variable-coefficient problems. we do not present the results here due to the space limitation. Instead, we want to focus on those behavior of our ELLAM schemes that comes solely from variable-coefficient problems.

The test problem is as follows: The domain is $(0,1)$, the initial time $t=0$, the final time $T=0.5$, $\Phi(x,t)=1$, $V(x,t)=(1+x)/3$, $D(x,t)=0.01$, $K(x,t)=0.2$. The source term $f(x,t)$ is given by

$$\begin{aligned} f(x,t) &= \frac{1}{3}G(x-V(x,t)t,t) - \frac{(1+x)\,t}{9}G_x(x-V(x,t)t,t) \\ &\quad + \frac{Dt(6-t)}{9}G_{xx}(x-V(x,t)t,t) \end{aligned} \tag{5.2}$$

where $G(x,t)$ is

$$G(x,t) = \frac{\exp(-Kt)\exp\left(-\pi x^2/(0.1+4\pi Dt)\right)}{\sqrt{10}\sqrt{0.1+4\pi Dt}}. \tag{5.3}$$

It is easy to verify that the exact solution $u(x,t)$ of this example is

$$u(x,t) = \frac{\exp(-Kt)\exp\left(-\pi(x-V(x,t)t)^2/(0.1+4\pi Dt)\right)}{\sqrt{10}\sqrt{0.1+4\pi Dt}} \tag{5.4}$$

whose initial condition is a Gaussian hill given by $\exp\left(-10\pi x^2\right)$. In the numerical tests, we use $u(x,t)$ to compute the boundary conditions. We use inflow flux and outflow Neumann boundary conditions, since these conditions often arise in real applications and many ELM have difficulties in treating these boundary conditions. Also, we choose the velocity $V(x,t)$ to be a linear function, since in most applications, $V(x,t)$ is usually given as a piecewise-linear numerical solution of an associated pressure equation to (1.4).

From the numerical experiments presented in the tables, we see that both our ELLAM schemes have second-order convergence in space and first-order convergence in the time. Scheme 2 further reduces the time truncation error present in Scheme 1. In Table 2, the error of Scheme 2 is much smaller than that of the Scheme 1. Because the time step Δt is relatively larger than the space mesh Δx in this case, so the time truncation error dominates the numerical solution. Since Scheme 2 approximates temporal part more accurately, so Scheme 2 should work much better than Scheme 1 as confirmed by Table 2. In Table 1, the errors with Scheme 1 and Scheme 2 are basically the same. Because the time step Δt is much smaller than the space mesh Δx, so the space error dominates the numerical solutions. Since the two schemes treat the space similarly, the errors with these two schemes should be about the same as confirmed by Table 1. In realistic applications, the feasible time step can not be very small due to the limitations of computational cost. Thus, Scheme 2 works better than Scheme 1 at the cost of computing two more terms that only affect the right-hand side of the discrete system.

TABLE I
Test for α.

Δt	Δx	L^2 error for Scheme 1	L^2 error for Scheme 2
1/500	1/5	2.136462E − 2	2.120703E − 2
1/500	1/10	4.824073E − 3	4.804058E − 3
1/500	1/20	1.179526E − 3	1.159527E − 3
1/500	1/40	3.022359E − 4	2.822362E − 4
		$\alpha = 2.0405$	$\alpha = 2.1410$

TABLE II
Test for β.

Δt	Δx	L^2 error for Scheme 1	L^2 error for Scheme 2
1/20	1/100	1.901269E −3	9.012794E − 4
1/40	1/100	9.561227E − 4	4.501249E − 4
1/80	1/100	4.844626E − 4	2.244622E − 4
1/160	1/100	2.501478E − 4	1.101484E − 4
		$\beta = 0.9743$	$\beta = 1.0743$

We conclude this paper by one more comment on our ELLAM schemes developed in this paper. In [13], the test functions are taken to vary exponentially along the characteristics for our ELLAM scheme for a constant-coefficient advective-diffusive-reactive transport equation. The derived scheme has a symmetric and positive-definite coefficient matrix on the interior nodes. The matrix may be symmetric on the boundary depending on which kind of boundary condition is specified [13]. However, for variable-coefficient equations the same test functions yield nonsymmetric matrix on each node due to the spatial variation of the reaction term. With our test functions in this paper, our ELLAM schemes still have a symmetric and positive-definite coefficient matrix (at least on the interior nodes). For those applications where the reaction does not dominate the physical process, our ELLAM schemes derived in this paper work very well. If the reactions dominate the physical process, our guess is that the schemes with the test functions defined in [13] might be better. The drawback for the test functions in [13] are that the coefficient matrix is nonsymmetric, and special care must be taken when defining the reaction coefficients especially for nonlinear problems. Since small spatial oscillations may arise in the numerical solutions, they can cause small negative concentrations to appear. For linear equations this does not cause problems, but when the reaction coefficient is a function of the concentration, a change of sign in the concentration can cause a change of sign in the reaction coefficient. This can cause incorrect sign in the exponent that defines the exponential change in the test functions, which can lead to numerical instability. We are currently conducting intensive research in this direction.

6. Acknowledgements

This research was supported in part by Office of Naval Research Contract No. 0014–88–K–0370, by National Science Foundation Grant No. DMS-8922865, by funding from the Institute for Scientific Computation at University of Wyoming, by funding from the Institute for Scientific Computation at Texas A&M University, by funding from DOE DE–ACO5–840R21400, Martin Marietta Subcontract SK966V, and by the funding from the Norwegian Research Council for Science and the Humanities.

References

1. Brooks, A. and Hughes, T.J.R.: 1982, 'Streamline Upwind Petrov-Galerkin Formulation for Convection Dominated Flows With Particular Emphasis on the Incompressible Navier-Stokes Equations', *Comp. Meth. Appl. Mech. Engrg.* **32**, pp. 199–259.
2. Celia, M.A., Herrera, I., Bouloutas, E.T. and Kinder, J.S.: 1989 'A New Numerical Approach for the Advective-Diffusive Transport Equation', *Numerical Methods for PDE's* **5**, pp. 203–226.
3. Celia, M.A., Russell, T.F., Herrera, I. and Ewing, R.E.: 1990, 'An Eulerian-Lagraingian Localized Adjoint Method for the Advection-Diffusion Equation', *Advances in Water Resources* **13**, pp. 187–206.
4. Celia, M.A. and Zisman, S.: 1990, 'An Eulerian-Lagrangian Localized Adjoint Method for Reactive Transport in Groundwater', *Computational Methods in Subsurface Hydrology, Proceedings of the Eighth International Conference on Computational Methods in Water Resources*, Venice, Italy, pp. 383–392.
5. Dahle, H.K., Espedal, M.S., Ewing, R.E. and Sævareid, O.: 1990, 'Characteristic Adaptive Sub-Domain Methods for Reservoir Flow Problems', *Numerical Methods for PDE's*, pp. 279–309.
6. Dahle, H.K., Ewing, R.E. and Russell, T.F.: to appear, 'Eulerian-Lagrangian Localized Adjoint Methods for a Nonlinear Convection-Diffusion Equation.
7. Demkowicz, L. and Oden, J.T.: 1986, 'An Adaptive Characteristic Petrov-Galerkin Finite Element Method for Convection-Dominated Linear and Nonlinear Parabolic Problems in Two Space Variables', *Comp. Meth. in Appl. Mech. and Eng.* **55**, pp. 63–87.
8. Douglas, Jr., J. and Russell, T.F.: 1982, 'Numerical Methods for Convection-Dominated Diffusion Problems Based on Combining the Method of Characteristics With Finite Element

or Finite Difference Procedures', *SIAM J. Numer. Anal.* **19**, pp. 871–885.
9. Espedal, M.S. and Ewing, R.E.: 1987, 'Characteristic Petrov-Galerkin Subdomain Methods for Two-Phase Immiscible Flow', *Comp. Meth. in Appl. Mech. and Eng.* **64** pp. 113–135.
10. Ewing, R.E. (ed.): 1983, *Research Frontiers in Applied Mathematics, Vol. 1*, SIAM, Philadelphia.
11. Ewing, R.E.: 1991, 'Operator Splitting and Eulerian-Lagrangian Localized Adjoint Methods for Multiphase Flow', *The Mathematics of Finite Elements and Applications VII* (MAFELAP 1990), (Whiteman, I., ed.), Academic Press, Inc., San Diego, California, pp. 215–232.
12. Ewing, R.E. and Wang, H.: 1991; 'Eulerian-Lagrangian Localized Adjoint Methods for Linear Advection Equations', *Proceedings of International Conference on Computational Engineering Science*, Melbourne, Australia, pp. 245–250.
13. Ewing, R.E. and Wang, H.: 1992, 'Eulerian-Lagrangian Localized Adjoint Methods for Reactive Transport in Groundwater', *IMA Preprint Series* **1014**, Institute for Mathematics and Its Applications, University of Minnesota; to appear in *IMA Volume in Mathematics and Its Applications* (Wheeler *et al.*, eds.), Springer Verlag, Berlin.
14. Herrera, I., Chargoy, L. and Alduncin, G.: 1985, 'Unified Formulation of Numerical Methods. 3', *Numerical Methods for PDE's* **1(4)**, pp. 241–258.
15. Herrera, I.: 1987, 'The Algebraic Theory Approach for Ordinary Differential Equations: Highly Accurate Finite Differences', *Numerical Methods for PDE's* **3(3)**, pp. 199–218.
16. Hughes, T.J.R. and Brooks, A.: 1982, 'A Theoretical Framework for Petrov-Galerkin Methods With Discontinuous Weighting Functions. Applications to the Streamline-Upwind Procedure' (Gallagher, ed.), *Finite Elements in Fluids* **4**, Wiley.
17. Johnson, C.: 1987, *Numerical solutions of Partial Differential Equations by the Finite Element Method*, Cambridge Univrsity Press, Cambridge.
18. Russell, T.F.: 1989, 'Eulerian-Lagrangian Localized Adjoint Methods for Advection-Dominated Problems. Numerical Analysis', *Proceedings of the 13th Dundee Conference on Numerical Analysis* (Griffiths, D.F. and Watson, G.A., eds.), *Pitmann Research Notes in Mathematics Series* **228**, Longman Scientific & Technical, Harlow, U.K., pp. 206–228.
19. Russell, T.F. and Trujillo, R.V.: 1990, 'Eulerian-Lagrangian Localized Adjoint Methods With Variable Coefficients in Multiple Dimensions. Computational Methods in Surface Hydrology', *Proceedings of the Eighth International Conference on Computational Methods in Water Resources*, Venice, Italy, pp. 357–363.
20. Varoglu, E. and Finn, W.D.L.: 1980, 'Finite Elements Incorporating Characteristics for One-Dimensional Diffusion-Convection Equation', *J. Comp. Phys.* **34**, pp. 371–389.
21. Wang, H., Ewing, R.E. and Russell, T.F.: 1992, 'ELLAM for Variable-Coefficient Convection-Diffusion Problems Arising in Groundwater Applications', *Computational Methods in Water Resources IX. Vol. I: Numerical Methods in Water Resources* (Russell, Ewing, Brebbia, Gray and Pinder, eds.), Computational Mechanics Publications and Elsevier Applied Science, London and New York, pp. 25–31.
22. Wang, H. and Lin, T. and Ewing, R.E.: 1992, 'ELLAM With Domain Decomposition and Local Refinement Techniques for Advection-Reaction Problems With Discontinuous Coefficients', *Computational Methods in Water Resources IX. Vol. I: Numerical Methods in Water Resources* (Russell, Ewing, Brebbia, Gray and Pinder, eds.), Computational Mechanics Publications and Elsevier Applied Science, London and New York, pp. 17–24.
23. Wang, H., Dahle, H., Ewing, R.E. and Espedal, M.S.: 1993, 'Eulerian-Lagrangian Localized Adjoint Methods for Convection-Diffusion Problems in Multidimensions', presentation at *SIAM Conference on Mathematical and Computational Issues in the Geosciences*, Houston, Texas, April 19–21.
24. Wang, H. and Ewing, R.E. and Russell, T.F.: submitted, 'Eulerian-Lagrangian Localized Adjoint Methods for Convection-Diffusion Equations and Their Convergence Analysis'.

THE COMMUNICATION PATTERNS OF NESTED PRECONDITIONINGS FOR MASSIVELY PARALLEL ARCHITECTURES

J.C. DíAZ *
Center for Parallel and Scientific Computing
The University of Tulsa
600 Sth. College Ave.
Tulsa, OK 74104-3189
diaz@babieco.mcs.utulsa.edu

and

K. E. JORDAN
Kendall Square Research
170 Tracer Lane
Waltham, MA 02154-1379
kirk@ksr.com

Abstract. Preconditioned Iterative methods are used to solve sparse linear systems from discrete convection dominated diffusion problems. The preconditionings are based on nested incomplete factorization with approximate tridiagonal inverses using a two color line ordering of the discretization grid. These preconditionings can be described in terms of *vector-vector to vector* operations of dimension equal to half the total of grid points. We discuss the communication patterns of these preconditionings for exploitation of massively parallel architectures.

Key words: Preconditioning, Massively Parallel, Communication Pattern

1. Model Problem.

Consider the non-self adjoint differential equation modeling steady state convection-diffusion.

$$\nabla \cdot (D\nabla u) - V \cdot \nabla u = f, \qquad in \quad [0,1]X[0,1],$$

where

$$D = \begin{bmatrix} a & 0 \\ 0 & b \end{bmatrix}, V = \begin{bmatrix} v_1 \\ v_2 \end{bmatrix},$$

subject to appropriate boundary conditions. Here D is the diffusion parameter and V is the velocity parameter.

Assume a square grid with n points on each side and $\Delta x = \Delta y = \frac{1}{n+1}$. The five-point upwinded discretization on such a mesh produces the following set of

* Research supported in part by the Oklahoma Center for Advancement of Science and Technology Grants RB9-008 (3748) and ARO-36 (3910), and supported by, or in part by the Army Research Office contract number DAAL03-89-C-0038 with the University of Minnesota Army High Performance Computing Research Center.

S. Gomez and J-P. Hennart (eds.), Advances in Optimization and Numerical Analysis, 207–218.
© 1994 *Kluwer Academic Publishers.*

equations

$$u_{i-1,j}(\frac{v_1\Delta x + a}{\Delta x^2}) + u_{i+1,j}(\frac{a}{\Delta x^2}) + u_{i,j+1}(\frac{b}{\Delta y^2})$$
$$+u_{i,j-1}(\frac{v_2\Delta y + b}{\Delta y^2}) - u_{i,j}(\frac{2a}{\Delta x^2} + \frac{2b}{\Delta y^2} + \frac{v_1}{\Delta x} + \frac{v_2}{\Delta y}) = f_{i,j}$$

where both $i, j = 1, \cdots n$. The above equations can be expressed as a linear system $Au = f$ where the matrix structure depends on the ordering of the grid. Several schemes for numbering the grid points can be considered.

The natural ordering yields a matrix

$$A = \begin{pmatrix} T_1 & U_1 & & & \\ L_2 & T_2 & U_2 & & \\ & L_3 & T_3 & U_3 & \\ & & \cdots & & \\ & & L_{n-1} & T_{n-1} & U_{n-1} \\ & & & L_n & T_n \end{pmatrix}$$

where the submatrices, T_i, L_i, U_i, are of order n by n, the matrices $T_i, i = 1, \cdots, n$ are tridiagonal, and the matrices $U_i, L_{i+1}, i = 1, \cdots, n-1$ are diagonal.

On the other hand, a zebra-like two color-line ordering allows array dimensions of $m = n\lceil\frac{n}{2}\rceil$. It yields a matrix

$$A = \begin{pmatrix} T_1 & U \\ L & T_2 \end{pmatrix} \tag{1}$$

where the submatrices T_1, T_2, U, and L are of order m, the submatrices T_1, T_2 are tridiagonal, and the submatrices U and L are bidiagonal. The bidiagonal submatrix U (L) has a main diagonal and an upper (lower) subdiagonal located n diagonals off the main diagonal.

2. Iterative Procedure.

The preconditioned conjugate gradient-type iterative algorithm has been found to be an effective method of solution for large systems of equations. Since the matrix is not symmetric, we use the conjugate residual algorithm because of its simplicity of implementation. The conjugate residual algorithm is described in the Figure 1.

The iterative algorithm is composed of four inner-products, four SAXPY vector–vector operations, a matrix–vector multiply $v = Az$, and the preconditioning application $Qz = r$.

The matrix vector multiply can be represented in terms of vector-vector operations. For matrices structure as in Equation 1, the matrix-vector multiply

$$\begin{pmatrix} T_1 & U \\ L & T_2 \end{pmatrix} \cdot \begin{pmatrix} z_1 \\ z_2 \end{pmatrix} = \begin{pmatrix} r_1 \\ r_2 \end{pmatrix} \tag{2}$$

is computed by:

$$r_1 = T_1 z_1 + U z_2, \tag{3}$$
$$r_2 = L z_1 + T_2 z_2, \tag{4}$$

x^0 be the initial guess,
$r^0 = f - Ax^0$,
z^0 solves $Qz^0 = r^0$, where Q a preconditioning,
$v^0 = Az^0$, and
$\sigma_0 = (v^0, v^0)$.
For $k = 0, 1, \cdots$ until convergence:

$$\begin{array}{ll}
\alpha_k & =\frac{(r^k, v^k)}{\sigma_k} \\
x^{k+1} & =x^k + \alpha_k z^k \\
r^{k+1} & =r^k - \alpha_k v^k \\
\text{if} & ((r^{k+1}, r^{k+1}) \leq \text{Tol}), \text{Stop} \\
Qz^{k+1} & =r^{k+1} \\
v^{k+1} & =Az^{k+1} \\
\beta_k & =\frac{(v^k, v^{k+1})}{\sigma_k} \\
z^{k+1} & =z^{k+1} - \beta_k z^k \\
v^{k+1} & =v^{k+1} - \beta_k v^k \\
\sigma_{k+1} & =(v^{k+1}, v^{k+1})
\end{array}$$

Fig. 1. Iterative Method

which takes the form of matrix vector multiplies by tridiagonal and bidiagonal matrices with arrays of dimension $m = n\lceil \frac{n}{2} \rceil$ per color.

It is highly desirable that the preconditioning application, $Qz = r$, also be representable in terms of vector-vector operations.

A preconditioning based on the idea of nested factorization with approximate inverses is described in detail by Leaf *et al* [4, 5]. A fully vectorizable formulation of the tridiagonal approximate-inverse preconditioner was suggested by Díaz and Macedo [2].

This algorithm was implemented for the Connection Machine by Díaz and Dutt [1]. The particular preconditioning yields a preconditioning application $Qz = r$ of the form

$$\begin{pmatrix} I_1 & 0 \\ P_2 L & I_2 \end{pmatrix} \cdot \begin{pmatrix} I_1 & P_1 U \\ 0 & I_2 \end{pmatrix} \cdot \begin{pmatrix} z_1 \\ z_2 \end{pmatrix} = \begin{pmatrix} P_1 r_1 \\ P_2 r_2 \end{pmatrix}$$

where P_1 and P_2 are tridiagonal. Hence, the equation $Qz = r$ is solved by

$$y_1 = P_1 r_1, \tag{5}$$
$$z_2 = P_2(r_2 - Ly_1), \tag{6}$$
$$z_1 = y_1 - P_1(Uz_2), \tag{7}$$

which takes the form of matrix–vector multiplies with tridiagonal and bidiagonal matrices having dimension $m = n\lceil \frac{n}{2} \rceil$ per color.

3. Implementational Issues.

A simple data structure is used. Both the sparse matrices and vectors have dimension $m = n\lceil \frac{n}{2} \rceil$. Tridiagonal matrices are stored by diagonals with zero padding as necessary.

$$T = \begin{pmatrix} d_1 & u_1 & & & & \\ l_2 & d_2 & u_2 & & & \\ & l_3 & d_3 & u_3 & & \\ & & \cdots & \cdots & \cdots & \\ & & & l_{m-1} & d_{m-1} & u_{m-1} \\ & & & & l_m & d_m \end{pmatrix}$$

Tridiagonal matrices are represented by the three arrays where 0's represent padding:

$$\begin{pmatrix} u \\ d \\ l \end{pmatrix} = \begin{pmatrix} 0 & u_1 & u_2 & \cdots & u_{m-2} & u_{m-1} \\ d_1 & d_2 & d_3 & \cdots & d_{m-1} & d_m \\ l_2 & l_3 & l_4 & \cdots & l_m & 0 \end{pmatrix}.$$

The product by a tridiagonal matrix, $z = Tr$, is computed by:

```
z = 0
z = z + cshift( u * r,  1)
z = z + d * r
z = z + cshift( l * r, -1)
```

Similarly,

$$L = \begin{pmatrix} d_1 & & f_1 & & & \\ & \cdots & & \cdots & & \\ & & \cdots & & \cdots & \\ & & & d_{m-n} & & f_{m-n} \\ & & & & \cdots & \\ & & & & & d_m \end{pmatrix},$$

is represented by the two arrays:

$$\begin{pmatrix} f \\ d \end{pmatrix} = \begin{pmatrix} 0 & \cdots & 0 & f_1 & \cdots & f_{m-n} \\ d_1 & \cdots & d_n & d_{n+1} & \cdots & d_m \end{pmatrix}.$$

The product $z = Lr$ is computed by:

```
z = 0
z = z + cshift( f * r , n)
z = z + d * r
```

And correspondingly

$$U = \begin{pmatrix} d_1 & & & & & \\ & \cdots & & & & \\ f_{n+1} & & d_{n+1} & & & \\ & \cdots & & \cdots & & \\ & & \cdots & & \cdots & \\ & & & f_m & & d_m \end{pmatrix},$$

is represented by the two arrays:

$$\begin{pmatrix} d \\ f \end{pmatrix} = \begin{pmatrix} d_1 & \cdots & d_{m-n} & d_{m+1-n} & \cdots & d_m \\ f_{n+1} & \cdots & f_m & 0 & \cdots & 0 \end{pmatrix}.$$

The matrix vector multiply, Equations 5 and 6, takes 20 vector-vector operations of dimension m, and has two shifts of order n and two shifts of order 1. The application of the preconditioning Equations 5, 6, and 7, can be completed in 28 vector-vector operations of dimension m, and Equation 7, has two shifts of order n and three shifts of order 1. The four inner-products take $8mlog_2m$ operations, and the four SAXPY operations take $16m$ operations. The total number of floating point operations for one iteration of the loop is $(64 + 8log_2m)m$. Each iteration of the loop has 5 shifts of order 1 and 4 shifts of order n.

4. Architectures.

The massively parallel architectures considered are two models of the Connection Machine (CM) manufactured by Thinking Machines, Inc. The two models considered were the CM2/200 and the CM5. The CM2 and CM200 are similar in design. The CM200 operates at 10 MHz clock versus the 7 MHz clock for the CM2, [6, 7]. The Connection Machine CM2/200 consist of thousands of bit-serial processors each with associated memory connected via a hypercube network. Each CM2/200 processor chip packages 16 bit-serial processors. The router node is hardware on the CM2/200 processor chip that handles interprocessor communications. Communication of processors not on the same chip is handled by messages routed to other CM2/200 processor chips over a network. The router nodes on all the processor chips are wired together to form the router network and the topology of this network is a boolean n-cube, a hypercube network. In addition to the bit-serial processors, there are floating-point ALU chips, one for every two CM2/200 processor chips. There is one floating-point ALU for every 32 bit-serial processors. A 32 bit word can be represented in a sliced wise fashion where each bit is stored in a different bit-serial processor-memory. Herein, we take the view that the floating-point ALU with two CM2/200 processor chips and associated memory form the essential floating point component. The floating-point ALU's are off-the-shelf chips based on the Weitek WTL3164. They have multiply and add units that can be chained to form vector multiply-add operations. The CM2/200 is programmed in the data parallel paradigm usually using CM Fortran, a Fortran 90 variant. Single instructions are issued to all processors giving the CM2/200 a SIMD style of computation.

A Connection Machine CM5 is a massively parallel, distributed memory computer like the CM2/200 but with a different network topology based on a "Fat Tree", see [8] for details. It may be configured with tens to thousands of processing nodes. Each node consists of a SPARC scalar chip set, four vector pipelines, 32 MB of memory, and a network interface. Nodes communicate with each other and with a variety of I/O devices via point to point data routing network (DR) and a multifunctions broadcast, combining, reduction network (CN). The CM5 may be programmed in the message passing style using the CMMD message passing library or may be programmed in the data parallel style using either CM Fortran or C*, a parallel

TABLE I
Peak Rates for CM configurations

Model	Number (Type) of Procs	Peak Rate per Proc	Total Peak Rate
CM2	256 Weitek	13.9 MFlops/WC	3.6 GFlops
CM5-sp	32 Sparc	7 MFlops/Sp	.23 GFlops
CM5–vu	32 Sparc × 4vu	32 MFlops/vu	4.1 GFlops

TABLE II
Array dimension = Physical Grid×Local Grid = P×L

n	CM2	CM5–vu	CM5Sp
128	256X32	128X64	32X256
256	256X128	128X512	32X1024
512	256X512	128X1024	32X4096
1024	256X2048	128X4096	32X16384

extension of the C language. The message passing version supports node level code written in Fortran 77 or C. The data parallel languages may be used to construct programs that are global in extent (they manage the resources of all processors in the system) or they may be used to create node level programs which communicate using the message passing library. Currently, the vector pipelines may be accessed either by writing data parallel code or by writing assembly level routines embedded in message passing programs. Node level programs written in Fortran 77 will not be able to take advantage of vector performance of the CM5 without the addition of such assembly level routines.

The program, was implemented on the CM2 and CM5. A sequencer of a CM2 was used. This consists of 256 Weitek chips on an interconnect network. The CM5–vu with 32 processors and a full complement of four vector units per processor was also used. For reference point, a CM5–sp, a CM5 with 32 Sparc processors without the vector units was also considered. The peak rates for these CM configurations are given in Table I.

Arrays are layout in memory on the physical grid which consist of the floating point processors and associated memories. On each processor's memory the arrays are blocked into the local grid (or subgrid). If L is the local grid and P the physical grid, then the elements $(1 + L * (p-1))$, $\cdots$, $L * p$, of an array are stored on the p-th (Processor + Memory) node: where p = $1, \cdots, P$. Table II gives the physical and local grid (or subgrid) layout of memory elements for each CM configuration used as a function of n.

TABLE III
Performance in Megaflops

n	CM2	CM5-sp	CM5–vu
128	39.30	14.87	32.45
256	52.00	15.20	96.42
512	71.59	12.26	172.07
1024	84.86	12.16	227.00

Fig. 2. **Performance in Megaflops**

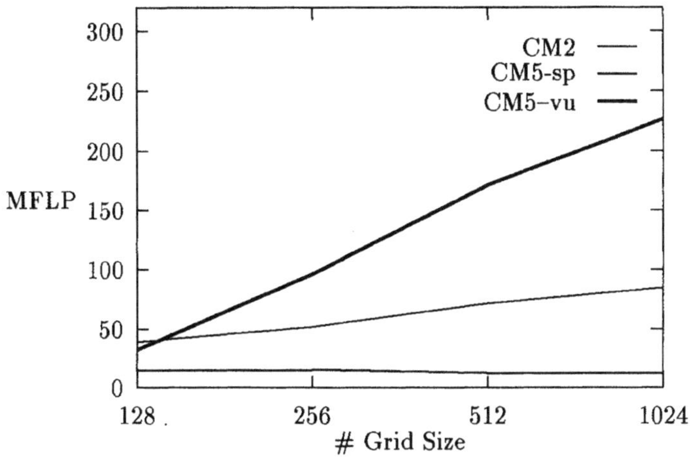

5. Results.

The results are presented in Table III and Figures 2 through 8. The performance rate measured in Megaflops for each configuration and problem size, is given in Table III. Figure 2 presents graphically the results of Table III.

The communication cost of the algorithm can be a significant part of the total cost. The algorithm discussed above has communications only on the shifts. Each shift of order n results in the movement of n elements from one processor to its immediate neighbor. For instance, if $n = 128$, in the CM2 with layout provided by Table II, a shift of order n results in the movement of the total contents of a local array across four processors. For $n = 128$ for the CM5–sp it is only half of the array stored in the inmediate neighboring processor. For the CM5–vu the movement is more complicated. Only half of the array stored in the inmediate neighboring processor needs to be brought in. However, each two vector units is

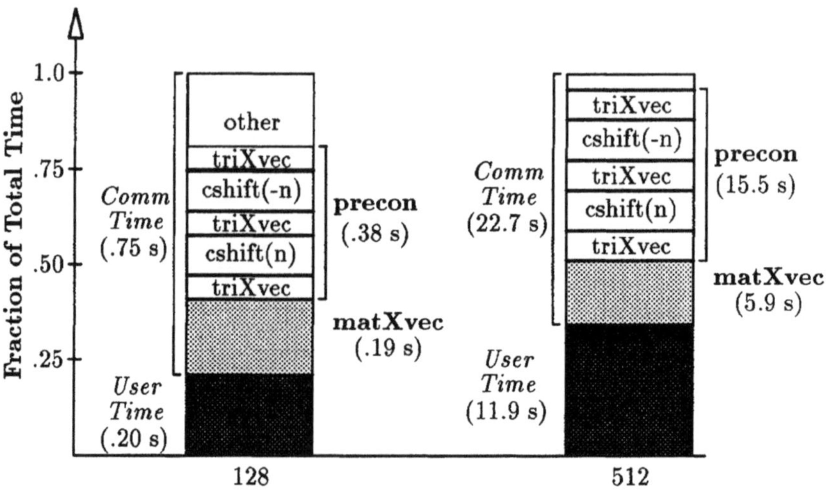

Fig. 3. **CM–2**

associated to one memory bank. Although the vector units associated with each Sparc chip have direct memory access, there is the possibility of banking conflicts occurring. For $n = 1024$ the movement is of the order one–half for the CM2, one-fourth for the CM5–vu and one sixteenth for the CM5–sp. Clearly, shift of order 1 results in only one memory element being moved across immediate processors. These communication costs are reflected in the Figures 3 through 8.

Figures 3 through 8 show the percent or fraction of time used by the different portions of the program. The following terms are used in Figures 3 through 8.

- *User Time* refers to the time the user is fully utilizing the cpu's.
- *CommTime* refers to the time spent using communication between processors.
- **matXvec** refers to the time spent by the routine performing the matrix vector multiply.
- **precon** refers to the time spent by the routine performing the matrix vector multiply.
- triXvec refers to the time spent by the routine performing the tridiagonal matrix times a vector.
- cshift(-n) refers to the time spent by performing shifts n places to the right.
- cshift(n) refers to the time spent by performing shifts n places to the left.

Figures 3 and 4 present the results for the CM–2. The CM5–sp, without vector units, are presented in Figures 5 and 6. And the results for the CM5 with vector units are presented in Figures 7 and 8.

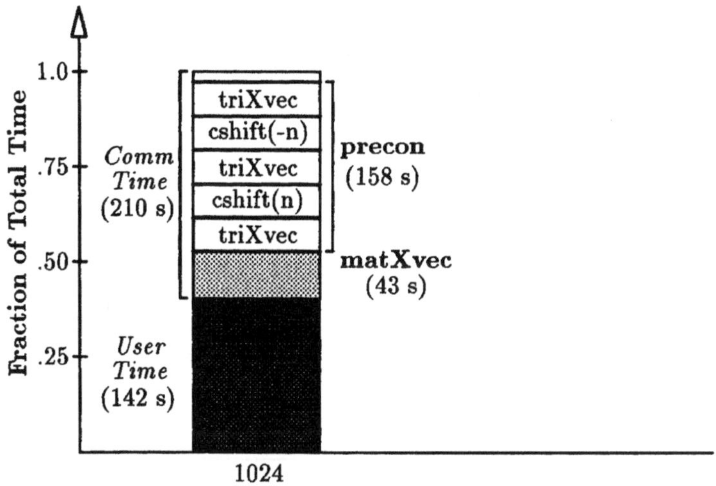

Fig. 4. **CM–2**

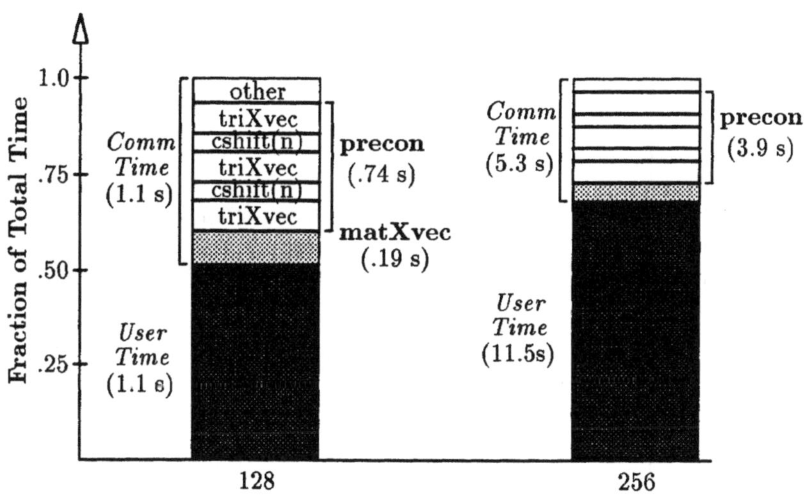

Fig. 5. **CM5 Sparc**

6. Conclusions.

The results reported for the CM5–sp without the vector units is for reference only. However, Table III and Figure 2 illustrate the power of the vector units. Figure 2 shows better performance by CM5–vu as the size of the problem grows. The performance is from 85% faster to 167% faster than the CM2. ¿From Table I, it can

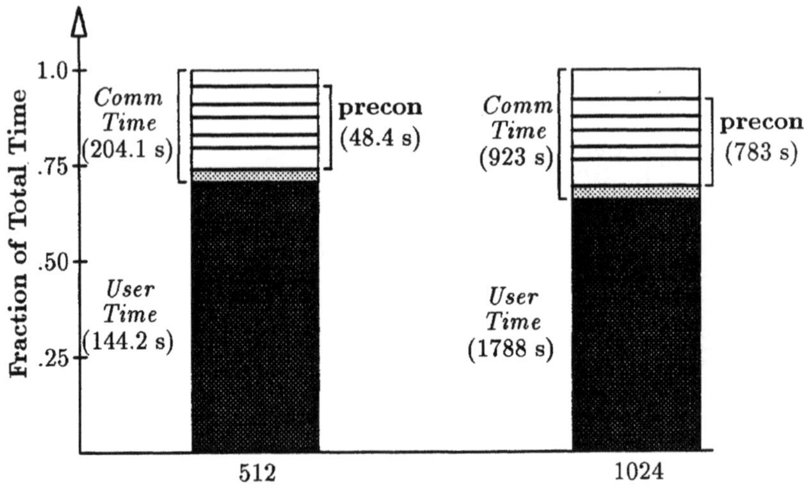

Fig. 6. **CM5 Sparc**

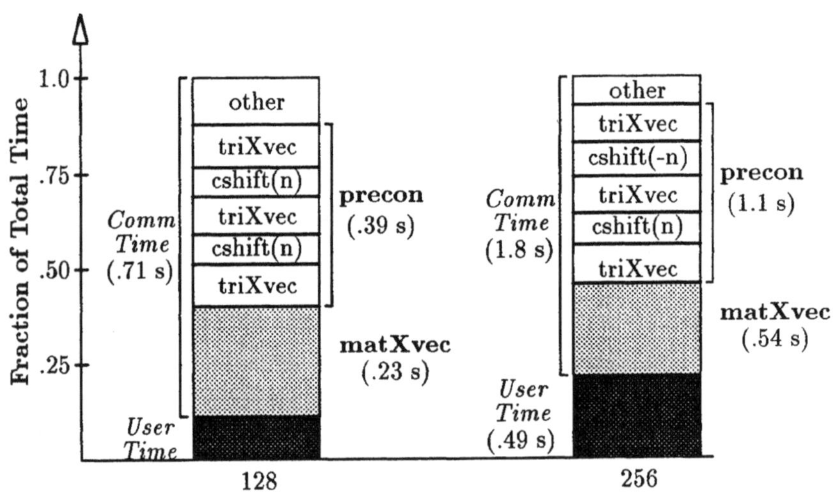

Fig. 7. **CM5–vu**

be seen that the peak performance of the CM5–vu is about 14% faster than that of the CM2. Therefore, the program shows a much better potential for the CM5–vu.

The comparison of the peak rate (Table I) and the actual performance (Table III) shows a significant discrepancy. For this reason, the cost of communications was examined. The figures 3 through 8 give a graphical representation of the communi-

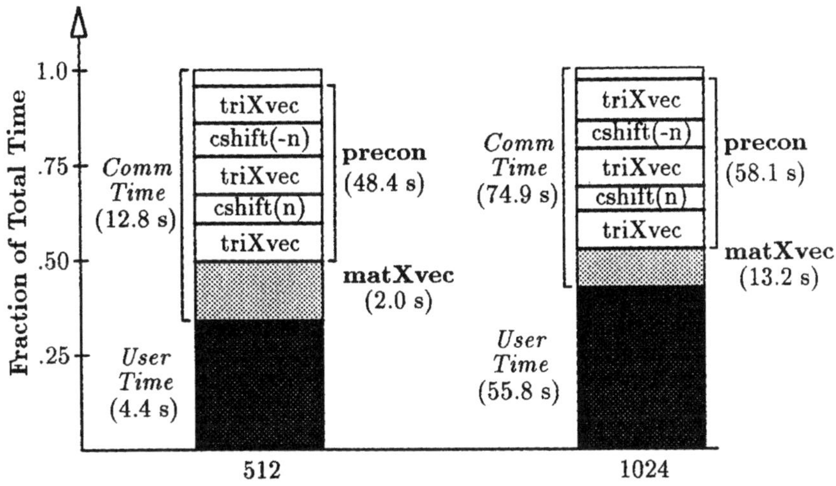

Fig. 8. **CM5–vu**

cation costs involved. The application of the preconditioning is the one that incurs the highest communication cost. A significant amount of this time is spent doing cshift operations of order n. The computation time on the vector units has been speeded up. Hence, the percent of time spent in communications is greater.

A representation of the arrays as two dimensional arrays may improve the performance by reducing the cshift operation from order n to order 1. This work is in progress and will be reported elsewhere.

Acknowledgements

The authors wish to thank Professor J. Hensley of The University of Tulsa for assistance in producing Figures 3 through 8.

References

1. Díaz, J. C. and A. Dutt: 1992, 'Experiences with Line Ordered-Nested Block Preconditionings for Nonsymmetric Systems on the CM2', Proceedings of the VII IMACS International Conference, on Computers Methods for PDEs, (eds. R. Vichnevetsky, D. Knight and G. Richter) New Brunswick, NJ. USA, June 22-24, pp. 219–223
2. Díaz, J. C. and Macedo, C. G., Jr.: 1989, 'Fully Vectorizable Block Preconditionings with Approximate Inverses for Non Symmetric systems of Equations', *Inter. J. of Numerical Methods in Engineering*, Vol. **27**, pp. 501–522
3. Macedo, C. G., Jr.: 1990, 'Parallel and Vector Algorithms for Numerical Modeling Using Adaptive Grid Techniques', *Ph. D. dissertation,* The Univ. of Oklahoma, Norman, Oklahoma
4. Leaf, G, K., M. Minkoff, and J. C. Díaz, Preconditioned iterative methods for partial differential equations, **Proceedings of the 6th IMACS International Symposium on Computer Methods for Partial Differential Equations** (eds. Vichnevetsky, Stepleman), June 23-26, 1987, Bethlehem, PA, pp. 551-555.

5. Leaf, G, K., M. Minkoff, and J. C. Díaz, 'Nested Factorization Preconditioners for Convective-Diffusion Problems in Three Dimensions,' in **Mathematics for Large Scale Scientific Computing**, (J. C. Díaz, ed.), Lecture Notes Series, Marcel Decker, 1989, pp. 217-263.
6. Connection Machine Model CM2 Technical Summary, Version 6.0, Thinking Machines Corporation, Cambridge, MA, Dec 1990.
7. Connection Machine CM200 Technical Summary, Thinking Machines Corporation, Cambridge, MA, June 1991.
8. CM5 Technical Summary, Thinking Machines Corporation, Cambridge, MA, Oct. 1991.

SMOOTHNESS AND SUPERCONVERGENCE FOR APPROXIMATE SOLUTIONS TO THE ONE DIMENSIONAL MONOENERGETIC TRANSPORT EQUATION

G. DONALD ALLEN
Department of Mathematics
Texas A&M University
College Station, TX 77843

Abstract. Over the past several years, numerous papers proposing a variety of projection methods for the solution of the linear one-dimensional monoenergetic transport have been published. For the characteristic form, many of these method reduce to standard methods for the solution of linear integral equations of the second kind. In this paper we propose a general framework for several projection methods based on function approximation of specified orders of approximation and/or smoothness. In addition, we illustrate how superconvergence (of the moment type) arises and how to obtain a desired order of superconvergence while maintaining other desired properties, such as smoothness.

Key words: integral equations, projection methods, interpolation methods, superconvergence

1. Introduction.

Over the past several years, numerous papers have been published on approximations to the solution of the discrete-ordinates transport equation in slab geometry. The list of contributers is long and includes, among others: Alcouffe et al. [1], Allen and Nelson [2], Gopinath et al. [6], Ganguly [19], Hennart et al. [7], Keller and Nelson [8], Larsen [9],[10], Lathrop [11], Lee [12], Menon et al. [13], Miller [9], Nelson [10], Neta [14],[15], Vaidyanathan [18], and Victory [14], [15],[19].

It is not our purpose to review this literature here, but rather to illustrate that many of the methods developed in these papers fit into a general classification which applies to a large class of linear integral equations of the second type with *smoothing* kernels. In addition we introduce several new methods of approximation that can be used in the approximate solution of linear integral equations. Generally of the projection type, they can exhibit desired properties of endpoint or gaussian point interpolation and/or orders of convergence and superconvergence. In particular, we propose to examine approximation methods for the characteristic form of this equation from the viewpoints of the order of **convergence**, the order of **superconvergence** and the order of **smoothness/interpolation**. To do this we introduce a classification scheme which will ultimately involve three indices. They are:

S. Gomez and J-P. Hennart (eds.), Advances in Optimization and Numerical Analysis, 219–232.
© 1994 *Kluwer Academic Publishers.*

r	:=	the order of convergence of the approximation,
s	:=	the order of superconvergence of the approximation,
ℓ	:=	the (differential) order of smoothness at partition points, or the number of interior interpolation points.

In most cases, the methods given are simply constructed.

2. Preliminaries.

The monoenergetic transport equation in slab geometry for the angular neutron flux, $\psi(x,\mu)$, has the form

$$\mu\frac{\partial\psi}{\partial x}(x,\mu)+\sigma(x)\psi(x,\mu)=c(x)\sigma(x)\int_{-1}^{1}k(x,\mu',\mu)\psi(x,\mu')d\mu'+q(x,\mu) \tag{1}$$

where μ denotes the cosine of the scattering angle, x denotes the spatial variable $\in[0,a]$, σ, c, and q are nonnegative and piecewise continuous on $[0,a]$, and $k(x,\mu',\mu)$ denotes the (nonnegative) scattering kernel. For our results k will require some piecewise smoothness conditions. The **discrete ordinates** approximation $\psi_i^e(x)$ to (1), determined by application of Gaussian quadrature, is given by

$$\mu_i\frac{d\psi_i^e}{dx}(x)+\sigma(x)\psi_i^e(x)=c(x)\sigma(x)\sum_{j=1}^{N}\omega_j k(x,\mu_j,\mu_i)\psi_j^e(x)+\tilde{q}_i(x) \tag{2}$$

An **interface point** $x\in[0,a]$ is a point where one or more of the functions c, k, or q is discontinuous. We assume that there are only a finite number of these. Denote by $C_N^r[0,a]$ the Banach function space of N-vector of r-times continuously differentiable functions on $[0,a]$ with the supremum norm.

Integrating (2) we obtain the formal class of integral equations to be considered:

$$\psi^e=K\psi^e+\tilde{K}q \tag{3}$$

where K and $\tilde{K}$ are r-smoothing (defined below) and

$$q\in C_N^r(J)=C_N^r([0,a]-\{\text{interface points}\}).$$

In the following we take $N=1$, and suppress writing it, with no loss in generality. Our first basic assumption on K is to assume that it is subcritical. This means that its spectral radius is less than one. Thus, there exists a unique solution ψ in L^2 (resp. $C^r(J_\pi)$) of $\psi=K\psi+g$ for each $g\in L^2$ (resp. $C^r(J_\pi)$). For a given operator M, the approximate solutions to (3) are defined by

$$\psi=KM\psi+\tilde{K}Mq. \tag{4}$$

The difference between the approximate and exact solution is

$$\psi-\psi^e=KM(\psi-\psi^e)+K(M-I)\psi^e+\tilde{K}(M-I)q. \tag{5}$$

is the quantity to be measured. Assuming $\|KM\| < 1$ (in whatever norm that is chosen) we have

$$(\psi - \psi^e) = [I - KM]^{-1}[K(M - I)\psi^e + \tilde{K}(M - I)q]. \tag{6}$$

Thus estimates of the norm of $M - I$ is the key to estimates of $\psi - \psi^e$. The type of questions we consider are as follows: What does the order of approximation of $(M-I)\psi^e$ and $(M-I)q$ imply about the difference $\psi - \psi^e$? The answer, as we shall see, depends upon M and upon which aspects of $\psi - \psi^e$, e.g. pointwise estimates or moments, are considered.

To define the projection method we consider **partitions** π of $[0, a]$ defined by the mesh points

$$0 = x_{1/2} < x_{3/2} < \cdots < x_{H+1/2} = a.$$

Each interval $C_m = (x_{m-1/2}, x_{m+1/2}), m = 1, \ldots, H$ is termed a **cell** and x_m denotes the midpoint of the cell C_m. Assume the interface points of the transport equation are included in mesh points of each partition. Define

$$\begin{aligned} h_m &= x_{m+1/2} - x_{m-1/2}, \quad m = 1, \ldots, H \\ h &= \max_{1 \le m \le H} h_m. \end{aligned}$$

In what follows we will consider sequences of partitions $\{\pi_h\}$ for which $h \to 0$. We will always assume that the sequence is **quasi-uniform**, which means there is a fixed constant c so that for each partition π_h in the sequence $h_m < ch$.

The compression of the operator is defined in the usual way. Given π_h, define the mapping

$$(\rho_{h,m} f)(x) := \left[\frac{2}{h_m}\right]^{1/2} f\left(\frac{h_m}{2}x + x_m\right), \quad m = 1, \ldots, H, \quad -1 \le x \le 1. \tag{7}$$

Clearly $\rho_{h,m}$ is a unitary transformation from $L_2(C_m)$ onto $L_2[-1, 1]$. The inverse $\rho_{h,m}^{-1}$, also a unitary transformation from $L_2[-1, 1]$ to $L_2(C_m)$, is given by

$$(\rho_{h,m}^{-1} f)(x) = \left[\frac{h_m}{2}\right]^{1/2} f\left(\frac{2}{h_m}(x - x_m)\right). \tag{8}$$

Let M be a bounded linear operator on $L_2[-1, 1]$; with norm denoted by $\|M\|_2$. For a given partition π_h, define the linear transformation M_h from $L_2[0, a]$ to $L_2[0, a]$ by

$$M_h = \sum_{m=1}^{H} \chi_{C_m} \rho_{h,m}^{-1} M \rho_{h,m} \chi_{C_m}. \tag{9}$$

Remark. Although (9) is defined for bounded linear operators M on $L_2[-1, 1]$, the same formula can be used for operators M on $C[-1, 1]$, replacing $\rho_{h,m}$ by $(\tilde{\rho}_{h,m} f)(x) = f\left(\frac{h_m}{2}x + x_m\right)$ and $\rho_{h,m}^{-1}$ by $(\tilde{\rho}_{h,m}^{-1} f) = f\left(\frac{2}{h_m}(x - x_m)\right)$. However as we observe in the next result, M_h will be a bounded operator from $C[-1,1]$ to $C(J_\pi)$, where $J_\pi = [0,a] - \{x_{1/2}, \dots, x_{H+1/2}\}$. The following result is elementary.

Theorem 1. Let π_h is a partition of $[0,a]$ with H subdivisions and cells $C_m, m = 1 \dots H$.

(i) If M is a bounded operator on $L_2[-1,1]$, then M_h is a bounded operator on $L_2[0,a]$ and $\|M_h\|_2 = \|M\|_2$. Moreover $\dim M_h = H \dim M$. (Here, $\dim M$ means the dimension of the range of M.) In particular, if M is a projection then so also is M_h.

(ii) If M is a bounded operator on $C[-1,1]$, then M_h is a bounded operator on $C(J_\pi)$, where $J_\pi = [0,a] - \{x_{1/2}, \dots, x_{H+1/2}\}$, and $\|M_h\|_{C(J_\pi)} = \|M\|_{C[-1,1]}$.

Examples. We need only define operators on some Banach function space based on the interval $[-1,1]$. The following examples, standard in the transport theory literature, are of the type M_h for various M. Let $[-1,1]$ be the reference interval and $C^r[-1,1]$ (or $L^2[-1,1]$) the reference space.

(1) Step Characteristic.

$$(M_0 f)(x) = \frac{1}{2} \int_{-1}^{1} f(y) dy \chi_{[-1,1]}(x),$$

where $\chi_{[-1,1]}(x)$ is the indicator function over $[-1,1]$.

(2) Diamond Difference.

$$(M_D f)(x) = \frac{1}{2}(f(-1) + f(1)) \chi_{[-1,1]}(x)$$

Note here that M_D is a bounded operator on $C[-1,1]$.

(3) Linear Discontinuous.

$$(M_{LD} f)(x) = \frac{1}{2}(f(-1) + f(1)) \chi_{[-1,1]}(x) + \frac{1}{2}(f(1) - f(-1))x$$

(4) Linear Moments.

$$(M_1 f)(x) = (M_0 f)(x) + \frac{3x}{2} \int_{-1}^{1} y f(y) dy$$

(5) Linear Characteristic.

$$(M_{LC}) f(x) = (M_0 f)(x) + \frac{f(1) - f(-1)}{2} x$$

(6) Quadratic. (See Gopinath et al. [6].)

$$(M_Q) f(x) = (M_{LC}) f(x) + (f(1) + f(-1) - 2 M_0 f(0)) \left(\frac{3}{4} x^2 - \frac{1}{4}\right)$$

(7) Quadratic Moments.

$$(M_2) f(x) = (M_1 f)(x) + \frac{5}{4}(3x^2 - 1) \int_{-1}^{1} (3y^2 - 1) f(y) dy$$

(8) Legendre Moments.

$$(P_r)f(x) := \sum_{m=0}^{r} \frac{2m+1}{2} p_m(x) \int_{-1}^{1} p_m(y) f(y) dy$$

where $p_0(x), \ldots p_r(x)$, denote the first r Legendre functions. Each of these is a linear projection. The projections M_0, M_1, M_2 and P_r above are orthogonal (on L^2). The others are not. However, some of them are *partially orthogonal*, a term to be defined below.

It is the set of operators M_h based on some M that will be used in the approximation (4). The basic equation is then

$$\psi = KM_h\psi + \tilde{K}M_hq. \tag{4'}$$

Our second basic assumption about K is that for the particular operators M and M_h under consideration, unique solutions to (4') exist in $C^r(J_{\pi_h})$, where $r \geq 0$ is an integer associated with M in a way described in the next section. Following similar arguments to those in Victory and Ganguly ([19], Lemma 1, p. 83), it is shown that a necessary condition for convergence of the approximate solutions to ψ^e is that M_h should approximate the identity I in the strong operator topology. Each of the operators in the above examples do this. The following theorem gives, in turn, necessary and sufficient conditions for this strong convergence.

Theorem 2. (i) Suppose that M is a bounded operator on $L_2[-1,1]$. Then $\lim_{h\to 0} M_h = I$ in the strong operator topology if and only if $M\chi_{[-1,1]} = \chi_{[-1,1]}$.
(ii) Let M be a bounded operator on $C[-1,1]$. Then for every $f \in C(J_\pi)$

$$\lim_{h\to 0} \|M_hf - f\|_\infty = 0$$

if and only if $M\chi_{[-1,1]} = \chi_{[-1,1]}$.

A proof be found in Allen and Nelson ([2], Theorem 3.2).

3. O-Type Operators.

Operators satisfying the condition of Theorem 2 alone do not provide sufficiently rapid convergence to have much practical value. By restricting the application of M_h to differentiable functions, finer types of approximation and hence faster types of convergence can be obtained. Let M be any finite rank bounded operator on $L_2[-1,1]$ (or $C[-1,1]$). Suppose that $r \geq 0$ is any integer. We say that M is of **O-Type** r, $r \geq 1$, if for each function $f \in C^r[0,a]$, the function

$$(M_h - I)f = O(h^r)$$

(pointwise) as $h \to 0$ over quasi-uniform partitions. We say that M is of **O-Type** 0 if for each $f \in C[0,a]$, the function

$$(M_h - I)f = o(1)$$

(pointwise) as $h \to 0$ over quasi-uniform partitions. Theorem 3, below, gives simple necessary and sufficient conditions for the construction of O-Type operators. Again, the proof, part of which applies the results of Bramble and Hilbert [3], can be found in Allen and Nelson [2].

Theorem 3. Suppose M is a bounded operator on $C[-1,1]$, and $r > 0$ is an integer. Then M is of O-Type r if and only if $M(f)(x) = x^j$, where $f(x) = x^j$, $j = 0, 1, \ldots, r-1$.

In Table I above we tabulate the operators above according to O-Type. These values are easly verified by direct application of Theorem 3.

TABLE I

Classification by O-Type

Method	O-Type
M_D	1
M_{LD}	2
M_{LM}	2
M_{LC}	2
M_Q	3
M_{QM}	3
P_{r-1}	r

As an application to integral equations we have a result that relates the order of the O-Type operator to the order of the solution. We need only define that an operator K is called r-**smoothing** if for each $f \in C^r(J_\pi)$ it follows that $Kf \in C^{r+1}(J_\pi)$, for any partition π.

Theorem 4. (Orders of Convergence.) Suppose K, $\tilde{K}$, are r-smoothing and $q \in C_N^r(J)$. Suppose that M is a projection of O-Type r and a bounded operator from $C_N^r[-1,1]$ to $C_N^r[-1,1]$. Then

$$(\psi - \psi^e)(x) = O(h^r) \tag{10}$$

uniformly for $x \in J_{\pi_h}$. If, in addition $M_0 M = M_0$, then

$$(\psi - \psi^e)(x) = O(h^{r+1}). \tag{11}$$

Thus, for example, the higher order $O(h^{r+1})$ is obtained if $M = P_{r-1}$. Various versions of this theorem have appeared for different approximation schemes. For this particular formulation, see Allen and Nelson [2]. For the Galerkin formulation in the Sobolev spaces W^k, see de Boor and Fix [5] and Chandler [4].

Remark. For the diamond difference method the second order estimate is obtainable, provided one order higher of smoothness, namely C^2, is assumed for K, $\tilde{K}$ and q. Proving this is quite simple. It is just a matter of showing that

$M_{D,h} - M_{0,h} = O(h^2)$ on C^2. Having done this, using the condition of the proposition below, proceed using (6) as follows:

$$\begin{aligned}(\psi - \psi^e) &= [I - KM_{D,h}]^{-1}[K(M_{D,h} - I)\psi^e + \tilde{K}(M_{D,h} - I)q] \\ &= [I - KM_{D,h}]^{-1}[K(M_{D,h} - M_{0,h} + M_{0,h} - I)\psi^e + \\ &\qquad \tilde{K}(M_{D,h} - M_{0,h} + M_{0,h} - I)q] \\ &= [I - KM_{D,h}]^{-1}[K(M_{D,h} - M_{0,h}) + K(M_{0,h} - I)\psi^e + \\ &\qquad \tilde{K}(M_{D,h} - M_{0,h}) + \tilde{K}(M_{0,h} - I))q].\end{aligned}$$

The $O(h^2)$ estimate follows for each piece. For other O-Type 1 operators M, however, the same result may hold, namely that the order of convergence is $O(h^2)$, rather than $O(h)$ as the theorem predicts. Necessary and sufficient conditions for this are given in the following

Proposition. Suppose that M is O-Type 1. Suppose that M is bounded on C^2. A necessary and sufficient condition that $M_h - M_{0,h} = O(h^2)$ on C^2 is that $Mx = 0$. More generally, suppose that M is bounded on C^{r+1} and O-Type r. A necessary and sufficient condition that a O-Type r operator M satisfy $M_h - P_{r-1,h} = O(h^{r+1})$ on C^{r+1} is that $Mp_r = 0$.
The proof of this result is elementary and is very similar to the methods used to prove Theorem 3.

4. P-Type Operators.

Usually, superconvergence is described in one of two ways. Either we say that an operator M of O-Type r is **superconvergent** if

$$(\psi - \psi^e)(\text{special points}) = O(h^{r+s})$$

or

$$\langle \varphi, (\psi - \psi^e) \rangle = O(h^{r+s}),$$

where $s \geq 1$ and φ is at least piecewise C^s. For our purposes we use the second definition, though there are many results, particularly in finite element methods, which prove the first condition. What we have observed is this: *For a particular operator M the order of superconvergence is linked to it orthogonal projective nature.* To see how this works, we say the O-Type r operator M is **P-Type** s, where $s > 0$ is an integer, if

$$P_{s-1}M = P_{s-1}.$$

This, of course, implies that

$$P_{s-1}^{\perp}M^{\perp} = M^{\perp}.$$

Here we use the notation $M^{\perp}$ for the operator $I - M$, even though M is not generally an orthogonal projection. If this equation does not hold for any integer $s > 0$, we say that M is P-Type 0. It is easy to see that for each $0 \leq s \leq r$, there exists an O-Type r, P-Type s operator M. For convenience such operators will be called **type**(r, s). If

$0 < s \leq r$ we say that M is **partially orthogonal** of order s. Note that the linear characteristic and quadratic methods are nontrivially partially orthogonal, yet are not orthogonal projections. The basic superconvergence result for such partially orthogonal operators is this:

Theorem 5. (Superconvergence.) Suppose that K and $\tilde{K}$ are r-smoothing and $q \in C^r(J)$. Suppose that M is bounded on $C^r(J)$ and is of type (r, s), $s > 0$. Then for each $\varphi \in C^s(J)$ we have

$$\langle \varphi, (\psi - \psi^e) \rangle = O(h^{r+s}).$$

Proof. Below, we omit the subscripted parameter h which properly appears each time the operator M or P_{s-1} is invoked. Again, we use the notation $M^\perp$ for the operator $I - M$, even though M may not be an orthogonal projection. Since $I - K$ is well posed, it follows that $I - KP_{s-1}$ The Fredholm alternative shows that $I - P_{s-1}K^*$ is also well posed. Since K^* is r-smoothing it follows that for any $\phi \in C^s(J)$ there is a $\chi \in C^s(J)$ such that $(I - P_{s-1}K^*)\chi = \phi$. Thus

$$\begin{aligned}
\langle \phi, \psi - \psi^e \rangle &= \langle (I - P_{s-1}K^*)\chi, \psi - \psi^e \rangle \\
&= \langle \chi, (I - KP_{s-1})(\psi - \psi^e) \rangle \\
&= \langle \chi, (I - KM + K(P_{s-1}M))(\psi - \psi^e) \rangle \\
&= \langle \chi, (I - KM)((\psi - \psi^e)) + \langle \chi, K(P_{s-1} - M)(\psi - \psi^e) \rangle \\
&= \langle -\chi, KM^\perp \psi^e + \tilde{K}M^\perp q \rangle + \langle \chi, KM^\perp(\psi - \psi^e) \rangle \\
&\quad + \langle \chi, -KP_{s-1}^\perp(\psi - \psi^e) \rangle
\end{aligned}$$

Since $P_{s-1}^\perp M^\perp = M^\perp$, it follows that the first term on the right hand side of the equality just above is

$$\langle -P_{s-1}^\perp K^* \chi, M^\perp \psi^e \rangle + \langle -P_{s-1}^\perp \tilde{K}^* \chi, M^\perp q \rangle$$

and since both ψ^e and q are in $C^r(J_\pi)$ this is $O(h^{r+s})$. Using this same argument and (10) gives both the second and third terms above to be $O(h^{r+s})$. This proves the theorem.

Similar results hold for Galerkin-Petrov methods (i.e. using MK instead of KM). Negative norm results such as those in this theorem can also be found in the literature, in particular see Chandler [4] and Sloan and Thomée [17]. Table II illustrates the order of superconvergence of the methods defined above.

5. I-Type Operators.

If an operator M is of O-Type r interpolates certain points in the interval $[-1, 1]$, we say that it is of **I-Type**. For example, the methods

$$(M_{LD}f)(x) = \frac{1}{2}(f(-1) + f(1))\chi_{[-1,1]}(x) + \frac{1}{2}(f(1) - f(-1))x$$

TABLE II

Classification by O- and P- Type			Order of
Method	O-Type	P-Type	Superconvergence
M_0	1	1	2
M_D	1	0	1*
M_{LD}	2	0	2*
M_{LM}	2	2	4
M_{LC}	2	1	3*
M_Q	3	1	4
M_{QM}	3	3	6
P_{r-1}	r	r	2r

*The order is one higher with additional smoothness assumed.

TABLE III

Classification by O-, P- and I_H-Type			
Method	O-Type	P-Type	I_H-Type
M_0	1	1	−1
M_D	1	0	−1
M_{LD}	2	0	0
M_{LM}	2	2	−1
M_{LC}	2	1	−1
M_Q	3	1	0
M_{QM}	3	3	−1
P_{r-1}	r	r	−1

and

$$(M_Q)f(x) = (M_{LC})f(x) + (f(1) + f(-1) - 2M_0 f(0))\left(\frac{3}{4}x^2 - \frac{1}{4}\right)$$

interpolate the endpoints ± 1. Thus, if $f \in C[0,a]$, both $M_{LD,h}$ and $M_{Q,h}$ are continuous functions for each partition. Our goal now to to further classify O- and P- type operators according to their interpolation properties. We illustrate two distinct sub-types.

I_H-Type Operators. Suppose that M is type (r,s) and $\ell = 0, 1, \ldots$. We say that M is I_H-Type(r, s, ℓ) if for $f \in C^r[-1,1]$, the derivatives

$$(Mf)^{(j)}(\pm 1) = f^{(j)}(\pm 1), \qquad j = 0, 1, \ldots \ell.$$

In this way the index ℓ denotes the (differential) order of cell edge smoothness of M_h. If M does not interpolate points we use the default index $\ell = -1$. Table III gives a tabulation of the usual set of operators according to their O-, P-, and I_H-Types. As is evident only M_{LD} and M_Q are interpolatory type larger than -1. The next result gives a partial answer to the existence and construction of I_H-Type operators in the sense that various orders are given. Further existence results can also be established.

Theorem 6. (Existence) For any given r, s, and $\ell \geq -1$ with $r \geq s + 2(\ell + 1)$, there exists an operator of I_H-Type (r, s, l).

Proof. The proof follows by construction. First, let us suppose that $r = s+2(\ell+1)$, and $\ell \geq 0$. Define

$$M = \underset{\substack{\uparrow \\ \text{Front End}}}{P_{s-1}} + \underset{\substack{\uparrow \\ \text{Back End}}}{\sum_{j=0}^{2\ell+1} c_j p_{s+j}(x)},$$

where the $p_{s+j}(x)$ denote the corresponding Legendre functions, and where the coefficients c_j depend on the function $f(x)$. The interpolatory conditions require that

$$\sum_{j=0}^{2\ell+1} c_j p_{s+j}^{(k)}(\pm 1) = f^{(k)}(\pm 1) - (P_{s-1} f)^{(k)}(\pm 1), \quad k = 0, \ldots, \ell.$$

This produces a linear system for the coefficients c_i, which we show to be nonsingular. Hence the solution is unique and moreover determines M as a projection of rank $s + 2(\ell + 1)$.

To establish the nonsingularity of the system, we suppose to the contrary, that there are constants $c_0, \ldots, c_{2\ell+1}$ for which

$$p(x) = \sum_{j=0}^{2\ell+1} c_j p_{r+j}$$

together with its first ℓ derivatives is zero at ± 1. This implies that $p(x)$ has the form

$$p(x) = (1 - x^2)^{\ell+1} q(x),$$

where $q(x)$ has degree $\leq s - 1$. Since $P_{s-1} p = 0$, we have with inner product on $[-1, 1]$ that

$$\langle p, q \rangle = \langle (1 - x^2)^{\ell+1} q, q \rangle = 0.$$

Thus $q(x) = 0$, as the above function is positive on $(-1, 1)$. This is a contradiction.

In the case that $r > s + 2\ell + 1$ and the *front end* can not be selected as P_{s-1} it is a routine construction to let P be some (nonorthogonal) projection to the first $r - (2\ell + 1) - 1$ powers which commutes with P_{s-1}. This provides the base type(r, s) operator. A final step, to show that the operator M is invariant on the first r powers, is elementary and is omitted.

Remark. Note that, P-Type operators do not interpolate points. Thus, the I_H-Type contributes to the order of convergence, but not to the order of superconvergence.

The following is a Mathematica script written to generate the matrices, $p_{s+j}^{(k)}(\pm 1)$.

```
inv/: inv[s_, l_] := Inverse[genarray[s, l]]

genarray/: genarray[s_, l_] :=
   Table[nn[i, (j + s) - 1], {i, 2*(l + 1)}, {j, 2*(l + 1)}]

nn/: nn[i_, j_] := mm[i, j, x] /. x -> (-1)^Floor[i - 1]

mm/: mm[i_, j_, x_] := D[l[j, x], {x, Floor[(i - 1)/2]}]

l/: l[s_, x_] := LegendreP[s, x]
```

Examples. Below are listed a few $\mathrm{I_H}$-Type(r,s,l) methods generated from this script.

I_H-Type(3,1,0):

$$\begin{aligned} Mf &= P_0f + c_1p_1(x) + c_2p_2(x) \\ \begin{pmatrix} 1 & 1 \\ -1 & 1 \end{pmatrix}\begin{pmatrix} c_1 \\ c_2 \end{pmatrix} &= \begin{pmatrix} f(1)-(P_0f)(1) \\ f(-1)-(P_0f)(-1) \end{pmatrix} \\ \begin{pmatrix} c_1 \\ c_2 \end{pmatrix} &= \frac{1}{2}\begin{pmatrix} 1 & -1 \\ 1 & 1 \end{pmatrix}\begin{pmatrix} f(1)-(P_0f)(1) \\ f(-1)-(P_0f)(-1) \end{pmatrix} \end{aligned}$$

Thus

$$Mf = P_0f + \frac{1}{2}(f(1)-f(-1))p_1(x) + (f(1)+f(-1)-2P_0f(1))\frac{1}{2}p_2(x),$$

which is the Quadratic method.

I_H-Type(5,1,1):

$$\begin{aligned} Mf &= (P_0f)(x) + c_1p_1(x) + c_2p_2(x) + c_3p_3(x) + c_4p_4(x) \\ \begin{pmatrix} 1 & 1 & 1 & 1 \\ -1 & 1 & -1 & 1 \\ 1 & 3 & 6 & 10 \\ 1 & -3 & 6 & -10 \end{pmatrix}\begin{pmatrix} c_1 \\ c_2 \\ c_3 \\ c_4 \end{pmatrix} &= \begin{pmatrix} f(1)-(P_0f)(1) \\ f(-1)-(P_0f)(-1) \\ f'(1) \\ f'(-1) \end{pmatrix} \\ \begin{pmatrix} c_1 \\ c_2 \\ c_3 \\ c_4 \end{pmatrix} &= \begin{pmatrix} \frac{3}{5} & -\frac{3}{5} & -\frac{1}{10} & -\frac{1}{10} \\ \frac{5}{7} & \frac{5}{7} & -\frac{1}{14} & \frac{1}{14} \\ -\frac{1}{10} & \frac{1}{10} & \frac{1}{10} & \frac{1}{10} \\ -\frac{3}{14} & -\frac{3}{14} & \frac{1}{14} & -\frac{1}{14} \end{pmatrix}\begin{pmatrix} f(1)-(P_0f)(1) \\ f(-1)-(P_0f)(-1) \\ f'(1) \\ f'(-1) \end{pmatrix} \end{aligned}$$

Thus

$$\begin{aligned} Mf = {} & (P_0f)(x) + [\frac{3}{5}(f(1)-f(-1)) - \frac{1}{10}(f'(1)+f'(-1))]p_1(x) \\ & + [\frac{5}{7}(f(1)+f(-1)-2(P_0f)(0)) - \frac{1}{14}(f'(1)-f'(-1))]p_2(x) \\ & + [-\frac{1}{10}(f(1)-f(-1)) + \frac{1}{10}(f'(1)+f'(-1))]p_3(x) \\ & + [-\frac{3}{14}(f(1)+f(-1)+2(P_0f)(0)) + \frac{1}{14}(f'(1)-f'(-1))]p_4(x) \end{aligned}$$

I_H-Type(4,2,0):

$$\begin{aligned} Mf &= (P_1f)(x) + [f(1) - f(-1) - ((P_1f)(1) + (P_1f)(-1))](\frac{3}{4}x^2 - \frac{1}{4}) \\ &\quad + [f(1) + f(-1) - ((P_1f)(1) - (P_1f)(-1))](\frac{5}{4}x^3 - \frac{3}{4}x) \end{aligned}$$

I_G-Type Operators. Suppose that M is type (r, s) and $\ell = 0, 1, \ldots$. We say that M a type(r, s) operator is I_G-Type(r, s, ℓ) if for $f \in C^r[-1, 1]$ M interpolates f at the ℓ Gaussian points in $[-1, 1]$. If M is not interpolatory, we use the default index $\ell = -1$. Because P-Type operators do not interpolate points the inequality $r \geq s + \ell$ obtains. For example, if $r = s + \ell$, then M may have the form

$$Mf = (P_{s-1}f)(x) + \sum_{j=1}^{\ell} c_j p_{s+j-1}(x),$$

where the c_k depend on f. Let $\{g_{\ell,k},\ k = 1, \ldots, \ell\}$ denote the Gaussian points corresponding to p_ℓ. The interpolatory conditions imply that

$$\sum_{j=1}^{\ell} c_j p_{s+j-1}(g_{\ell,k}) = f(g_{\ell,k}) - (P_{s-1}f)(g_{\ell,k}), \quad k = 1, \ldots, \ell. \tag{12}$$

Existence and uniqueness of such approximations depends on the singularity of this matrix system. We have resolved this question in a few specific cases described below in

Theorem 7. (Existence.) (i) If $\ell = s - 1$ and ℓ is even, the system (9) is nonsingular and yields a projection M of I$_G$-Type$(s + \ell, s, \ell)$.
(ii)If $\ell \neq 1$ is odd and if the interpolating polynomials are contiguously labeled Legendre polynomials, there are no I$_G$-Type$(s + \ell, s, \ell)$ operators, for any s.

Proof. (i) As in the proof of Theorem 6 we assume the system matrix, in this case $p_{s+j-1}(g_{s-1,k})$, $1 \leq j, k \leq s - 1$, is singular. For a nontrivial set of coefficients $c_1, \ldots c_{s-1}$, the polynomial $p(x) = \sum_{j=1}^{s-1} c_j p_{s+j-1}$ is zero at the gaussian points $g_{s-1,j}$. This implies that $p(x)$ factors as $p(x) = p_{s-1}(x)q(x)$, where $\deg q \leq s - 1$. We know that $q(x) = P_{s-2}q + cp_{s-1}$. Since $P_{s-1}p = 0$, it follows that $c = 0$. Thus $q(x) = P_{s-2}\,q$. If $s - 2 = 0$, then q is a constant and $p = cp_{s-1}$ which is a contradiction. So, $s - 2 > 0$ and $q = P_{s-3}q + dp_{s-2}$. Since $\langle p_{s-1}q, x\rangle = 0$, it follows that $d = 0$ and hence p_{s-1} is orthogonal to $xq(x)$, establishing that $\deg q \leq s - 3$, or what is the same thing that $P_{s-3}q = q$.

Continue this process reducing the degree of q, using successively higher powers until $\deg q = 0$ is established. This completes the proof of (i).

For (ii) it is easy to see that if $\ell > 1$ is odd, then one of the gaussian points is 0 and so also is at least one of the columns of the system matrix, yielding a singular system, making the interpolation impossible.

Remark. With somewhat more effort, it is possible to show that in the case $\ell = 2$, all values of $s > 2$ can be selected. The proof uses some special properties of the

Legendre polynomials. For larger even ℓ, the result is completely unknown and is related to the open question concerning the duplication of zeros of different orthogonal polynomials corresponding to the same measure. Yet, the system has been checked for a large number of ℓ values and the system matrix was always nonsingular. If in (ii) we take $\ell = 1$, the Legendre polynomial used for the interpolation must be even. Such methods may also be termed "midpoint methods" because they, in effect, interpolate at the cell midpoint. Because these operators can be defined in so many different ways, it is difficult to make general qualitative statements about them.

Examples. There do not seem to be any operators of this type in the literature. One reason may be that for many problems superconvergence at gaussian nodes is often obtained, thus giving marginal reason to require exactness at these points. Yet, the gaussian nodes were selected only as an example. Any set of points for which the system matrix in (12) is nonsingular could be selected. For a reference on superconvergence at gaussian nodes see, for example, the paper by Richter [16].

I_G-Type(3,2,1):

$$Mf = (P_1 f)(x) - 2[f(0) - (P_1 f)(0)](\frac{3}{2}x^2 - \frac{1}{2}).$$

I_G-Type(3,1,1):

$$Mf = (M_{LC} f)(x) - 2[f(0) - (M_{LC} f)(0)](\frac{3}{2}x^2 - \frac{1}{2}).$$

I_G-Type(5,3,2).

$$Mf = (P_2 f)(x) + ap_3(x) + bp_4(x),$$

where

$$\begin{aligned} a &= \frac{-3\sqrt{3}}{4}[f(\frac{1}{\sqrt{3}}) - (P_2 f)(\frac{1}{\sqrt{3}}) - f(-\frac{1}{\sqrt{3}}) + (P_2 f)(-\frac{1}{\sqrt{3}})] \\ b &= \frac{-9}{7}[f(\frac{1}{\sqrt{3}}) - (P_2 f)(\frac{1}{\sqrt{3}}) + f(-\frac{1}{\sqrt{3}}) - (P_2 f)(-\frac{1}{\sqrt{3}})] \end{aligned}$$

References

1. R.E. Alcouffe, E.W. Larsen, W.F. Miller, Jr. and B.R. Wienke, 'Computational efficiency of numerical methods for the multigroup discrete-ordinates transport equation. The slab geometry case', *Nucl. Sci. Eng.* 71 (1979) pp. 111-127.
2. G. D. Allen and P. Nelson, Jr., 'On Generalized Finite difference Methods for Approximating solutions to Integral Equations', in *Advances in Numerical Partial Differential equations and Optimization*, Proceedings of the Fifth Mexico-United States Workshop, S. Gomez, et. al., eds. (1989), pp. 112-140.
3. J.H. Bramble and S. R. Hilbert, 'Estimation of Linear Functionals on Sobolev Spaces with Application to Fourier Transforms and Spline Interpolation', *SIAM J. Numer. Anal.*, 7 (1970) pp. 112-124.
4. G. A. Chandler, 'Superconvergence for second kind integral equations', in *The Application and Numerical Solution of Integral Equations*, R. S. Anderssen et.al., eds., Sijthoff & Noordhoff, Alphen aun den Rijn, 1980, pp. 300-304.
5. C. de Boor and G. J. Fix, 'Spline approximation by quasi-interpolants', *J. Approx, Th.* 8 (1973), pp. 19-45.

6. D.V. Gopinath, A. Natarajan, and V. Sundarazaman, 'Improved interpolation schemes in anisotropic source-flux iteration techniques', *Nucl. Sci. Eng.*, 75 (1980) pp. 181-184.
7. J.P. Hennart, E. del Valle, F. Serrano, and J. Valdés, 'Discrete- ordinates equations in slab geometry: A generalized nodal finite element formalism', *Int. Top Mtg, Advances in Reactor Physics, Mathematics and Computers*, Paris, April 1987, CEC/OECD, pp. 1283-1288.
8. H.B. Keller and P. Nelson, 'Closed linear one-cell functional spacial approximations: consistency implies convergence and stability', *Transport Theory and Statistical Physics,* 17 (1988) 191-208.
9. E.W. Larsen and W.F. Miller, Jr., 'Convergence rates of spatial difference equations for the discrete-ordinates neutron transport equations in slab geometry', *Nucl. Sci. Eng.*, 73 (1980), pp. 76-83.
10. E.W. Larsen and P. Nelson, Jr., 'Finite-difference approximation approximations and superconvergence for the discrete-ordinates equations in slab geometry', *SIAM J. Numerical Analysis*, 19 (1982), pp. 334-348.
11. K.D. Lathrop, 'Spatial differencing of the transport equation: positivity vs. accuracy', *J. Comp. Phys.*, 4 (1969), pp. 475-498.
12. S.M. Lee and R. Vaidyanathan, 'Comparison of the order of approximation in several spatial difference schemes for the discrete-ordinates transport equation in one-dimensional plane geometry', *Nucl. Sci. Eng.*, 76 (1980), pp. 1-9.
13. S. V. G. Menon and D. C. Sahni, 'Convergence of Discrete Ordinates Iteration Scheme', *Transport Theory and Statistical Physics,* 14(3) (1985), pp. 353-372.
14. B. Neta and H.D. Victory, Jr., 'The convergence analysis for sixth-order methods for solving discrete ordinates slab transport equations', *Numer. Funct. Anal. and Optimiz.*, 5(1) (1982), pp. 85-126.
15. ____________, 'A new fourth-order finite-difference method for solving discrete- ordinates slab transport equations', *SIAM J. Numerical Analysis,* 20 (1983), pp. 94-105.
16. G. A. Richter, 'Superconvergence for piece-wise polynomial Galerkin approximations for Fredholm integral equations of the second kind', *Numer. Math.* 31 (1978) pp. 63-70.
17. Ian H. Sloan and Vidar Thomée, 'Superconvergence of the Galerkin Iterates for Integral equations of the Second Kind, *Journal of Integral Equations* 9 (1985), pp. 1-23.
18. R. Vaidyanathan, 'A finite moments algorithm for particle transport problems, *Nuclear Sci, Engrg.*, 71 (1979), pp. 46-54.
19. H.D. Victory, Jr., and K. Ganguly, 'On finite-difference methods for solving discrete-ordinates transport equations', *SIAM J. Numer. Anal.* 23 (1986) pp. 78-108.

EXPERIMENTS WITH THE POWER AND ARNOLDI METHODS FOR SOLVING THE TWO-GROUP NEUTRON DIFFUSION EIGENVALUE PROBLEM

JÉRÔME JAFFRÉ
INRIA, B.P. 105, 78153 Le Chesnay Cédex, France

and

JEAN-LOUIS VAUDESCAL
INRIA and Université Paris-Dauphine, Place du Mal. de Lattre de Tassigny, 75775 Paris Cédex 16, France

Abstract. The algebraic solution to the two-group neutron diffusion problem is investigated. It is a generalized nonsymmetric eigenvalue problem for which the dominant eigenvalue – which is real – and the corresponding eigenvector are sought. We present comparisons, for this problem, of the Arnoldi method with the power method, both combined with Chebyshev acceleration.

Key words: Nodal methods, power method, Arnoldi's method, Chebyshev's acceleration, generalized eigenvalue problem, multigroup neutron diffusion.

1. Introduction

A standard model for the simulation of civil nuclear plants is the two-group neutron diffusion model. The neutron fluxes are the solution of a nonsymmetric generalized eigenvalue problem. The dominant eigenvalue of this problem describes the general behaviour of the plant : when larger than one, the plant is super-critical, when smaller than one, it is under-critical and when equal to one, it is self-sustained. This problem is usually solved numerically with the power method. The speed of convergence of this method is proportional to the ratio of the largest eigenvalue to the next to the largest one. Therefore, the power method is rather slow when this ratio is close to one, which is the case in practical applications, especially when refining the mesh. On the other hand, another way to solve unsymmetric eigenvalue problems is the Arnoldi method [4]. We make here some comparisons between these two methods when they are combined with Chebyshev's acceleration and show some advantages of the Arnoldi method.

2. Algebraic structure of the eigenvalue problem

Using standard notations for nuclear engineering, the static distribution of neutrons is governed by the following set of equations :

$$\begin{cases} -\mathrm{div}(D_1\overrightarrow{\mathrm{grad}}\Phi_1) + (\Sigma a_1 + \Sigma_R)\Phi_1 = \frac{1}{\lambda}(\nu\Sigma f_1\Phi_1 + \nu\Sigma f_2\Phi_2)\ , \\ -\mathrm{div}(D_2\overrightarrow{\mathrm{grad}}\Phi_2) + \Sigma a_2\Phi_2 = \Sigma_R\Phi_1\ . \end{cases}$$

S. Gomez and J-P. Hennart (eds.), Advances in Optimization and Numerical Analysis, 233–244.
© 1994 *Kluwer Academic Publishers.*

The unknowns are the real functions Φ_1 and Φ_2 of the space variable and are respectively the neutron flux of the fast group and that of the thermal or slow group. The coefficients in the equation are D_1, D_2, the diffusion coefficients, $\Sigma a_1, \Sigma a_2$, the macroscopic absorption cross sections, Σ_R, the macroscopic slowing-down cross section of group 1, and $\nu\Sigma f_1, \nu\Sigma f_2$ the macroscopic fission cross sections.

The first equation represents the conservation of the fast neutrons. The source term on the righthand side is due to fission. In the second equation describing conservation of the slow neutrons, the source term represents the production of slow neutrons produced by slowing down fast neutrons. This is why the same term appears in the lefthand side of the equation. Other effects taken into account in the equations are diffusion and absorption.

In addition to these equations, the neutron flux is normalized with respect to the given total power of the reactor.

As a practical example, we consider the case of nuclear cores of Pressurized Water Reactor (PWR) type. A nuclear core is almost a cylindar whose cross section is made out of a rectangular grid of assemblies. For commodity of calculation, the core is completed by a so-called reflector, in order to compute in a rectangular box, as shown in figure 1.

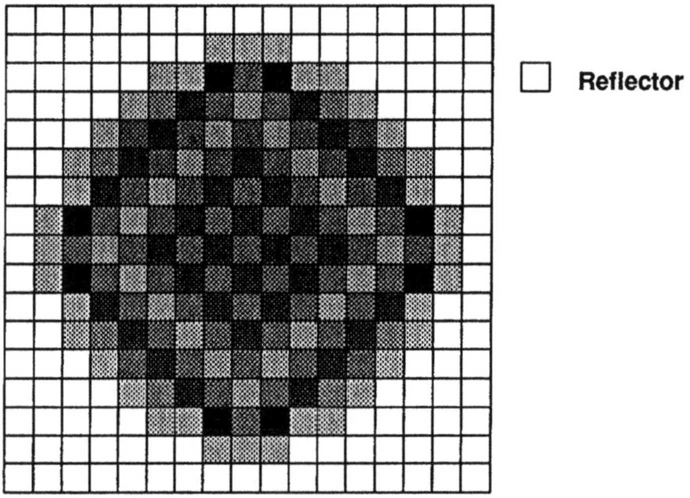

Fig. 1. Cross section of a PWR nuclear core

Boundary conditions are of Robin's type :

$$D_i \frac{\partial \Phi_i}{\partial n} + \alpha_i \Phi_i = 0 \quad , i = 1, 2.$$

The problem is discretized with a nodal method of lowest order. It is a nonconforming finite element method based on a mixed-hybrid formulation [2, 3]. Conservation of neutrons is written inside each discretization cell and continuity of the current is enforced at the interelement boundaries. There are two types of unknowns for the flux, cell unknowns (one per cell) and face unknowns (one per face) (see fig. 2).

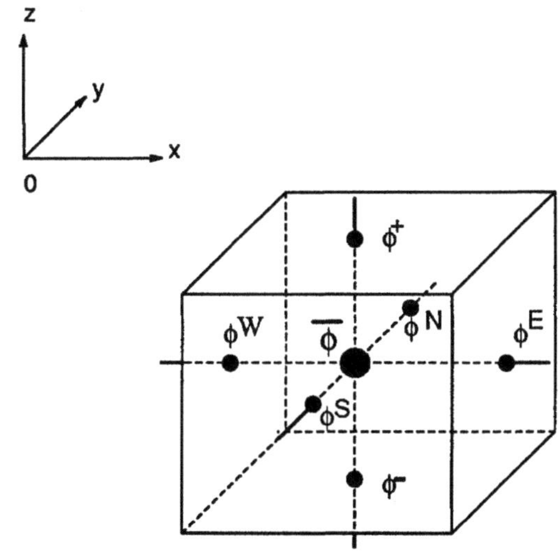

Fig. 2. Flux unknowns in a 3-D discretized problem

In usual practice, only the dominant eigenpair (λ_0,Φ_0) is sought. Under some assumptions, existence and uniqueness of the solution can be proved. It can also be shown that Φ_0 is positive and that all other eigenvalues in the spectra have a strictly smaller modulus than λ_0 (see [8], or [1] for the case with a finite difference approximation).

In matrix form, the problem takes the form of a generalized nonsymetric eigenvalue problem

$$A\Phi = \frac{1}{\lambda} B\Phi \, ,$$

where

$$A = \begin{pmatrix} A_{11} & 0 \\ -\Sigma_R & A_{22} \end{pmatrix} , \quad B = \begin{pmatrix} \nu\Sigma f_1 & \nu\Sigma f_2 \\ 0 & 0 \end{pmatrix} , \quad \Phi = \begin{pmatrix} \Phi_1 \\ \Phi_2 \end{pmatrix} .$$

The block A_{11} denotes the matrix representing the discretized analog of the continuous operator $-\mathrm{div} D_1 \overrightarrow{\mathrm{grad}} + \Sigma a_1 + \Sigma_R$ and A_{22} that for the continuous operator $-\mathrm{div} D_2 \overrightarrow{\mathrm{grad}} + \Sigma a_2$. The blocks $\nu\Sigma f_1$ and $\nu\Sigma f_2$ in matrix B denote now the discretized fission operators while Σ_R in matrix A denotes the discretized slowing down operator. The sparsity structures of matrices A and B are shown in fig. 3.

Note that the matrix B is low rank. However the matrix A is nonsingular, the diagonal blocks A_{11} and A_{22} being positive definite. Therefore the eigenvalue problem that we solve is actually

$$A^{-1}B\Phi = \lambda\Phi.$$

In practice, a coarse mesh calculation requires one $x - y$ cell per assembly and 20 z-intervals which gives a total of about 40000 unknowns. Of course, finer meshes are also used.

The power method is usually used to solve the eigenvalue problem. When using fine meshes or when solving inverse problems, the computation cost becomes prohibitive. Therefore it is useful to investigate alternatives to the power method such as the Arnoldi method [4].

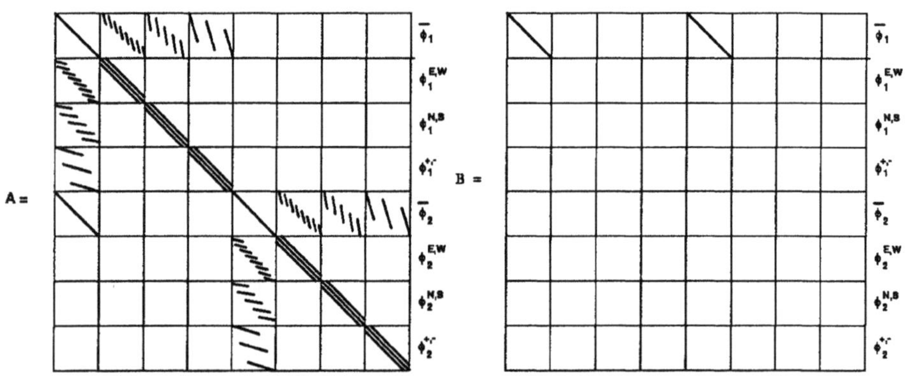

Fig. 3. Structures of diffusion and fission matrices

3. The power method and the Arnoldi method

In this section, we recall the two algorithms that we will consider. First, the power method :

Given an initial vector $\Phi^{(0)} := B\Psi^{(0)}$.
For $n = 1, \ldots$ do (outer iterations)
  solve $A\widetilde{\Phi}^{(n)} = \Phi_1^{(n-1)}$
  $\Psi^{(n)} := B\widetilde{\Phi}^{(n)}$
  $\lambda^{(n)} := \dfrac{(\Psi^{(n)}, \Psi^{(n)})}{(\Psi^{(n-1)}, \Psi^{(n)})}$
  if convergence is reached then
    set $\lambda_0 = \lambda^{(n)}$ and stop
  else
    $\Phi_1^{(n)} := \dfrac{\Psi^{(n)}}{\lambda^{(n)}}$
  endif

One standard stopping criteria is the following one [6]. Define the sequences :

$$\lambda_{min} = \min_{\Psi \neq 0} \frac{\Psi^{(n)}}{\Psi^{(n-1)}} \quad , \lambda_{max} = \max_{\Psi \neq 0} \frac{\Psi^{(n)}}{\Psi^{(n-1)}} \ .$$

For a given tolerance ε, the convergence is said to be reached when

$$\frac{| \lambda_{max} - \lambda_{min} |}{2\lambda^{(n)}} < \varepsilon \ .$$

Now, for the Arnoldi method, we construct an orthonormal basis $\{v_j\}_{j=1,..,m}$ for the Krylov subspace of dimension m and an upper Hessenberg matrix $H_m = \{h_{ij}\}_{i,j=1,..,m}$ using the following algorithm :

Given an initial vector v^1 of unit norm and m the dimension of the Krylov subspace,
For $j = 1$ until m do (outer iterations)
  solve $Aw = Bv^j$
    For $i = 1$ until j do
    $h_{ij} = (w, v^j)$
    $w = w - h_{ij} v^i$
  $h_{j+1,j} = (w, w)$
  $v^{j+1} = \dfrac{w}{h_{j+1,j}}$

Then, we solve the approximate eigenvalue problem of dimension m :

$$(H_m - \lambda_m I) y_m = 0.$$

The eigenvalues of this smaller problem are approximations of the original problem. The closer the eigenvalue is to the extremities of the spectra, the more accurate its approximation is. More details can be found in [4].

The Arnoldi method is more efficient than the power method in terms of number of iterations since the approximate eigenvalue problem involves all the iterates while the power method "forgets" the previous iterates at each iteration.

Another important difference concerning the generalized eigenvalue problem is that it has been observed experimentally that the Arnoldi method requires for each outer iteration a more accurate solution to the linear system than the power method. Therefore, some care must be taken concerning the solution of the linear systems. We used a preconditioned conjugate gradient algorithm.

4. Chebychev's acceleration

Usually the power method is combined with Chebyshev's acceleration [7] and this new algorithm can be written as follows :

Given $\Phi^{(0)}$, set $\alpha = 0.5, \beta = 0$ and $\overline{\sigma} = 1$
for $n = 1, \ldots$ do

if $n > 1$ then

$$\text{set } \gamma = \cosh^{-1}\left(\frac{2}{\overline{\sigma}} - 1\right), \quad \alpha = \frac{\cosh\left((n-1)\gamma\right)}{\cosh\left(n\gamma\right)}, \quad \beta = \frac{\cosh\left((n-2)\gamma\right)}{\cosh\left(n\gamma\right)}$$

endif

solve $A\Phi^{(n)} = B\Phi^{(n-1)}$

calculate $\overline{\sigma}$, $\alpha_1 = \dfrac{4.\alpha}{\overline{\sigma}.\lambda^{(n)}}$

$\Phi^{(n)} = \alpha_1\Phi^{(n)} - 2\alpha\Phi^{(n-1)} - \beta\Phi^{(n-2)}$

if convergence then

$\lambda_0 = \lambda^{(n)}$ stop

else

$$\Phi^{(n)} = \frac{\Phi^{(n)}}{\lambda^{(n)}}$$

endif

In this algorithm the number $\overline{\sigma}$ denotes the dominant ratio $\dfrac{\lambda_1}{\lambda_0}$ where λ_1 is the next to the largest eigenvalue λ_0. The practical difficulty is that, since the eigenvalues are unknown, one can only estimate $\overline{\sigma}$.

A standard estimate for $\overline{\sigma}$ is obtained from the following result [7]:

$$\frac{\lambda_{max}^{(n)} - \lambda_{min}^{(n)}}{\lambda_{max}^{(n-1)} - \lambda_{min}^{(n-1)}} \longrightarrow \overline{\sigma} \text{ when } n \to \infty.$$

The Arnoldi method provides another way to estimate $\overline{\sigma}$ which can be calculated from the values of the two largest eigenvalues obtained by this method. Therefore, one can start with a few iterations of the Arnoldi method until a suitable estimation of $\overline{\sigma}$ is obtained and then switch to Chebyshev's accelerations as described above. In [5] this is referred to as the Arnoldi-hybrid method. However, when we are calculating only one eigenvalue, there is no need to restart Arnoldi's process after the Chebychev acceleration iterations, as opposed to what should be done when calculating several eigenvalues.

5. Numerical experiments

5.1. Some remarks on the solution of linear systems

The linear systems associated with the nodal approximation that have to be solved at each outer iteration are symmetric positive definite. Two options are available : either solve the full linear system with both face and cell unknowns or solve a reduced system obtained by eliminating the cell unknowns.

Since the matrices of the full systems are positive definite, the reduced system is better conditionned that the full one. Figure 4 illustrates this fact on a calculation in a realistic case. The comparisons are made without preconditioning, with diagonal preconditioning and with tridiagonal preconditioning.

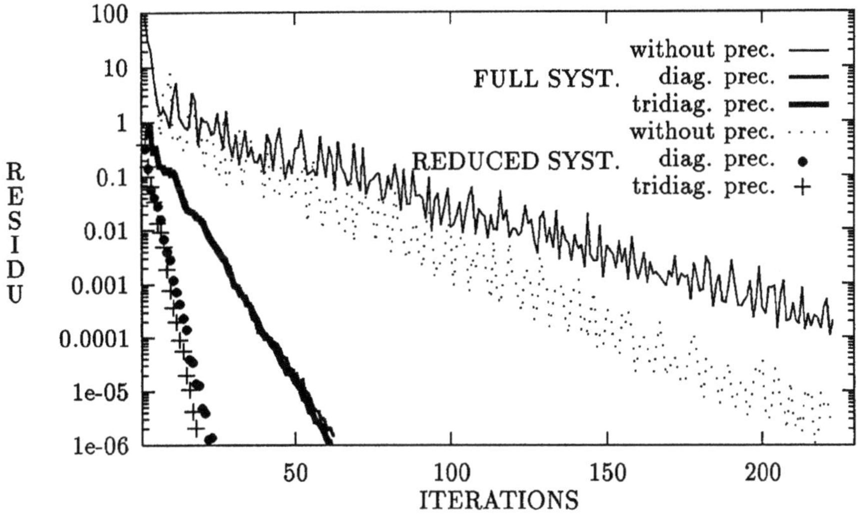

Fig. 4. Convergence of the preconditioned conjugate gradient for full and reduced systems

We remark that the difference in terms of iterations between diagonal and tridiagonal preconditioners is very small. Several other preconditioners have been tried without significant improvement in the number of iterations [8]. Therefore, in terms of computation cost, diagonal preconditioning, which is the cheapest preconditioner, is the most efficient to obtain a given accuracy.

5.2. Comparisons between the power method and Arnoldi's method

The test case that we consider corresponds to real data given by Electricité de France, and coming from a PWR core. The geometry of the cross-section is shown in figure 1. A coarse discretization has one cell per assembly in the $x-y$ plane and 20 intervals in the z direction.

We compare the power method and the Arnoldi method when using the same solver for the linear systems : the conjugate gradient method with diagonal preconditioning. However, much fewer iterations are necessary for the power method (5 iterations) than for the Arnoldi method (15 iterations).

To allow fair comparisons, the same residual norm has been used for both methods that is, $||(A - \lambda B)\Phi||$, even though it is not the one usually used for the power method.

In table I, we show results obtained when increasing m, the size of the Krylov subspace, which is often called the number of Arnoldi iterations. For each m, the residual, the CPU time on a Cray 2 and the calculated dominant eigenvalue λ_0 are shown. In table II, we give the same quantities for the power method – p is the number of power iterations. In addition, we show values of the tolerance ε only used with this method. Enginners are usually satisfied when $\varepsilon = 10^{-5}$. One can see that

m	$\|\|(A - \lambda_0 B)\Phi\|\|$	CPU(s)	λ_0
10	1.1×10^{-2}	3.47	1.03317
15	3.43×10^{-3}	5.33	1.03324
20	3.04×10^{-4}	7.16	1.03325
23	7.51×10^{-5}	8.32	1.03325

TABLE I
Results for the Arnoldi method (15 inner iterations per group) depending on the dimension m of the Krylov space (coarse mesh)

p	tolerance ε	$\|\|(A - \lambda_0 B)\Phi\|\|$	CPU(s)	λ_0
37	1×10^{-2}	7.43×10^{-2}	4.9	1.03252
96	1×10^{-3}	8.65×10^{-3}	7.17	1.03324
163	1×10^{-4}	8.69×10^{-4}	15.96	1.03325
235	1×10^{-5}	8.26×10^{-5}	26.32	1.03325

TABLE II
Results for the power method (5 inner iterations) depending on the number p of power iterations (coarse mesh

the power method needs $p = 235$ iterations and the Arnoldi method needs $m = 23$ iterations to obtain a residu of 8×10^{-5}, which corresponds to the usual accuracy that engineers require for this type of calculation. Comparisons of CPU time show that the Arnoldi method is three times as fast as the power method.

In tables III and IV, we show the effects of refinement for both methods. Each cell has been divided into 8 equal cells. In this case, comparing the line $p = 253$ for the power method to the line $m = 25$ for the Arnoldi method, that is for roughly equal residus, we observe that the Arnoldi method is twice as fast as the power method.

5.3. Comparisons when using Chebyshev's acceleration

Table V gives results obtained with the coarse mesh when using Chebychev's acceleration. As we see, Chebychev's acceleration cut the computational cost by more than half for the power method and by more than 2/3 for the Arnoldi method. We observe that now the Arnoldi hybrid method is twice as fast as the accelerated power method.

In figure 5 we show the convergence rate of the power method, of the Chebychev iterations initialized by the power method and in the Arnoldi hybrid method after

m	$\|\|(A - \lambda_0 B)\Phi\|\|$	CPU(s)	λ_0
10	1.13×10^{-2}	50.34	1.03182
15	2.88×10^{-3}	76.11	1.03197
20	3.13×10^{-4}	101.7	1.03198
25	1.81×10^{-5}	127.7	1.03198
30	6.01×10^{-6}	154.32	1.03198

TABLE III
Results for the Arnoldi method (15 inner iterations) depending on the number m of the Krylov space (refined mesh)

p	tolerance ε	$\|\|(A - \lambda_0 B)\Phi\|\|$	CPU(s)	λ_0
47	1×10^{-2}	$2.08.10^{-2}$		1.03171
108	1×10^{-3}	$2.6.10^{-3}$	109	1.031979
180	1×10^{-4}	$2.8.10^{-4}$	172	1.03198
253	1×10^{-5}	$2.87.10^{-5}$	223	1.03198

TABLE IV
Results for the power method (5 inner iterations) depending on the number p of power iterations (refined mesh)

10 Arnoldi iterations. Even though we use the same Chebychev iterations for both accelerated methods, the rate of convergence differs only because of the starting point. This could be due to the fact that, at the starting point, the vector provided by the Arnoldi method is a better approximation of the true eigenvector than that provided by the power method.

Table VI and figure 6 give corresponding results for the refined mesh. The Arnoldi hybrid method is no longer significantly faster than the accelerated power method. The reason is that it is now more difficult to solve the larger systems and methods

	$\|\|(A - \lambda_0 B)\phi\|\|$	CPU(s)	λ_0
Arnoldi ($m = 23$)	7.51×10^{-5}	8.32	1.03325
Power ($p = 235$)	8.26×10^{-5}	26.32	1.03325
Power + Chebychev ($p = 81$)	7.17×10^{-5}	10.9	1.03325
Arnoldi($m = 10$) + Chebychev ($p = 35$)	7.06×10^{-5}	5.09	1.03325

TABLE V
Comparison of the power method and the Arnoldi method with or without acceleration (coarse mesh)

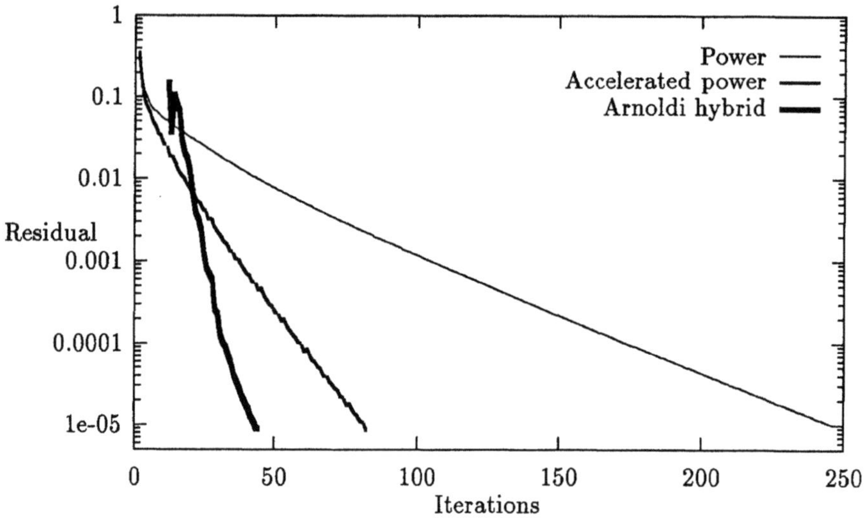

Fig. 5. Convergence curves for various methods (coarse mesh)

like multigrid should be used. We observe that the convergence of the accelerated power method is not as smooth as for the coarse mesh for reasons that are not understood for the moment. For the Arnoldi hybrid method the angle in the convergence curve can be removed by increasing the required precision in the solution of the linear systems, which cannot be done without increasing the computation cost of the method.

	$\|\|(A - \lambda_0 B)\phi\|\|$	CPU(s)	λ_0
Arnoldi ($m = 25$)	1.81×10^{-5}	127.7	1.03198
Power ($p = 235$)	2.87×10^{-5}	223	1.03198
Power + Chebychev ($p = 81$)	2.03×10^{-5}	82	1.03198
Arnoldi($m = 10$) + Chebychev ($p = 49$)	1.15×10^{-5}	70.2	1.03198

TABLE VI
Comparison of the power method and the Arnoldi method with or without acceleration (refined mesh)

However these results show that the number of iterations has not increased even though the dimension of the matrix is 8 times larger. This fact illustrates the efficiency of the Chebychev acceleration.

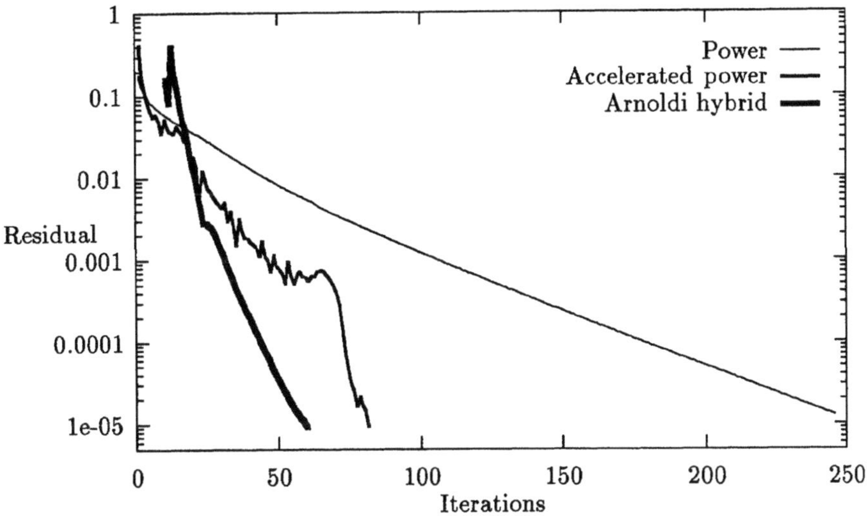

Fig. 6. Convergence curves for various methods (refined mesh)

6. Conclusion

These numerical experiments show that for this problem, the Arnoldi method, though more efficient than the power method, does not provide very important gains. The gain in number of iterations is large but not for the overall computation time. The reason is that the problem that we solve is not a standard eigenvalue problem but a generalized eigenvalue problem for which 90% of the computation time is spent in solving linear systems. Therefore, most efforts must be given to efficient algorithms for the solution of linear systems. This question is crucial for the Arnoldi method since it requires accurate solutions for the linear systems.

This conclusion remains valid when Chebyshev's acceleration, which provides important gain in efficiency, is used.

However, a big advantage of the Arnoldi method is that it can calculate several eigenvalues and eigenvectors without increasing significantly the computation time. This fact could be of interest for engineering purposes.

References

1. R. Dautray and J. L. Lions. *Mathematical Analysis and Numerical Methods for Science and Technology, Vol.4.* Spinger-Verlag, 1990.
2. J. P. Hennart. A general family of nodal schemes. *SIAM J. Sci. Stat. Computing*, 3:264–287, 1986.
3. J. P. Hennart J. Jaffre J.E Roberts. A constructive method for deriving finite elements of nodal types. *Numerische Mathematik*, 53:701–738, 1988.
4. Y. Saad. Variation on Arnoldi's Method for computing eigenelements of large unsymmetric matrices. *Linear Algebra and its Aplications*, 34:269–295, 1980.

5. Y. Saad. Chebyshev Acceleration Techniques for solving Nonsymmetric Eigenvalue Problems. *Mathematics of Computation*, 42:576–588, 1984.
6. R.S Varga. On estimating Rates of Convergence in Multigroup Diffusion Problems. Technical Report WAPD-TM 41, Bettis Atomic Power Laboratory, 1957.
7. R.S Varga. *Matrix Iterative Analysis*. Prentice Hall, 1962.
8. J.-L. Vaudescal. *Résolution Numérique de l'Equation de la Diffusion Neutronique Multigroupe*. Thèse de l'Université Paris-Dauphine, 1993.

COMPUTATIONAL STUDY OF A FREE-BOUNDARY MODEL

L. H. JUÁREZ, P. SAAVEDRA and M. SALAZAR
Dept. de Matemáticas, Universidad Autónoma Metropolitana, Apdo. Postal 55-534, México, 09340, D.F. México.

Abstract. We present computational results for different data sets of a free boundary model problem treated by P. Saavedra and L.R. Scott [12]. This model is used to study the numerical approximation of free boundary fluid-flow problems. Through a heuristical algorithm the order of convergence is estimated and compared with the theoretical value. The nonlinear problem is solved using Newton's method and a functional iteration method, the performance of both methods is compared. The rate of convergence for Neumann boundary conditions is also presented.

Key words: Free-Boundary, Flow problems, Finite element method, Newthon-Rhapson

Introduction

Recently, several articles on fluid-flow phenomena with free boundaries have been published. The models that simulate these phenomena are expressed by a system of nonlinear partial differential equations in domains where a part of the boundary has to be determined as an unknown of the problem. Their solution can not be found analytically, it must be approximated by a numerical method. Finite element methods are preferred over other numerical methods mainly for their flexibility. The nonlinear system of equations that results from the discretization is solved by the Newton-Raphson method because of its second order rate of convergence. There exist few articles with theoretical results on this subject (H. Okamoto [6], V. Pukhnachev [10], V.A. Solonnikov [15]) but there are several with numerical results that solve very complex problems (C. Cuvelier [1], Kistler and Scriven [2], Kruyt and Cuvelier [3], Y. Lin [4], Orr and Scriven[7], Ryskin and Leal [11], Saito and Scriven [13]). This subject is developing rapidly but there are still important issues that have to be settled. Recently W.G. Pritchard, L.R. Scott and S.J. Tavernier [9], in their study of a viscous free-surface flow over a perturbed inclined plane, noticed the importance of the choice of boundary conditions in modeling practical flows and its effect on the well-posedness of the associated mathematical model. They also mention that for certain values of the parameters, large values of Reynold's number or small values of the surface tension, the radius of convergence for Newton's method becomes quite small.

Eventhough the finite element method is the numerical method most widely used, few studies have been published on the convergence of this method when it is applied to a free boundary problem for viscous flows. The only ones we know in this direction are the articles of J.A. Nietsche [5] and P. Saavedra-R. Scott [12]. This last work presents an error analysis for the numerical approximation of an abstract model that has some of the main features of free boundary fluid-flow problems. The analysis

S. Gomez and J-P. Hennart (eds.), Advances in Optimization and Numerical Analysis, 245–260.
© 1994 *Kluwer Academic Publishers.*

has been done in the case when a finite element method is applied to the variational formulation of the model. Piecewise linear polynomials are used and optimal order of convergence is proved.

In this work we present computational results related to the problem treated in [12] for different data values. The order of convergence is estimated through a heuristical algorithm and it is compared with the theoretical value obtained in [12]. We also take advantage of the simplicity of the model to study the dependence of the convergence of Newton-Raphson's method with respect to the parameters. Its performance is compared with a function iterative method, which was used in [12] to prove the existence of a solution. Finally Neumann boundary conditions are also studied and the computational results show that for small values of the data the order of convergence is similar to the Dirichlet case, eventhough it has not been rigorously proved.

The structure of this paper is as follows: In the first section the problem treated by [12] is introduced and given its weak formulation. In section 2 the discrete problem and its solution are presented. In 3 numerical examples are given and the performance of Newton's method is analysed and finally, in section 4, Neumann boundary conditions are studied.

1. Problem Formulation

We shall make use of the Lebesgue and Sobolev spaces $L_p(\Omega)$ and $W_p^m(\Omega)$, respectively, $1 \leq p \leq \infty$, $m \in \mathcal{N}$, defined for an open set, Ω, in $\Re^n$, $n = 1$ or 2. These spaces are provided with the usual norms $||.||_{L^p(\Omega)}$ and $||.||_{W_p^m(\Omega)}$, respectively. $\mathring{W}_p^1(\Omega)$ are those functions in $W_p^1(\Omega)$ wich vanish on the boundary of Ω in the generalized sense.

Consider a function $\gamma \in W_\infty^1(0,1)$ such that $||\gamma||_{W_\infty^1(0,1)} < 1/2$; let $g \in W_p^2(\Omega^*)$, where $\Omega_\gamma \subset \Omega^* = [0,1] \times [0,3/2]$ for all γ satisfying the first hypothesis, and suppose $g(x,y) = 0$ for $y \geq 1/2$. Related to γ, consider the following sets:

$$\begin{aligned} \Omega_\gamma &= \{(x,y) |\ 0 < x < 1, \quad 0 < y < 1 + \gamma(x)\ \}, \\ \Gamma_\gamma &= \{(x,y) |\ 0 < x < 1, \quad y = 1 + \gamma(x)\ \}. \end{aligned}$$

We are interested in the approximation of the solution of the following problem: Find γ and u such that

$$\begin{aligned} \Delta u &= 0 \quad in\ \Omega_\gamma, \\ u &= g \quad on\ \partial\Omega_\gamma, \\ \frac{-s\gamma''(x)}{[1+\gamma'(x)^2]^{1/2}} &= \frac{\partial u(x, 1+\gamma(x))}{\partial \vec{n}}, \quad \forall x \in [0,1], \\ \gamma(0) &= \gamma(1) = 0. \end{aligned} \tag{1.1}$$

The parameter $s > 0$ plays the role of the "surface tension" and $\vec{n}$ is the outward normal vector to the boundary $\partial\Omega_\gamma$ of Ω_γ. Note that u is a scalar field and that in the balance of forces on the boundary we are using a nonlinear term which is not the curvature of γ.

The weak formulation of this problem is the following: Find $\gamma \in \mathring{W}_\infty^1(0,1)$ and $u \in g \bigoplus \mathring{W}_p^1(\Omega_\gamma)$ such that

$$\begin{aligned} a_\gamma(u,v) &= 0, & \forall v \in \mathring{W}^1_q(\Omega_\gamma) \\ b(\gamma,\chi) &= a_\gamma(u, E_\gamma\chi), & \forall \chi \in \mathring{W}^1_1(0,1). \end{aligned} \tag{1.2}$$

where $a_\gamma(.,.)$ and $b(.,.)$ are defined by

$$\begin{aligned} a_\gamma(u,v) &= \int_{\Omega_\gamma} \nabla u \cdot \nabla v \, dx\, dy, \\ b(\gamma,\chi) &= s\int_0^1 \gamma'(x)\, \chi'(x)\, dx, \end{aligned}$$

and $E_\gamma\chi$ is the extension of $\chi \in \mathring{W}^1_1(0,1)$ to Ω_γ in such a way that

$$E_\gamma\chi|_{\Gamma_\gamma} = \chi \text{ and } \quad E_\gamma\chi|_{\partial\Omega-\Gamma_\gamma} = 0.$$

It is proved in [12] that problem (1.2) admits a unique solution $(\gamma, u) \in \mathring{W}^1_\infty \times g \bigoplus \mathring{W}^1_p(\Omega_\gamma)$ for small values of the norm of g, provided that $p \in (2, P)$ for some $P > 2$.

The discrete problem

We approximate the solution of this problem using a finite element method. Denote $\Omega_0 = [0,1] \times [0,1]$ and let π_h^0, $0 < h \leq h_o < 1$, be a quasi-uniform triangulation of Ω_0 and denote by π_h^γ the triangulation obtained by transforming the vertices n_i of π_h^0 via the mapping $(\xi,\eta) \to (x,y)$ from Ω_0 to Ω_γ defined by

$$x = \xi,\ y = (1+\gamma(\xi))\eta. \tag{2.1}$$

More precisely, let F^γ denote the continuous, piecewise-affine interpolant of the mapping (2.1) with respect to the mesh π_h^0, and let π_h^γ be the image of π_h^0 with respect to F^γ. Denote by Ω_h^γ the image of Ω_0 with respect to F^γ. Corresponding to the triangulation π_h^γ, we define the following spaces:

$$\begin{aligned} V_h^\gamma &= \{v \in C^0(\Omega_h^\gamma) \quad | \, v|_K \in P^1(K)\ \forall K \in \pi_h^\gamma\}, \\ \mathring{V}_h^\gamma &= \{v \in V_h^\gamma |\ v(n_i) = 0 \quad \forall n_i \in \partial\Omega_h^\gamma\}. \end{aligned}$$

Let $\xi_i \in [0,1], 1 \leq i \leq l_h$, denote all ξ-coordinates such that $(\xi_i, 1)$ is a vertex of a triangle $K \in \pi_h^0$. Denote $I_i = [\xi_i, \xi_{i+1}]$ and decompose $[0,1] = \bigcup_{i=1}^{l_h-1}[I_i]$, and associate with this mesh the following discrete space:

$$S_h = \{\gamma \in C^0(0,1) \quad | \, \gamma|_{I_i} \in P^1(I_i); \quad \gamma(0) = \gamma(1) = 0\}.$$

For functions $v \in W_p^1(\Omega_h^\gamma)$ with $p > 2$ we shall use the usual piecewise-linear interpolant $I_h^\gamma : W_p^1(\Omega_h^\gamma) \to V_h^\gamma$, namely

$$I_h^\gamma v = \sum_{k=1}^{NN} v(n_k) v_k^h,$$

where $\{v_k^h : 1 \leq k \leq NN\}$ is the usual Lagrange basis of V_h^γ defined by

$$v_k^h(n_j) = \delta_{kj}, \quad 1 \leq k, j \leq NN,$$

and NN is the number of vertices of the triangulation π_h^γ. We will write $g_h^\gamma = I_h^\gamma g$. Finally we define the discrete extension E_h of a function $\chi \in S_h$ as:

$$E_h\chi(x) = \begin{cases} \sum_{j=2}^{l_h-1} \chi(\xi_j)\ v_j^h(x) & if\ x \in \Gamma_\gamma; \\ 0 & if\ x \notin \Gamma_\gamma, \end{cases}$$

where v_j^h are the basis functions of V_h^γ associated with the nodes $(\xi_j, 1+\gamma(\xi_j))$.

Associated with these spaces we define the following discrete problem: Find $\gamma_h \in S_h$ and $u_h \in g_h^{\gamma_h} \bigoplus \overset{\circ}{V}_h^{\gamma_h}$ such that

$$\begin{aligned} a_{\gamma_h}(u_h, v_h) &= 0, & \forall v_h \in \overset{\circ}{V}_h^{\gamma_h}, \\ b(\gamma_h, \chi) &= a_{\gamma_h}(u_h, E_h\chi), & \forall \chi \in S_h. \end{aligned} \tag{2.2}$$

Theorem.

Suppose (γ_h, u_h) is a solution of the discrete problem. If the small solution $(\gamma, u) \in V_\varepsilon$ of problem (1.2) for $p > 2$ satisfies $(\gamma, u) \in W_\infty^2 \times W_p^2(\Omega_0)$ and ε is sufficiently small, then there is a constant $C < \infty$ and a strictly positive h_0 such that for every h that satisfies $0 < h \le h_0$

$$||\gamma - \gamma_h||_{W_\infty^1(0,1)} + ||\hat{u} - \hat{u}_h||_{W_p^1(\Omega_0)} \le C\ h\ (||\gamma||_{W_\infty^2(0,1)} + ||\hat{u}||_{W_p^2(\Omega_0)}\ ), \tag{2.3}$$

where (γ_h, u_h) is the small-norm solution of the discrete problem (2.2), $\hat{u}$ is the image of the function u when (2.1) is applied, and

$$V_\varepsilon = \{(\gamma, \hat{u}) \in \overset{\circ}{W}{}_\infty^1(0,1) \times W_p^1(\Omega_0) |\ ||\gamma||_{W_\infty^1(0,1)} < 1/2, ||\hat{u}||_{W_p^1(\Omega_0)} < \varepsilon\}.$$

The nonlinear system of equations associated to problem (2.2) is the following

$$\begin{aligned} F_1(\vec{w}, \vec{\gamma}) &= A(\vec{\gamma})\vec{w} + \vec{b}(\vec{\gamma}) = 0 \\ F_2(\vec{w}, \vec{\gamma}) &= sB\vec{\gamma} - D(\vec{\gamma})\vec{w} - \vec{d}(\vec{\gamma}) = 0 \end{aligned} \tag{2.4}$$

where $\vec{u} = \vec{w} + \vec{g}$ and

$$\begin{aligned} A_{kj}(\vec{\gamma}) &= \int_{\Omega_h^\gamma} \nabla v_k^h \cdot \nabla v_j^h\ dx\ dy, \\ b_k(\vec{\gamma}) &= \sum_{j=1}^{NN} g(n_j) \int_{\Omega_h^\gamma} \nabla v_j^h \cdot \nabla v_k^h\ dx\ dy, \\ B_{kj} &= \int_0^1 \chi_k'(x)\ \chi_j'(x)\ dx, \\ d_j(\vec{\gamma}) &= \sum_{k=1}^{NN} g(n_k) \int_{\Omega_h^\gamma} \nabla v_k^h \cdot \nabla E_h\chi_j\ dx\ dy, \\ D_{kj}(\vec{\gamma}) &= \int_{\Omega_h^\gamma} \nabla v_j^h \cdot \nabla E_h\chi_k\ \ dx\ dy. \end{aligned}$$

This system can be solved by different methods. In this work a functional iteration method and the Newton- Raphson method are presented.

1.1. FUNCTIONAL ITERATION METHOD

Let $\vec{\gamma}_0 = 0$ and $\vec{u}_0$ be the solution of the discrete problem associated with the following problem:

$$\begin{aligned} \Delta u &= 0 \quad in\ \Omega_0, \\ u &= g \quad on\ \partial\Omega_0. \end{aligned} \tag{2.5}$$

Given $(\vec{w}_i, \vec{\gamma}_i)$, the vector $\vec{\gamma}_{i+1}$ is the solution of the system

$$sB\ \vec{\gamma}_{i+1} = D(\vec{\gamma}_i)\vec{w}_i + \vec{d}(\vec{\gamma}_i)$$

and $\vec{w}_{i+1}$ is the solution of the linear system

$$A(\vec{\gamma}_{i+1})\ \vec{w}_{i+1} = \vec{b}(\vec{\gamma}_{i+1}).$$

This iterative procedure converges to the solution of problem (2.4) with a rate of convergence K equal to $K = C\varepsilon/s$, where C and ε are defined in expression (2.3).

1.2. NEWTON'S METHOD

Given $\vec{\gamma}_0 = 0$ and $\vec{w}_0$ the solution of the homogeneous problem associated to (2.5) and ${}^t\vec{X}_i = [\vec{w}_i, \vec{\gamma}_i]$, $\vec{Y}_i$ is the solution of

$$DF(\vec{X}_i)\ \vec{Y}_i = -\vec{F}(\vec{X}_i)$$

where $DF(\vec{X}_i)$ is the Jacobian matrix of F

$$DF(\vec{X}_i) = \begin{pmatrix} \frac{\partial \vec{F}_1}{\partial \vec{w}} & \frac{\partial \vec{F}_1}{\partial \vec{\gamma}} \\ \frac{\partial \vec{F}_2}{\partial \vec{w}} & \frac{\partial \vec{F}_2}{\partial \vec{\gamma}} \end{pmatrix}$$

$\vec{X}_{i+1}$ is obtained by

$$\vec{X}_{i+1} = \vec{Y}_i + \vec{X}_i.$$

In the computational examples, for both methods the convergence of the iterative process is determined in terms of the relative error between two successive approximations. The iterative procedure is said to have converged when

$$||\vec{u}_{i+1} - \vec{u}_i||_{l^\infty(\Omega_0)}/||\vec{u}_{i+1}||_{l^\infty(\Omega_0)} \leq \rho \quad \text{and} \quad ||\vec{\gamma}_{i+1} - \vec{\gamma}_i||_{l^\infty(0,1)}/||\vec{\gamma}_{i+1}||_{l^\infty(0,1)} \leq \rho, \tag{2.6}$$

where ρ is a fixed real positive number.

The first method has the advantage that the systems of equations have a very simple structure, unfortunately its convergence rate is only linear. Newton-Raphson's method is preferred in this type of problems for its second order rate of convergence. In this case the position of the free boundary and the field variable are simultaneously calculated through an iterative procedure which involves solving a non symmetric sparse linear system of equations. A big disadvantage of this method is that the Jacobian of $\vec{F}$ must be calculated at each iteration. The assemblage of the Jacobian is not an easy task and can not be done using standard finite element codes. Also it has been reported for other free boundary problems that the radius of convergence for this method seemed to become quite small for large values of the Reynolds number or for small values of the surface tension, see [9]. One of the goals of this study is to compare, for this simple case, the performance of both methods with respect to the variation of the parameters: the norm of g and the surface tension s and to analyse if it is worth using the Newton's method.

Examples

In this section we present numerical results for two different cases. The order of convergence is estimated in each case for different values of the parameters. Also the performance of Newton's method and the function iteration method are compared.

We use the following heuristical algorithm to estimate the order of convergence. Suppose γ exactly describes the free boundary of problem (1.2). Let $\xi_i \in I = (0,1)$, and let h be the length of the discretization step of I; if r is the order of convergence in the norm L^∞, then for some positive constant C, independent of h, we have

$$\gamma(\xi_i) - \gamma_h(\xi_i) \approx Ch^r.$$

Similarly, if the length were $2h$

$$\gamma(\xi_i) - \gamma_{2h}(\xi_i) \approx C(2h)^r$$

and hence

$$\gamma_{2h}(\xi_i) - \gamma_h(\xi_i) \approx Ch^r(2^r - 1).$$

In the same way we get

$$\gamma_{4h}(\xi_i) - \gamma_{2h}(\xi_i)\ C(2h)^r(2^r - 1),$$

which implies

$$2^r \approx \frac{\gamma_{4h}(\xi_i) - \gamma_{2h}(\xi_i)}{\gamma_{2h}(\xi_i) - \gamma_h(\xi_i)}.$$

Define a function r_1 applied to the spaces S_h that estimates the local order of convergence associated to the node ξ_i then, using the last expression we define

$$r_1(\xi_i) \approx \frac{ln[\frac{\gamma_{4h}(\xi_i)-\gamma_{2h}(\xi_i)}{\gamma_{2h}(\xi_i)-\gamma_h(\xi_i)}]}{ln\ 2}.$$

If we define

$$R_1 = \min_{1 \le i \le l_h} r_1(\xi_i)$$

then R_1 is a reasonable estimation of the order of convergence.

It is also interesting to estimate the order of convergence in the norm L^∞ when u is approximated by u_h. In this case, we recall that bound (2.3) is essentially an "energy" estimate, lower norm estimates (L^2, L^∞, etc) have not been attemped and, may not be very easy to prove. Nevertheless, it could be expected that we would get an extra power of h when we considered only function value errors instead of energy value errors; for example for the linear elliptic problems, L. R. Scott [14] proved that if $u \in H^1(\Omega_0) \bigcap W^2_\infty(\Omega_0)$ then there exists a positive constant C independent of h such that

$$||u - u_h||_{L^\infty(\Omega_0)} \leq C\ h^2\ lnh\ ||u||_{W^2_\infty(\Omega_0)}. \tag{3.1}$$

This last result encouraged us to introduce a variable R_2 that estimates the error in the L^∞ norm in the approximation of u. First, as in the case of γ we introduce the function r_2

$$r_2(n_i) \approx \frac{ln[\frac{u_{4h}(n_i)-u_{2h}(n_i)}{u_{2h}(n_i)-u_h(n_i)}]}{ln\ 2},$$

where $h = \sup\ \{\ \text{diam}\ (K) : \forall K \in \pi_h^{\gamma_h}\ \}$.

Then we calculate R_2 as

$$R_2 = \min_{1\leq i\leq n_i} r_2(n_i).$$

In the following numerical experiments, we use a tolerance $\rho = 10^{-5}$ to stop the iterative procedures. We obtained results for $h = 1/10, 1/20, 1/40$. In order to illustrate how the solution changes for different values of h and of the parameters, the values of u_h and γ_h are shown for $n_i = (1/2, 1 + \gamma_h(1/2))$ and for $\xi = 1/2$.

In this first example we present, in Table 1, the estimated orders of convergence R_1 and R_2 for different values of the norm of g. The surface tension is taken equal to one. This table also presents the required number of iterations for the Functional Iterative method NFIM and for Newton's method NNM. In both cases the number of iterations is independent of h. Table 2 shows, for $A = 8$ and $h = 1/20$, the effect of the surface tension on the solution and on the number of iterations required to obtain convergence. Figure 1 shows the free boundary for different values of the norm of g when $h = 1/40$ and Figure 2 illustrates the effect of the the surface tension on the free surface. In example 2 the tables are similar to those of example 1, Table 4 presents the effect of the surface tension on the free boundary when $A = 7$ and $h = 1/20$.

1.3. Example 1

$$g(x,y) = \begin{cases} A(y - 1/2)\ sin\pi x, & y \leq 1/2, \\ 0 & y \geq 1/2. \end{cases}$$

Table 1

A	$\|\|g\|\|_{W_2^1(\Omega^*)}$	NFIM	NNM	h	R_1	$\gamma_h(\xi_i)$	R_2	$u_h(n_i)$
2	1.38	6	4	1/40		.025695		-.191960
				1/20	1.99	.025887	1.97	-.192340
				1/10		.026657		-.193842
8	5.52	10	5	1/40		.086658		-.701007
				1/20	1.97	.087241	1.96	-.702040
				1/10		.089575		-.706094
16	11.04	15	5	1/40		.146868		-1.280383
				1/20	1.96	.147820	1.96	-1.281617
				1/10		.151632		-1.286409

Table 2

s	SIM	NM	$\gamma_h(\xi_i)$	$u_h(n_i)$
0.10		6	.392773	-.440529
0.125		6	.349584	-.470817
0.25	21	6	.234603	-.561612
0.5	15	5	.147820	-.640808
1.0	10	5	.087241	-.702040
2.0	8	5	.048576	-.743827
4.0	6	4	.025887	-.769362
15.0	5	4	.007276	-.790877

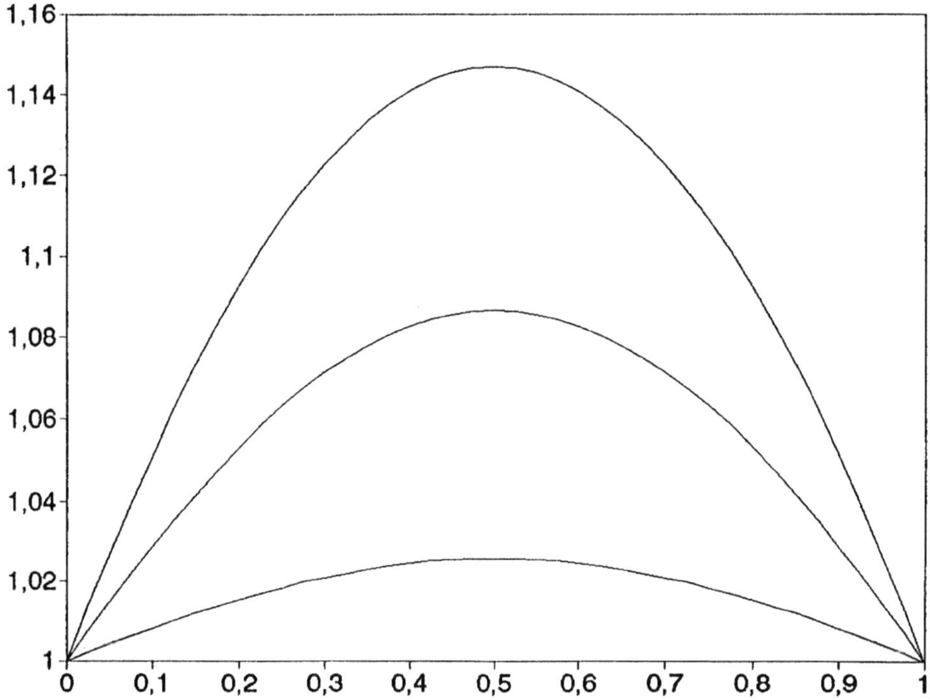

Fig.1 Variation of γ_h with respect to the norm of g.

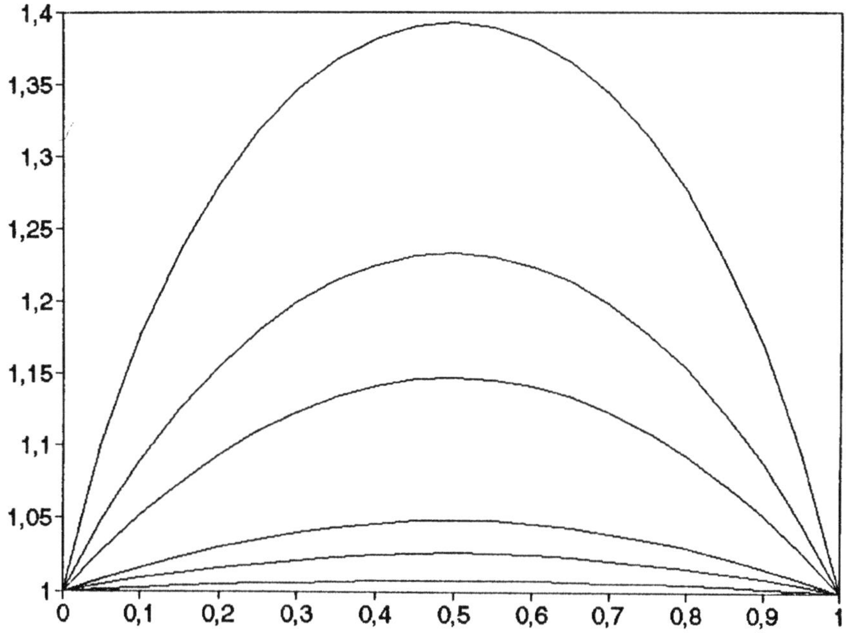

Fig.2 Variation of γ_h with respect to s.

Example 2.

$$g(x,y) = \begin{cases} \frac{A(1-x)\ x^2\ (y-1/2)^2\ \exp^{y+2}}{\cos y}, & y \leq 1/2; \\ 0 & y \geq 1/2. \end{cases}$$

Table 3

A	$\|\|g\|\|_{W_2^1(\Omega^*)}$	NFIM	NNM	h	R_1	$\gamma_h(\xi_i)$	R_2	$u_h(n_i)$
1	0.37	3	3	1/40		-.006712		.047913
				1/20	1.99	-.006767	1.98	.048018
				1/10		-.006986		.048431
7	2.59	8	4	1/40		-.053767		.359313
				1/20	2.02	-.054290	1.98	.360318
				1/10		-.056410		.364288
14	5.18	15	5	1/40		-.136988		.810164
				1/20	2.08	-.139051	2.04	.813981
				1/10		-.147794		.829657

Table 4

s	NFIM	NM	$\gamma_h(\xi_i)$	$u_h(n_i)$
0.5	14	5	-.139051	.406990
1	8	4	-.054290	.360317
1.5	7	4	-.034154	.349875
3.0	6	4	-.016222	.340827
6.0	5	3	-.007921	.336698
12.0	4	3	-.003915	.334720

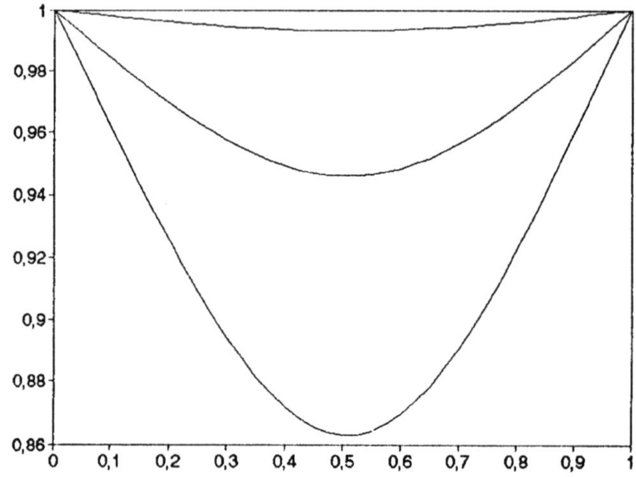

Fig.3 Variation of γ_h with respect to the norm of g.

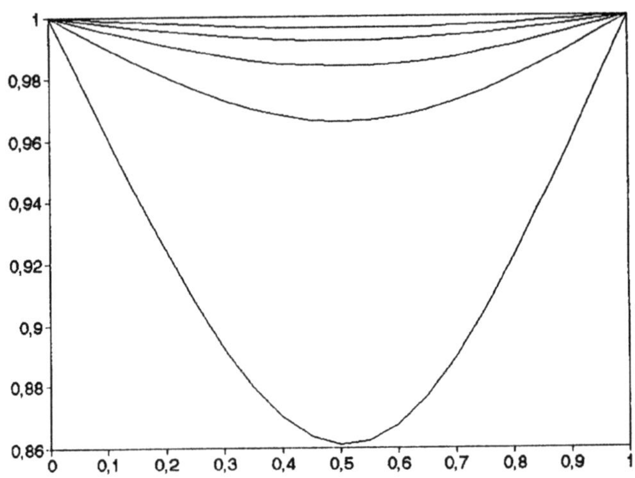

Fig.4 Variation of γ_h with respect to s.

The examples show that the estimated order of convergence for u_h and γ_h are in accord with the error bounds for elliptic problems. They also show that, as was expected, as the norm of g increases we get bigger solutions with bigger norm, and that the required number of iterations in both iterative methods also increase. If the surface tension remains constant, the required number of iterations in both iterative methods is independent of the mesh size. When it varies, we observe that as it increases there is more opposition from the free boundary to change with respect to the horizontal and vice-versa. In this case, the number of iterations needed to attain the bound (2.6) increases dramatically in the function iteration method as the surface tension decreases. Newton's method is more stable with respect to the parameters s and the norm of g. The required number of iterations is sensibly less than for the functional iterative method. This shows that even if the cost of Newton's method is higher it is worth its use.

For values $s < 0.1$ for example 1 and $s \leq 0.3$ for example 2, Newton's method diverges, we apply a continuation method in order to see if the convergence depended on the initial value, but we could not ameliorate the results. Eventhough we did not have an expression for the convergence rate of this method we can conclude that the radius of convergence depends on the value of s. As $s \rightarrow 0$, the radius also tends to zero.

Neumann conditions for the free boundary

We consider this case because it has some of the features of the problem considered by Pukhnachev [10] in his theoretical analysis of the flow of a liquid constrained only partly in a container, and that was the example that motivated us to study the numerical approximation of free boundary problems. Pukhnachev's work is a good example of the case when a part of the boundary of the domain filled by the liquid is an interface with another of much smaller density, and for which surface tension plays a significant role in determining the shape of the free surface. Instead of the Dirichlet boundary conditions for the curve γ we impose the following conditions:

$$\gamma'(0) = \gamma'(1) = 0, \qquad \int_0^1 \gamma(x)\, dx = 0. \tag{4.1}$$

The last equation expresses that the total volume of the liquid in the container remains constant. In order to give the weak formulation of this problem, we define the following spaces:

$$\bar{W}^1_\infty(0,1) = \{\gamma \in W^1_\infty(0,1) \mid \int_0^1 \gamma(x)\, dx = 0\},$$

and the space of test functions:

$$\bar{W}^1_1(0,1) = \{\chi \in W^1_1(0,1) \mid \int_0^1 \chi(x)\, dx = 0\}.$$

The weak formulation is: Find $\gamma \in \bar{W}^1_\infty(0,1)$ and $u \in g \bigoplus \mathring{W}^1_p(\Omega_\gamma)$ such that

$$\begin{aligned} a_\gamma(u,v) &= 0 && \forall v \in \mathring{W}^1_q(\Omega_\gamma), \\ b(\gamma,\chi) &= a_\gamma(u, E_\gamma \chi) && \forall \chi \in \bar{W}^1_1(0,1). \end{aligned} \tag{4.2}$$

The same argument as in [12] is used to prove existence and uniqueness of a small solution of problem (4.2). In this case V_ε is the following set

$$V_\varepsilon = \{(\gamma, u) \in \bar{W}^1_\infty(0,1) \times W^1_p(\Omega_0) \mid \|\gamma\|_{W^1_\infty(0,1)} < \frac{1}{2}, \|u\|_{W^1_p(\Omega_o)} < \varepsilon\}.$$

and the mapping T is defined in the following way: Given $(\gamma, u) \in V_\varepsilon$

$$T(\gamma, u) = (T_1(\gamma, u), T_2(\gamma, u)) = (\bar{\gamma}, \bar{u})$$

where $(\bar{\gamma}, \bar{u})$ is determined as follows: First, $\bar{\gamma} \in \bar{W}^1_\infty(0,1)$ is found satisfying

$$b(\bar{\gamma}, \chi) = a_\gamma(u, E_\gamma(\chi)) \qquad \forall \chi \in \bar{W}^1_1(0,1),$$

then $\bar{u} - g \in \mathring{W}^1_p(\Omega_{\bar{\gamma}})$ solves

$$a_{\bar{\gamma}}(\bar{u}, v) = 0 \qquad \forall v \in \mathring{W}^1_q(\Omega_{\bar{\gamma}}). \tag{4.3}$$

In order to prove that T is a contraction we need to make a new hypothesis on the function g : $g(0, y) = g(1, y) = 0 \; \forall y \in (0,1)$. This guarantees that $\hat{g}^1 - \hat{g}^2 \in \mathring{W}^1_p(\Omega_0)$ where $\hat{g}^i$ is the transformation of g from Ω_{γ_i} to Ω_0. This condition is essential to prove that $\hat{u}_1 - \hat{u}_2 \in \mathring{W}^1_p(\Omega_0)$, where u_i is the solution of the equation (4.3).

For the discretisation of problem (4.2) we use the same discrete spaces to approximate the functions u, g, $E_\gamma \chi$, to approximate $\gamma \in \bar{W}^1_1(0,1)$ we use the following space:

$$\bar{S}_h = \{\gamma \in C^0(0,1) \mid \gamma|_{I_i} \in P^1(I_i) \quad \int_0^1 \gamma(x)\, dx = 0\}.$$

It is not difficult to prove that a basis of $\bar{S}_h$ is $\bar{U} = \{\chi_j\}_{j=1}^{l_h-1}$, where the functions χ_j are defined in the following way:

$$\chi_j(x) = \psi_j(x) - \int_0^1 \psi_j(x)\, dx, \quad j = 1, \dots, l_h - 1,$$

with $\psi_j \in U$ elements of the standard Lagrange basis of the space of piecewise linear polynomials defined in $[0,1]$.

The discrete problem associated with problem (4.2) is the following: Find $(\gamma_h, u_h) \in \bar{S}_h \times g_h^{\gamma_h} \bigoplus \mathring{V}_h^{\gamma_h}$ such that

$$\begin{aligned} a_{\gamma_h}(u_h, v_h) &= 0 & \forall v_h \in \mathring{V}_h^{\gamma_h}, \\ b(\gamma_h, \chi_h) &= a_{\gamma_h}(u_h, \bar{E}_h \chi_h) & \forall \chi_h \in \bar{S}_h, \end{aligned} \tag{4.4}$$

with

$$\bar{E}_h \chi = \sum_{k=1}^{l_h-1} \chi_h(\xi_k) v_k^h.$$

It can be proved that problem (4.4) admits a unique small solution using the same arguments as in the other cases. The computational algorithm to solve this problem is similar to the others. The only necessary modification is in the calculus of γ_h.

Suppose (γ_h^i, u_h^i) has been calculated, then γ_h^{i+1} is calculated by $\gamma_h^{i+1} = \sum_{j=1}^{l_h-1} d_j\chi_j \in \bar{S}_h$,

$$b(\gamma_h^{i+1}, \chi_j) = a_{\gamma_h}(u_h, \bar{E}_h\chi_j) \qquad \forall \chi_j \in \bar{U},$$

this is equivalent to solving the following system of linear equations:

$$\sum_{j=1}^{l_h-1} b_{kj} d_j = f_k, \qquad k = 1, 2 \ldots, l_h - 1, \tag{4.5}$$

with

$$\begin{aligned} b_{kj} &= s\int_0^1 \chi_k'(x)\chi_j'(x)dx, \\ f_k &= \sum_{m=1}^{l_h-1} \chi_k(\xi_m) \sum_{j=1}^{NN} u_h(n_j) \int_{\Omega_h^{\gamma_h^i}} \nabla v_j^h \cdot \nabla v_m^h \, dx\, dy \quad k = 1, \ldots, l_h - 1. \end{aligned}$$

Any direct method can be used to solve (4.5), because b_{kj} is a tridiagonal, symmetric and diagonally dominant matrix.

For this case, an error bound similar to (2.3) has not yet been proved. The main difficulty, as was explained in [12], is that we can no longer assure that the difference between the extension of χ_h and its interpolant $J_h E\chi_h$ is in $\mathring{W}_q^1(\Omega_0)$ which is essential to apply a similar argument as in the Dirichlet case. Nevertheless, the following computational results show that the order of convergence could probably be the same. In the following examples $\xi_i = 0.0$ was chosen because it is at this point where the value of γ varies more with respect to h and the value of u_h and γ_h in this point better illustrate how the solution changes with respect to h. In Figure 5 we show the variations of γ_h with respect to the norm of g and finally in Figure 6 the effect of the surface tension on γ_h and u_h. Observe that the form of the free boundary is different from the other examples; this is due to the conditions (4.1) that are imposed in this case.

1.4. EXAMPLE 3

$$g(x, y) = \begin{cases} A\ (y - 1/2)\ \exp^{2+x}\ cos\pi x\ sin\pi x, & y \le 1/2; \\ 0 & y \ge 1/2. \end{cases}$$

Table 5

A	$\|\|g\|\|_{W_2^1(\Omega^*)}$	No. Iter.	h	R_1	$\gamma_h(\xi_i)$	R_2	$u_h(n_i)$
4	27.6	7	1/40		.035138		.423158
			1/20	2.01	.036309	1.90	.422333
			1/10		.041069		.418530
6	41.4	10	1/40		.057667		.635125
			1/20	2.0	.059605	1.87	.633377
			1/10		.067521		.625501
9	62.1	21	1/40		.106519		.946146
			1/20	2.0	.110727	2.05	.941029
			1/10		.129324		.917074

1.5. Example 4

$$g(x,y) = \begin{cases} 6(y-1/2)\exp^{2+x} \ cos\pi x \ sin\pi x, & y \le 1/2; \\ 0 & y \ge 1/2. \end{cases}$$

$h = 1/20$.

Table 6

s	No.Iter.	$\gamma_h(\xi_i)$	$u_h(n_i)$
0.7	15	.101138	.629120
0.85	12	.074449	.632471
1.0	10	.059605	.633377
2.0	5	.026227	.633090
4.0	4	.012476	.632101
8.0	3	.006099	.631489

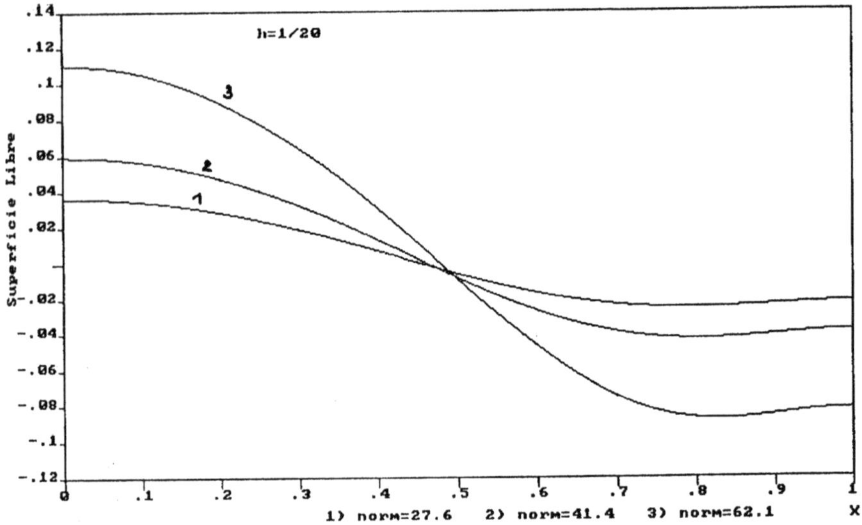

Fig. 5. Variation of the Free Surface with respect to the norm of g.

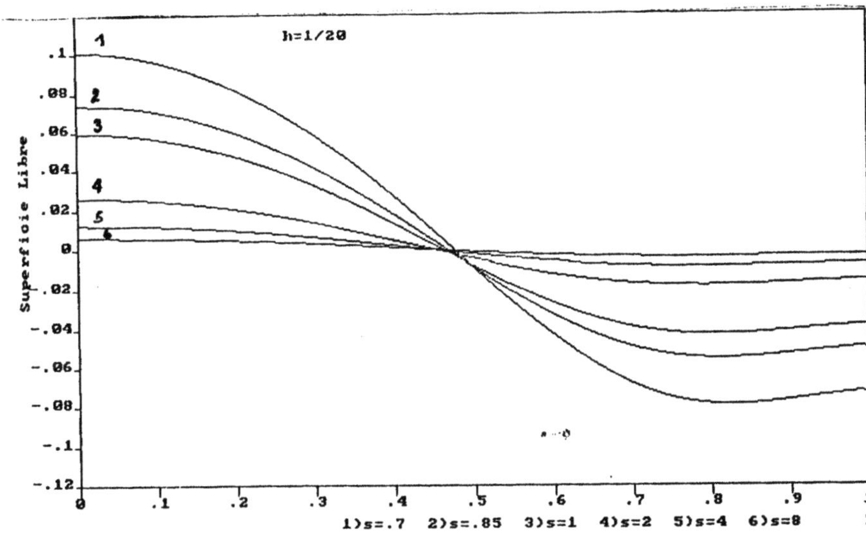

Fig.6. Variation of the Free Surface with respect to s.

Even if we have not proved an error bound for this case, the numerical results show that this must be similar to the Dirichlet case. The computational results encourage us to extend this study to more complicated problems such as the Stokes equations.

Acknowledgements

The authors wish to thank Professor L.R. Scott for his valuable comments concerning this work.

References

1. C. Cuvelier, *On the numerical solution of a capillary free boundary problem governed by the Navier-Stokes equations,* Springer lecture notes in Physics,**141** (1980), pp 123-137.
2. C. Cuvelier and R.M. Schulkes, *Some Numerical Methods for the Computation of Capillary Free Boundaries governed by the Navier-Stokes Equations,* Siam Review, **32. 3.** (1990), pp 355-423.
3. S.F, Kistler and L.E. Scriven, *Coating flow theory by finite Element and asymtotic analysis of the Navier-Stokes system,* International Journal for Numerical Methods in fluids, **4**, (1984), pp 207-229.

4. N.P. Kruyt, C. Cuvelier, A.Segal and J. Van der Zanden, *Total linearization methods for solving viscous free boundary flow problems by the finite element method*, International Journal for Numerical methods in Fluids, **8**, (1988), pp 357-363.
5. Y.Y.Lin, *Numerical solutions for flow in a partially filled rotating cylinder*, SIAM J. Sci. Stat. Comp.,**7**, No. 2, (1976).
6. J.A. Nietsche, *Free boundary problems for Stokes flow and finite element method*, "Ecuadiff 6", Lecture Notes in Math, Springer Verlag, **1192**, (1986), pp 327-332.
7. H. Okamoto, *Stationary free boundary problems for circular flow with or without surface tension*, Lecture Notes in Nu. App. Anal., **5**, (1982), pp 223-257.
8. F.M.Orr and L.E. Scriven,*Rimming flow: Numerical simulation of steady, viscous, free surface flow with surface tension*, J. Fluid Mech., **84** (1976), pp 145-165.
9. W.G. Pritchard, L.R. Scott and S.J. Tavernier, *Viscous free surface flow over a perturbed inclined plane*, Report no. AM83 (1991), Dept. Math., Penn State Univ., Philos. Trans. Soc. London, submitted
10. V.V. Pukhnachev, *Hydrodinamic free boundary problems*, "Nonlinear Partial Differential Equations and their Applications", College de France Seminar, **III**, Pitman, Boston, (1982), pp 301-308.
11. G. Riskin and L.G. Leal, *Numerical solution of free boundary problems in fluid mechanics*, Part 1-3. J. Fluid Mech., **148**, (1984), pp 1-43.
12. P. Saavedra and L.R. Scott, *Variational formulation of a model free boundary problem,* Math. of Computation, **57**, 196 (1991), pp. 451-475.
13. H.Saito and L.E. Scriven, *Study of coating flow by the finite element method*, J. Comp. Phys, **42**, (1981), pp 53-76.
14. L.R. Scott, *Optimal L^∞ estimates for the finite element method on irregular meshes*, Math. Comp., **30**, (1976), pp 681-697.
15. V.A. Solonnikov, *On the Stokes equations in domains with non-smooth boundaries and on viscous incompressible flow with a free surface*, "Nonlinear Partial Differential equations and their Applications, College de France Seminar, Volume III", Pitman, Boston, (1982), pp 340-423.

NUMERICAL APPROXIMATION TO A CLASS OF WEAKLY SINGULAR INTEGRAL OPERATORS *

M. LEVET
Université de Saint Etienne 23 rue du Dr. Paul Michelon 42023 Saint Etienne (France)

and

M. TELIAS
Universidad de Chile Casilla 170/3 correo 3 Santiago de Chile
mtelias@cec.uchile.cl

Abstract. In this work, we are interested in developing numerical approximations for a certain class of weakly singular integral operators. Firstly, we define the notion of singularity we are concerned with and the type of integral operators to be approximated. Secondly, we construct an approximation and we prove its uniform convergence. Thirdly, we recall some type of convergences weaker than uniform convergence and we find a priori verifiable condition for the convergence of numerical approximations. Finally, we give a practical method for the resolution of the numerical problem and a numerical illustration.

1. Setting the problem

For any metric spaces E and F we denote $C^0(E,F)$ the set of all continuous functions from E into F. In particular, let $B = C^0([0,1],\mathbb{C})$ be the Banach space of complex valued continuous functions defined on $[0,1]$ with the norm of uniform convergence. We are concerned with the approximation of the linear bounded integral operator T defined on B by

$$(Tb)(t) = \int_0^1 K(t,s)b(s)ds \qquad \forall b \in B, \forall t \in [0,1], \tag{1}$$

where $K : [0,1] \times [0,1] \to \mathbb{C}$ is a *weakly singular* kernel, in the following sense (cf. [2]):

$$\begin{aligned} K(t,s) = G(|t-s|)H(t,s) \qquad & \forall (t,s) \in [0,1]\times[0,1], \\ H \in C^0([0,1]\times[0,1],\mathbb{C}), & \\ G \in C^0(]0,1],\mathbb{R}) \cap L^1(0,1), & \end{aligned} \tag{2}$$

$\exists 0 < \delta < 1$ such that $G \geq 0$ and G nonincreasing on $]0,\delta]$,

Where $L^1(0,1)$ is the set of real valued Lebesgue integrable functions over [0,1]. It is well known that if K satisfies (2) then T is a compact operator in $L(B)$, and

* Partially supported by Grant 0879-89 Fondecyt Chile

S. Gomez and J-P. Hennart (eds.), Advances in Optimization and Numerical Analysis, 261–272.
© 1994 *Kluwer Academic Publishers.*

this, for any value of G at zero. Thus, we can suppose, without loss of generality, that $G(0) = 0$. If we define G_r and G_l on $[0,1] \times [0,1]$ by

$$G_r(t,s) = \begin{cases} 0 & \text{if } 0 \leq s \leq t \\ G(|t-s|) & \text{if } t \leq s \leq 1 \end{cases} \qquad \text{(r for right)}$$

$$G_l(t,s) = \begin{cases} G(|t-s|) & \text{if } 0 \leq s \leq t \\ 0 & \text{if } t \leq s \leq 1 \end{cases} \qquad \text{(l for left)}$$

then, $G = G_r + G_l$. Using the symmetry of $G_t(s) \equiv G(|t-s|)$ respect to t we conclude that

$$G_r(t,t+s) = G_r(0,s) \qquad \forall t \in [0,1], \forall s \in [0,1-t],$$

$$G_l(t,t-s) = G_l(1,1-s) \qquad \forall t \in [0,1], \forall s \in [0,t],$$

Thus we have

$$T = T_r + T_l$$

with T_r and T_l linear compact integral operators on B with kernels $K_r \equiv G_r H$ and $K_l \equiv G_l H$, respectively. Moreover, from (2), we get

$$\forall t \in [0,1], \forall s \in [0,t], G_r(t,s) = 0,$$

$$G_r(0,\cdot) \in C^0(]0,1], I\!R) \cap L^1(0,1),$$

$$\exists 0 < \delta \leq 1 \quad \textit{such that} \quad G_r(0,\cdot) \geq 0 \textit{ and } G_r(0,\cdot) \textit{ non increasing on }]0,\delta[,$$

$$\forall t \in [0,1], \forall s \in [0,1-t], G_r(t,t+s) = G_r(0,s),$$

and

$$\forall t \in [0,1], \forall s \in [t,1], G_l(t,s) = 0,$$

$$G_l(1,\cdot) \in C^0([0,1[, I\!R) \cap L^1(0,1),$$

$$\exists 0 < \delta \leq 1 \quad \textit{such that} \quad G_l(1,\cdot) \geq 0 \textit{ and } G_l(1,\cdot) \textit{ increasing on }]1-\delta,1[,$$

$$\forall t \in [0,1], \forall s \in [0,t], G_l(t,t-s) = G_l(1,1-s),$$

This allows us to consider the approximation of the following linear compact integral operator (instead of that defined in (1) and (2)):

$$(Tb)(t) = \int_0^1 k(t,s)b(s)ds \qquad \forall b \in B, \forall t \in [0,1] \tag{3}$$

with $k : [0,1] \times [0,1] \to \mathbb{C}$ such that

(4.1)
$$k(t,s) = g(t,s)h(t,s) \qquad \forall (t,s) \in [0,1] \times [0,1],$$

(4.2)
$$h \in C^0([0,1] \times [0,1], \mathbb{C}), \tag{4}$$

(4.3) the function g verifies

i.

$$\forall t \in [0,1], \forall s \in [0,t], \quad g(t,s) = 0,$$

ii.

$$g(0,\cdot) \in C^0(]0,1], \mathbb{R}) \cap L^1(0,1),$$

iii.

$$\exists\ 0 < \delta \leq 1 \ \textit{such that}\ g(0,\cdot) \geq 0 \ \textit{and}\ g(0,\cdot)\ \textit{nonincreasing on}\ ]0,\delta],$$

iv.

$$\forall t \in [0,1], \forall s \in [0,1-t], g(t,t+s) = g(0,s).$$

We are concerned with the following two problems associated with this operator:

(1) Finding some nonzero eigenvalue of T and the corresponding maximal invariant subspaces.

(2) Find $x \in B$ such that:

$$(T - zI)x = f \quad z \in \rho(T) : \textit{resolvent set of } T \quad I : \textit{identity in } B$$

We are led to solve

(1') Finding some nonzero eigenvalue of T_n and the corresponding maximal invariant subspaces.

(2') Find $x_n \in B$ such that:

$$(T_n - zI)x_n = f \qquad z \in \rho(T_n) : \textit{resolvent set of } T_n$$

Where T_n is, at least, a strongly stable (cf Chatelin [4]) approximation of T, that is:

For λ, eigenvalue of algebraic multiplicity equal to m, and for Γ_λ a Jordan curve included in $\rho(T)$ isolating λ, we have:

i. $T_n \xrightarrow{p} T$

ii. $\forall z \in \Gamma_\lambda$ and n large enough: $z \in \rho(T_n)$ and $(T_n - zI)^{-1}$ is bounded in n.

iii. For n large enough $\dim M_n = m$

M_n = Maximal invariant subspace associated with the group of eigenvalues of T_n isolated by Γ_λ.

This property is not a priori verifiable, in a next section we will look for verifiable sufficient conditions.

2. The Main Result

Here, we will propose a uniform approximation T_n^R (R for Rabinowitz) of T based on the application of the singular function product integration theory proposed in [3].

Since we will have to use two subscripts to label our variables, we will assume their dependence on n implicitly. Let

$$0 = s_0 < s_1 < \ldots < s_{n-1} < s_n = 1,$$

be a partition of $[0,1]$ and define $h_i = s_i - s_{i-1}, \forall i = 1, 2, \ldots, n$.

Divide now each subinterval $[s_{i-1}, s_i]$ using a grid $\{x_{i,j}\}_{j=0}^{m_i+1}$ such that

$$s_{i-1} = x_{i,0} \leq x_{i,1} < x_{i,2} < \ldots < x_{i,m_i-1} < x_{i,m_i} \leq x_{i,m_i+1} = s_i$$

where $m_i \geq 1, \forall i = 1, 2, \ldots, n$. We define

$$h_{i,j} = x_{i,j+1} - x_{i,j}, \quad \forall j = 0, 1, 2, \ldots, m_i.$$

For each $t \in [0,1]$ we consider the knots $\{s_i(t)\}_{i=0}^{n}$ defined by $s_i(t) = t + s_i, \forall i = 0, 1, \ldots, n$ and the grid $\{x_{i,j}(t)\}_{j=0}^{m_i+1}$ defined by $x_{i,j}(t) = t + x_{i,j}, \forall j = 0, 1, \ldots, m_i + 1$. Let $P_n^g(t, \cdot)$ be the piecewise polynomial function defined on $[0, 1+t]$ by

$$P_n^g(t,s) = \begin{cases} 0 & \text{if } 0 \leq s < t \\ L_1^g(t,s) & \text{if } s = t \\ L_i^g(t,s) & \text{if } s_{i-1}(t) < s \leq s_i(t) \end{cases}$$

where

$$L_i^g(t,s) = \sum_{j=1}^{m_i} l_{i,j}(t,s) g(t, x_{i,j}(t)) \qquad \forall s \in [s_{i-1}(t), s_i(t)]$$

and

$$l_{i,j}(t,s) = \prod_{k=1, k\neq j}^{m_i} \frac{s - x_{i,k}(t)}{x_{i,j}(t) - x_{i,k}(t)}.$$

In this definition, we have extended by zero the function $g(t, \cdot)$ out of $[0,1]$. If we define k_n by

$$k_n(t,s) = P_n^g(t,s) h(t,s)$$

then, extending by zero the functions $h(t, \cdot)$ and b out of $[0,1]$, we obtain the following approximation of T:

$$(T_n^R b)(t) = \int_0^1 k_n(t,s) b(s) ds$$

$$= \sum_{i=1}^{n} \sum_{j=1}^{m_i} w_{i,j}^b(t) g(t, x_{i,j}(t)) \qquad \forall b \in B, \forall t \in [0,1] \tag{5}$$

where

$$w_{i,j}^b(t) = \int_{s_{i-1}(t)}^{s_i(t)} l_{i,j}(t,s) h(t,s) b(s) ds \tag{6}$$

Now, we state our main result.

Theorem

If the grid points $x_{i,j}$ satisfy

(7.1)
$\exists 0 < d < 1$ such that

$$\forall n \in I\!N, \forall i = 1, 2, ..., n, \forall j = 1, 2, ..., m_i - 1, \;\; h_{i,j} \geq dh_i$$

(7.2)
$\exists M \geq 1$ such that

$$\forall n \in I\!N, M_n \equiv \max\{m_i : i = 1, 2, ..., n\} \leq M \tag{7}$$

(7.3)

$$\forall n \in I\!N, \forall i = 1, 2, ..., n, \;\; if \;\; h_{i,0} \neq 0 \;\; then \;\; h_{i,0} \geq dh_i$$

(where d is the constant of (7.1)),
(7.4) $\exists R > 0$ such that

$$\forall n \in I\!N, \forall i = 1, 2, ..., n, \;\; if \;\; h_{i,0} = 0 \qquad then \;\; Rh_{i-1} \geq h_i$$

(7.5)

$$H_n \equiv \max\{h_i : i = 1, 2, ..., n\} \to 0, \;\; as \;\; n \to \infty$$

and, if the kernel k verifies properties (4), then the operator T_n^R, defined in (5), converges uniformly to the operator T, defined in (3).

Proof : We first recall the following result :

Lemma
If the points $x_{i,j}$ verify properties (7.1) and (7.2) then $\exists L > 0$ such that

$$\forall n \in I\!N, \forall i = 1, 2, ..., n, \forall j = 1, 2, ..., m_i, \forall t \in [0, 1],$$

$$\forall s \in [s_{i-1}(t), s_i(t)], \qquad |l_{i,j}(t, s)| \leq L.$$

Proof : Cf. [3].
Let $\epsilon > 0, b \in B$ be such that $|b| = 1$. Let $t \in [0, 1]$ be given. Then

$$|\int_0^1 k_n(t, s)b(s)ds - \int_0^1 k(t, s)b(s)ds|$$

$$= |\sum_{i=1}^{n} \sum_{j=1}^{m_i} w_{i,j}^b(t)g(t, x_{i,j}(t)) - \int_0^1 h(t, s)g(t, s)b(s)ds|.$$

Since $g(0, \cdot) \in L^1(0, 1)$ then $\exists \beta \in]0, 1[$ such that

$$2\beta \leq \delta \;\; (\;\; where \;\; \delta \;\; is \;\; given \;\; in \;\; (4.3)(iii)),$$

$$\int_0^\beta |g(0, s)|ds < \frac{\epsilon}{6|h|} \min\{\frac{1}{4L\alpha M}, 1\},$$

where $\alpha = max\{\frac{1}{d}, R\}$, and $|h|$ denotes the uniform norm of h over $[0,1] \times [0,1]$. We define, for each $\beta \in]0,1[$ and for each $t \in [0,1]$, the function $g_\beta(t,\cdot)$ by

$$g_\beta(t,s) = \begin{cases} 0 & \text{if } 0 \le s \le t+\beta \\ g(t,s) & \text{if } t+\beta < s \le 1 \end{cases}$$

We note that if $t+\beta \ge 1$ then $g_\beta(t,s) = 0, \forall s \in [0,1]$. Hence, writing $w_{i,j}$ for $w^b_{i,j}$ we obtain

$$|\int_0^1 k_n(t,s)b(s)ds - \int_0^1 k(t,s)b(s)ds| \le A_1 + A_2 + A_3,$$

where

$$A_1 = |\sum_{i=1}^{n}\sum_{j=1}^{m_i} w_{i,j}(t)(g(t,x_{i,j}(t)) - g_\beta(t,x_{i,j}(t)))| \tag{8}$$

$$A_2 = |\sum_{i=1}^{n}\sum_{j=1}^{m_i} w_{i,j}(t) g_\beta(t,x_{i,j}(t)) - \int_0^1 h(t,s) g_\beta(t,s) b(s) ds| \tag{9}$$

$$A_3 = |\int_0^1 h(t,s)(g_\beta(t,s) - g(t,s))b(s)ds|. \tag{10}$$

As before, we have extended $g_\beta(t,\cdot)$ by zero out of $[0,1]$.

We prove that terms (9), (10) and (11) are arbitrarily small when n is large enough, uniformly in those $b \in B$ such that $|b| = 1$ and in $t \in [0,1]$

3. Projection Type Approximations

We remark that, in fact, T_n^R is only a semidiscretization of T, because analytical integral computations are involved.

In what follows, we will define approximations of T, based on T_n^R verifying weaker convergence properties which are directly verifiable:

(a) Quasi compact convergence (cf. [1]).

$$T_n \xrightarrow{qc} T \Longleftrightarrow T_n \xrightarrow{p} T$$

$$|(T_n - T)T_n| \overset{n\to\infty}{\longrightarrow} 0$$

(b) Collectively Compact convergence (cf Anselone)

$$T_n \xrightarrow{qc} T \Longleftrightarrow \begin{cases} T_n \xrightarrow{p} T \\ \cup_{n\in I\!N}(T_n - T)\{b \in B, |b| \le 1\} \quad \text{is relatively compact in B.} \end{cases}$$

(c) Uniform convergence

$$T_n \xrightarrow{||} T \Longrightarrow \sup_{b\in B, |b|\le 1} |T_n b - Tb| \overset{n\to\infty}{\longrightarrow} 0$$

we have the following properties

$$T_n \xrightarrow{||} T \Rightarrow \; T_n \xrightarrow{cc} T \Rightarrow \; T_n \xrightarrow{qc} T \Rightarrow \; T_n \xrightarrow{s.s.} T$$

Let $(\pi_n)_{n\in\mathbb{N}}$ be a sequence of finite rank projections such that $\pi_n \xrightarrow{p} I$ (I: identity in B).

We define the following operators

$T^{RP}_{n,m} = \pi_m T^R_n$ Projection Type

$T^{RS}_{n,m} = T^R_n \pi_m$ Sloan Type

$T^{RG}_{n,m} = \pi_m T^R_n \pi_m$ Galerkin Type

we prove the following properties for these operators

$$T^{RP}_{n,m} \xrightarrow{||} T, \; T^{RS}_{n,m} \xrightarrow{qc} T, \; T^{RG}_{n,m} \xrightarrow{qc} T, \qquad n, m \to \infty$$

4. Product Integration Type Approximation

T^R_n can be written as:

$$(T^R_n b)(t) = \sum_{i=1}^{m} \int_{s_{i-1}}^{s_i} \sum_{j=1}^{m_i} l_{i,j}(t,s)\underline{g(t, x_{i,j}(t))h(t,s)b(s)}\, ds$$

The underlined part can be numerically approximated. To do this we define:

$$(T^A_m(b))(t) = \int_0^1 g(t,s)P_m(t,s)ds$$

where

$$P_m(t,s) = \begin{cases} L_{1,m}(t,s) & \text{if } s = 0 \\ L_{i,m}(t,s) & \text{if } s_{i-1} < s \le s_i \end{cases}$$

$$L_{i,m}(t,s) = \sum_{j=1}^{m} l_{i,j}(t,s)h(t,x_{i,j})b(x_{i,j}) \qquad \forall s \in [s_{i-1}, s_i]$$

$\{s_i\}_{i=0}^m, \{x_{i,j}\}_{j=0}^{m_i+1}$ are the same we used in the definition of T^R_n. With the grid conditions from that definition we have

$$T^A_m \xrightarrow{cc} T$$

Thus, we can define:

$$(T^{RA}_{n,m} b)(t) = \int_0^1 P^g_n(t,s)P_m(t,s)ds$$

and

$$T^{R,F}_{n,m} = \pi_m T^{RA}_{n,m}$$

we have $T^{RA}_{n,m} \xrightarrow{||} T^A_m \quad \forall n$ and then

$$T^{RA}_{n,m} \xrightarrow{qc} T \qquad n,m \to \infty$$

$$T^{RF}_{n,m} \xrightarrow{qc} T \qquad n,m \to \infty$$

Because, in general, if $\{T_m\}_{m\in \mathbb{N}} \xrightarrow{qc} T$ and $\{T_{n,m}\}_{n,m\in \mathbb{N}} \xrightarrow{||} T_m$ uniformly in n, then $T_{n,m} \xrightarrow{qc} T$.

5. Defect Correction Method

Let D be a non empty subset of $(B, ||)$. Given $F : D \longrightarrow B$, we suppose that there is $\xi \in D$ such that $F(\xi) = 0$.

We say that $G : B \longrightarrow B$ is a local approximate inverse of F if and only if the sequence:

$$\xi_0 = G(0) \qquad \xi_{k+1} = G(0) + (1 - GF)(\xi_k) \qquad k \geq 0$$

is well defined for every k, and converges to ξ. Sufficient conditions for that are, for instance:

(i) $\exists \rho > 0$ such that $U = \{x/|x - \xi| \leq \rho\} \subset D \quad F(U) \subset DomG \quad G(0) \in U \quad , 1 - GF$ is a contraction on U.

(ii) GF is linear, bounded and $r_\sigma(1 - GF) < 1$ (r_σ stands for spectral radius).

We propose here a local approximated inverse operator in the case of the equation $(T - zI)b = f$ and for the computation of eigenvalues and maximal invariant subspaces.

For the equation we consider:

$$G(b) = (T_n - zI)^{-1}(b + f) \in B$$

then,

$$1 - G = (T_n - zI)^{-1}(T_n - T)$$

and for n large enough $r_\sigma(1 - GF) < 1$.

The iteration is

$$u^0 = (T_n - zI)^{-1} f$$

$$u^{k+1} = u^k - (T_n - zI)^{-1}((T - zI)u^k - f) \qquad k \geq 0 \ (Atkinson) \qquad (A)$$

if $z \in \rho(T_n)\backslash\{0\}$ we can write:

$$(T_n - zI)^{-1} = \frac{1}{z}((T_n - zI)^{-1} T_n - I)$$

if we define

$$R^B_n(z) = \frac{1}{2}((T_n - zI)^{-1} T - I)$$

the iteration is

$$u^0 = R^B_n(z) f$$

$$u^{k+1} = u^k - R_n^B(z)((T - zI)^{-1}u^k - f) \qquad k \geq 0 \ (Brakhage) \qquad (B)$$

For the spectral problem we want to compute in $X = B^m$, $\phi = (\phi_1, ..., \phi_m)$ a basis of M, maximal invariant subspace associated with λ, eigenvalue of multiplicity m.

Let $\chi_n \in X^*$ be such that $< \phi, \chi_n >= I_m$ where I_m is the identity in B^m. Then ϕ is an isolated zero of

$$F(x) = \underline{T}x - x < \underline{T}x, \chi_n >$$

where $\underline{T}x = (Tx_1, ..., Tx_m) \qquad (x_1, ..., x_m) \in X$.

The derivative of F is

$$D_xF : X \longrightarrow X$$

$$u \longrightarrow \underline{T}u - u < \underline{T}x, \chi_n > -x < \underline{T}x, \chi_n >$$

Let $\phi^{(n)}$ be a approximate solution, we define

$$P_nx = \phi^{(n)} < x, \chi_n > \qquad x \in X$$

$$G_nx = (\underline{T}x - \underline{P}x)\underline{T}_nx - x < \underline{T}_n\phi^{(n)}, \chi_n >$$

$$\Sigma_n = G_n^{-1}(I - \underline{P}_n) \qquad (11)$$

Σ_n defined by (11) is an approximate inverse of D_xF.

This approximation gives the following iterations:

$$I \begin{cases} \phi^0 = \phi^{(n)} \\ \phi^{k+1} = \phi^k - \Sigma_n(F_n(\phi_k)) \quad k \geq 0 \end{cases}$$

$$II \begin{cases} \phi^0 = \phi^{(n)} \\ \phi^{k+\frac{1}{2}} = \underline{T}\phi^k(\theta^k)^{-1} \\ \phi^{k+1} = \phi^{k+\frac{1}{2}} - \Sigma_n(\underline{T}\phi^{k+\frac{1}{2}} - \phi^{k+\frac{1}{2}}\theta^k) \quad k \geq 0 \end{cases}$$

$$III \begin{cases} \phi^0 = \phi^{(n)} \\ \phi^{k+1} = \phi^k - \delta^k \quad k \geq 0 \end{cases}$$

where $\theta^k =< \underline{T}\phi^k, \chi_n >$ and δ^k is obtained from the iteration

$$\delta_0^k = \Sigma_nF_n(\phi^k), \qquad \delta_j^k = \delta_0^k + (I - \Sigma_nD_xF(\phi^k))\delta_{j-1}^k \qquad j = 1, ..., \nu(k)$$

For n large enough but fixed and $\nu(k)$ large enough, these 3 methods are well defined and they converge to ϕ.

Furthermore, method III has superlinear convergence.

If $T_n \xrightarrow{qc} T$ we have the above results. The eigenvalue λ is approximated by the average of the eigenvalues of θ^k:

$$\lambda^k = \frac{1}{m}tr < \underline{T}\phi^k, \chi_n >$$

6. Fredholm Discretizations

We consider

$$(Tb)(t) = \int_0^1 k(t,s)b(s)ds \quad \forall b \in B \quad \forall t \in [0,1]$$

T weakly singular, bounded, compact.

Let $\Pi_n : B \longrightarrow B$ be the interpolation projection, continuous, piecewise linear,

$$\Pi_n b = \sum_{i=1}^{n} < b, e_i^{(n)*} > e_i^{(n)} \quad \forall b \in B$$

$\{e_i^{(n)}\}_{i=1}^n$ is the canonical basis of $B_n = \Pi_n B$ associated with the grid $t_i^{(n)} = \frac{(i-1)}{(n-1)}$ $i = 1, ..., n$.

$e^{(n)*}$ is defined by

$$< b, e_i^{(n)*} >= b(t_i^{(n)}) \quad \forall b \in B \quad i = 1, ..., n$$

The approximations T_n^{RP}, T_n^{RS}, and T_n^{RG} are defined by

$$(T_n^{RP} b)(t) = \sum_{i=1}^{n} \int_0^1 P_n^g(t_i^{(n)}, s)h(t_i^{(n)}, s)b(s)ds \quad e_i^{(n)}(t)$$

$$(T_n^{RS} b)(t) = \sum_{i=1}^{n} b(t_i^{(n)}) \int_0^1 P_n^g(t,s)h(t,s)e_i^{(n)}(s)ds$$

$$(T_n^{RG} b)(t) = \sum_{i=1}^{n}\sum_{j=1}^{n} b(t_j^{(n)}) \int_0^1 P_n^g(t_i^{(n)}, s)h(t_i^{(n)}, s)e_i^{(n)}(s)ds \quad e_i^n(t)$$

where the points $\{t_i^{(n)}\}$ are used to define the piecewise polynomial P_n^g.

The approximations $T_{n,m}^{RA}$ and $T_{m,n}^{RF}$ have the same form:

$$(T_{n,n}^{RA} b)(t) = \sum_{j=1}^{n} w_j^{(n)}(t)h(t, t_j^{(n)})b(t_j^{(n)})$$

$$(T_{n,n}^{RF} b)(t) = \sum_{i=1}^{n}\sum_{j=1}^{n} w_j^{(n)}(t)h(t_i^{(n)}, t_j^{(n)})b(t_j^{(n)})e_i^{(n)}(t)$$

Only T^{RF} and T^{RG} are completely discretized. For the numerical computations we use $\Pi_N B$ in place of B with $N >> n$.

7. Example

$$(Tb)(t) = \int_0^1 \ln(1 - \cos 2\pi|t-s|)b(s)ds$$

simple eigenvalue $\lambda_0 = -\ln 2$, $\phi =$ cte. semisimple eigenvalues $\lambda_k = -\frac{1}{k}$ $\quad k \in \mathbb{N}$.

$$\phi_k^{(1)}(t) = \cos 2\pi kt \qquad \phi_k^{(2)}(t) = \sin 2\pi kt$$

The equation $(T - zI)b = -\frac{3}{2}\cos 4\pi t \quad z \in \rho(T)$ has a solution

$$b(t) = \frac{3}{2z+1}\cos 4\pi t$$

equation	(method A)	(method B)
$T_{n,m}^{RP}$ discretization	16*	9*
$T_{n,m}^{RS}$ discretization	16*	9*
$T_{n,m}^{RG}$ discretization	19*	11*
residual	$.2 \times 10^{-12}$	$.1 \times 10^{-12}$
$\frac{r_{k-1}}{r_k}$	.17	.041

$*$: number of iterations to obtain the residual

$$\lambda = -1$$

Spectral Problem	(method I)	(method II)
$T_{n,m}^{RP}$ discretization	10*	5*
$T_{n,m}^{RS}$ discretization	9*	5*
$T_{n,m}^{RG}$ discretization	10*	7*
residual	$.34 \times 10^{-12}$	$.74 \times 10^{-12}$
$\frac{r_{k-1}}{r_k}$	.88	.0099

References

1. AHUES: 1987
'A class of Strongly Stable Operator Approximations'
J. Austral. Math. Soc. Ser. B Vol. no. **28**, pp. 435-442

2. AHUES, M. , DURAN, M. , LEVET, M. and TELIAS, M.
'On the Numerical Approximation of Singular Integral Operators'
submitted to the Journal of Australian Math. Society Series B

3. RABINOWITZ, Ph. and SLOAN, I. H.: 1984
'Product integration in the Presence of Singularity'
SIAM J. Numer. Analysis Vol. no. **21**, pp. 149 - 166.

4. CHATELIN, F.: 1983
'Spectral Approximation of Linear Operators'
Ac. Press. New York

5. LEVET, M.: 1990
'Numerical Approximation of Weakly Singular Integral Operators'
Engineer Thesis Universidad de Chile

DIRECTIONAL SECOND DERIVATIVE OF THE REGULARIZED FUNCTION THAT SMOOTHES THE MIN–MAX PROBLEM

CRISTINA GÍGOLA
ESFM-COFAA, Admon. ITAM

and

SUSANA GÓMEZ
Instituto de Investigaciones en Matemáticas Aplicadas y en Sistemas

1. Introduction

When solving the min–max problem

$$\min_x \max\Big(f_i(x), \quad i = 1, \dots m\Big) = \min_x \varphi(x), \tag{1}$$

where the $f_i : \mathbb{R}^n \to \mathbb{R}$ are continuously differentiable functions, as an unconstrained problem, special methods have to be used because the non differentiability of the max function φ. One possible way to solve this problem, is to generate differentiable approximations with local minima coinciding with those of problem (1). Using this idea in earlier papers the authors proposed a smoothing function called the Regularized Function [1] and [2], and developed first and second order numerical methods [3] for solving problem (1), using these approximations. Convergence for the convex problem when using a first order method was given in ref [1].

To get the Regularized Function we need to present an equivalent formulation of problem (1), see [1]

$$\varphi(x) = \sup_{u \in U} \sum_{i=1}^{m} u_i f_i(x) = \sup_{u \in U} u^T f(x) \tag{2}$$

where U is the convex set defined as

$$U = \left\{ u \in \mathbb{R}^m \,\middle|\, \sum_{i=1}^{m} u_i = 1 \quad u_i \geq 0 \quad i = 1, \dots m \right\}$$

The Regularized Function is then defined as

$$\varphi_v(x) = \sup_{u \in U} \left(u^T f(x) - \frac{1}{2} \|u - v\|^2 \right) \tag{3}$$

where $v \in U$ is a dual vector of parameters. Different values of vector v generate different smooth approximations to the function φ.

S. Gomez and J-P. Hennart (eds.), Advances in Optimization and Numerical Analysis, 273–285.
© 1994 *Kluwer Academic Publishers.*

The methods proposed in [1] and [3] generate sequences $\{x^k, v^k\}$ that converge to a stationary pair (x^*, v^*) (a Kuhn–Tucker point x^* and its associated Lagrange multiplier v^*) of the original problem (1), that is, (x^*, v^*) satisfy the following conditions

$$\begin{aligned} &\sum_{i=1}^{m} v_i^* \nabla f_i(x^*) = 0 \\ &\sum_{i=1}^{m} v_i^* = 1 \\ &v_i^* \geq 0 \qquad i = 1, \dots m \\ &v_i^* \Big(\varphi(x^*) - f_i(x^*)\Big) = 0 \qquad i = 1, \dots m \end{aligned} \tag{4}$$

At every point (x, v), we must evaluate the regularized function φ_v at x, so we need to compute the value of u for which the supremum is attained,

$$u(x, v) = \arg\sup_{u \in U} \left(u^T f(x) - \frac{1}{2} \|u - v\|^2 \right) \tag{5}$$

The solution of this Quadratic Programming problem can be computed explicitly

$$u_i(x, v) = \begin{cases} f_i(x) + v_i - \delta \text{ if } & f_i(x) + v_i \geq \delta \\ 0 & \text{otherwise} \end{cases} \tag{6}$$

where δ is such that $u \in U, i.e.$

$$\sum_{i=1}^{m} \Big(f_i(x) + v_i - \delta \Big)^+ = 1$$

and $(\cdot)^+ = \max(0, \cdot)$. In section 2 an explicit expression for δ will be given.

The functions $u_i(x, v)$ are continuous, and so φ_v has first continuous derivatives see [1],

$$\nabla_x \varphi_v(x) = \sum_{i=1}^{m} u_i(x, v) \nabla f_i(x) \tag{7}$$

$$\nabla_v \varphi_v(x) = u(x, v) - v \tag{8}$$

In order to prove convergence to a local minimum of the original problem, second order information of the regularized function is needed.

But in general, second continuous derivatives of φ_v do not exist at some points, because $u(x, v)$ is not differentiable. However we will show that φ_v has second directional derivatives along any direction.

The characteristics of the points where non differentiability is present, will be given in Section 2. In Section 3 an expression of the directional second derivative of

φ_v is obtained. Also second order optimality conditions for the Regularized Function will be given.

These results will allow us to modify the Regularized Function with a penalty parameter in order to assure convergence to a local minimum of the original problem, with the Regularization method. A paper in preparation will have these results, ref [12].

2. Differentiability of φ_v

The first derivative in variable x of the regularized function

$$\varphi_v(x) = \sup_{u \in U} \left(u^T f(x) - \frac{1}{2}\|u - v\|^2 \right) \tag{9}$$

has the following form, see ref [1],

$$\nabla_x \varphi_v(x) = \sum_{i=1}^{m} u_i(x, v) \nabla f_i(x) \tag{10}$$

where $u_i(x, v)$ is as in (6).

In order to get a directional second derivative of φ_v, we need first to prove that $u(x, v)$ has a directional derivative in all directions.

To accomplish this, we first note that $u(x, v)$ is the solution of the following constrained quadratic programming problem, for x and v fixed

$$\begin{aligned} &\sup_{u \in \mathbb{R}^m} \left(u^T f(x) - \frac{1}{2}\|u - v\|^2 \right) \\ &\text{subject to } \sum_{i=1}^{m} u_i = 1 \\ &\qquad\qquad u_i \geq 0 \quad i = 1, \dots m \end{aligned} \tag{11}$$

The lagrangean function is then formed as

$$L(u, \delta, \rho) = u^T f(x) - \frac{1}{2}\|u - v\|^2 - \delta\left(\sum_{i=1}^{m} u_i - 1 \right) + \sum_{i=1}^{m} \rho_i u_i \tag{12}$$

here δ and ρ are the associated Lagrange multipliers. The Kuhn-Tucker conditions for optimality in problem (11) can then be written as

$$\left[\begin{array}{l} f_i(x) - (u_i - v_i) - \delta + \rho_i = 0 \qquad i = 1, \dots m \\ \sum_{i=1}^{m} u_i = 1, u_i \geq 0 \qquad i = 1, \quad m \\ \rho_i u_i = 0 \qquad i = 1, \dots m \\ \rho_i \geq 0 \qquad i = 1, \dots m \end{array} \right. \tag{13}$$

In order to have a characterization of the solution function $u(x, v)$ and to understand some of its properties we define the following sets.

$$N(x) = \{i | u_i > 0\} \tag{14}$$

$$NM(x) = \{i | \rho_i > 0\} \tag{15}$$

$$NC(x) = \{i | \rho_i = 0, u_i = 0\} \tag{16}$$

Note that $N(x)$ is the set of non active constraints and $NC(x)$ is the set of active constraints that do not satisfy strict complementarity at the optimum of (11), when we fix x and v.

The explicit expression of $u(x, v)$, that is, the optimal solution of the quadratic problem (11), at point x, is then obtained as

$$u_i(x, v) = \max \left(0, f_i(x) + v_i - \delta(x)\right) \tag{17}$$

where $\delta(x)$ is the non differentiable function

$$\delta(x) = \frac{\sum_{i \in N(x)} (f_i(x) + v_i) - 1}{|N(x)|} \tag{18}$$

which represent the optimal Lagrange multiplier associated to the equality constraint of problem (11), at point x, and $|N(x)|$ is the cardinality of the $N(x)$ set.

The non differentiability in x, of $\delta(x)$, can be easily seen from the fact that in a neighbourhood of x, the cardinality of $N(x)$ could change, that is NC may be a non empty set.

We present now an example to show the non differentiability of $u(x, v)$

2.1. Example

Let us take the functions

$$\begin{aligned} f_1(x) &= 6 - 16x \\ f_2(x) &= 6 - 40x + 40x^2 \\ f_3(x) &= -4 + 4x \qquad \text{for } x \geq 0 \end{aligned} \tag{19}$$

which correspond to the problem of Demyanov–Malozemov [4], moving in a direction $d = (-2, 6)^T$ from the point $(-1, 1)^T$.

If we consider the dual parameter $v = (0, 1, 0)^T \in U$ where $U = \{u \in \mathbb{R}^3 | u_1 + u_2 + u_3 = 1, u_1 \geq 0, u_2 \geq 0, u_3 \geq 0\}$, we can calculate explicitly its dual variables $u_i(x, v)$ for $i = 1, 2, 3$ and $\delta(x)$ and the regularized function. In Fig. 1 the dual variables and their dependence on variable x are illustrated.

$$\varphi_v(x) = \begin{cases} 400x^4 - 480x^3 + 184x^2 - 40x + 6 & 0 \le x \le \frac{1}{10} \\ 5 - 16x & \frac{1}{10} \le x \le \frac{9}{20} \\ 100x^2 - 106x + \frac{101}{4} & \frac{9}{20} \le x \le \frac{11}{20} \\ -5 + 4x & \frac{11}{20} \le x \le \frac{12}{20} \\ 400x^4 - 880x^3 + 724x^2 + 260x + 31 & \frac{12}{20} \le x \le \frac{11+\sqrt{21}}{20} \\ 6 - 40x + 40x^2 & x \ge \frac{11+\sqrt{21}}{20} \end{cases} \tag{20}$$

$$\begin{cases} u_1(x,v) = -20x^2 + 12x \\ u_2(x,v) = 20x^2 - 12x + 1 \\ u_3(x,v) = 0 \end{cases} \quad 0 \le x \le \frac{1}{10} \tag{21}$$

$$\begin{cases} u_1(x,v) = 1 \\ u_2(x,v) = 0 \\ u_3(x,v) = 0 \end{cases} \quad \frac{1}{10} \le x \le \frac{9}{20} \tag{22}$$

$$\begin{cases} u_1(x,v) = \frac{11}{2} - 10x \\ u_2(x,v) = 0 \\ u_3(x,v) = -\frac{9}{2} + 10x \end{cases} \quad \frac{9}{20} \le x \le \frac{11}{20} \tag{23}$$

$$\begin{cases} u_1(x,v) = 0 \\ u_2(x,v) = 0 \\ u_3(x,v) = 1 \end{cases} \quad \frac{11}{20} \le x \le \frac{12}{20} \tag{24}$$

$$\begin{cases} u_1(x,v) = 0 \\ u_2(x,v) = 6 - 22x + 20x^2 \\ u_3(x,v) = -5 + 22x - 20x^2 \end{cases} \quad \frac{12}{20} \le x \le \frac{11+\sqrt{21}}{20} \tag{25}$$

$$\begin{cases} u_1(x,v) = 0 \\ u_2(x,v) = 1 \\ u_3(x,v) = 0 \end{cases} \quad x \ge \frac{11+\sqrt{21}}{20} \tag{26}$$

$$\delta(x) = \begin{cases} 20x^2 - 28x + 6 & 0 \le x \le \frac{1}{10} \\ 5 - 16x & \frac{1}{10} \le x \le \frac{9}{20} \\ -6x + \frac{1}{2} & \frac{9}{20} \le x \le \frac{11}{20} \\ -5 + 4x & \frac{11}{20} \le x \le \frac{12}{20} \\ 20x^2 - 18x + 1 & \frac{12}{20} \le x \le \frac{11+\sqrt{21}}{20} \\ 6 - 40x + 40x^2 & x \ge \frac{11+\sqrt{21}}{20} \end{cases} \tag{27}$$

Remember that from the first condition of (13)

$$\rho_i(x) = \max\left(0, -f_i(x) - v_i + \delta(x)\right) \qquad \text{for } i = 1,2,3 \tag{28}$$

In Figs. 1 and 2, at point marked 1, $NC(x)$ is an empty set

$$\rho_1 = 0 \qquad \rho_2 > 0 \qquad \rho_3 = 0$$

$$u_1 > 0 \qquad u_2 = 0 \qquad u_3 > 0$$

at such point $u_i(x,v), i = 1,2,3$ are all differentiable. The cardinality of $N(x)$ doesn't change when we move around x.

At the point marked 2, NC is a non empty set

$$\rho_1 > 0 \qquad \rho_2 = 0 \qquad \rho_3 = 0$$

$$u_1 = 0 \qquad u_2 = 0 \qquad u_3 > 0$$

at such point, $u_2(x,v)$ and $u_3(x,v)$ are non differentiable. We don't have strict complementarity in the second active constraint $u_2 = 0$. The cardinality of $N(x)$ changes because u_2 will become positive in a neighbourhood of x.

Although $u(x,v)$ is a non differentiable function, we will show that it has directional derivatives along any direction.

Taking into account that $u(x,v)$ is the solution of a quadratic programming problem with equality and inequality constraints, we can characterize this function using optimality results from the literature. The first useful result is in a theorem given by Fiacco and Mc. Cormick [5], where conditions for the solution point and Lagrange multipliers differentiability with respect to small perturbations , when strict complementary is satisfied, are given.

Later on, Bigelow and Shapiro [6], extended these results without considering the strict complementarity condition, but as discussed by Sargent [7], some extra formalization was needed. Jittomtrum [8], gave these formalized results, and we will use them to prove directional differentiability for the solution function $u(x,v)$ of the inequality constrained problem (11).

Consider the following perturbation of the quadratic programming problem (11)

$$P(t) = \sup_{u \in U} \left(u^T f(x+tp) - \frac{1}{2}\|u - v\|^2\right) \tag{29}$$

for x and p fixed, where x is considered a perturbation parameter and p is any direction in $\mathbb{R}^n$. Note that $u(t) = u(x+tp, v)$ is the solution of (29)

According to theorem 3 given in Jittontrum [8, p. 132] we need (11) to satisfy the following conditions:

a. The functions defining $P(t)$ are twice continuously differentiable at (u,t) in a neighbourhood of $(u^*, 0)$, where $u^* = u(0)$

b. The linear independence condition of the active constraints at the solution $u^* = u^*(x,v) = u(0)$

c. Second order sufficiency optimality condition in (11) or (29) at $t = 0$.

In our case we have that:

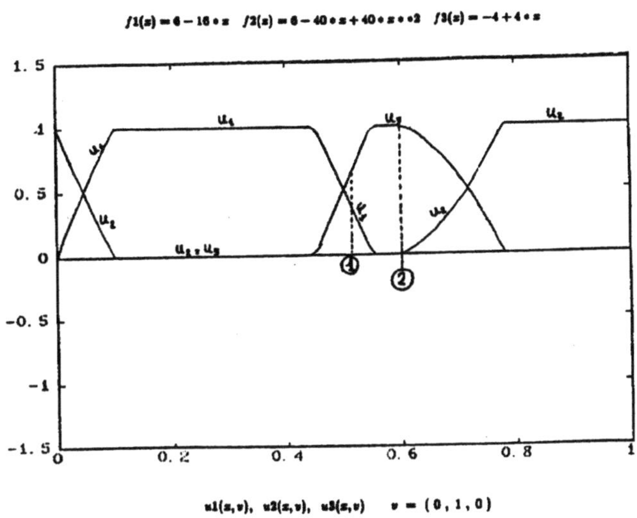
f1(x) = 6 − 16 * x f2(x) = 6 − 40 * x + 40 * x * * 2 f3(x) = −4 + 4 * x
1.5
1
0.5
0
-0.5
-1
-1.5
0
0.2
0.4
0.6
0.8
1
u1(x,v), u2(x,v), u3(x,v) v = (0,1,0)

Fig. 1

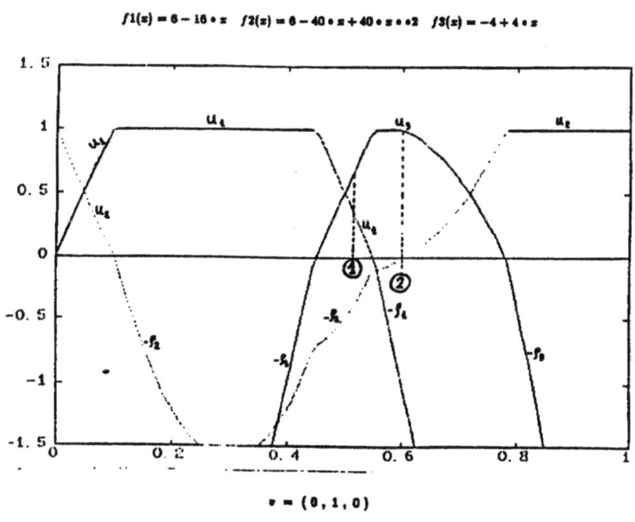
f1(x) = 6 − 16 * x f2(x) = 6 − 40 * x + 40 * x * * 2 f3(x) = −4 + 4 * x
1.5
1
0.5
0
-0.5
-1
-1.5
0
0.2
0.4
0.6
0.8
1
v = (0,1,0)

Fig. 2

a. is satisfied because we assume $f_i \in \mathbb{C}^2(\mathbb{R}^n)$ for all $i = 1, \ldots m$, and the objective function of problem $P(t)$ is quadratic in u.

b. is satisfied because the Jacobian of (19) for the active constrains is an $n \times n$ matrix.

$$\begin{bmatrix} & & -1 \\ & I_{n-1,n-1} & -1 \\ & & -1 \\ 1\ldots & 1 & 1 \end{bmatrix} \tag{30}$$

with rank n, because at least one bound (inequality constraint) is nonactive ($u_i > 0$ for some index i) in order to satisfy the equality constraint $\sum u_i = 1$.

c. is satisfied because the objective function of problem (19) is a strictly concave function.

From theorems 3 and 4, in Jittontrum [8, p. 13] we then have that problem $P(t)$ has a unique continuous solution $u(t), \delta(t), \rho(t)$ for $t \geq 0$. Furthermore the right-handed derivative $\dot{u}(0), \dot{\delta}(0), \dot{\rho}(u)$ exist where

$$\begin{array}{ll} \dot{u}(0) = \lim\limits_{t \to 0+} \frac{u(t)-u(0)}{t} & \dot{u}(0) = (\dot{u}_1(0), \ldots \dot{u}_m(0))^T \\ \dot{\delta}(0) = \lim\limits_{t \to 0+} \frac{\delta(t)-\delta(0)}{t} & \dot{\delta}(0) \in \mathbb{R} \\ \dot{\rho}(0) = \lim\limits_{t \to 0+} \frac{\rho(t)-\rho(0)}{t} & \dot{\rho}(0) = (\dot{\rho}_1(0), \ldots \dot{\rho}_m(0))^T \end{array} \tag{31}$$

These derivatives are the solution of the following inequality system, obtained deriving (with respect to the parameter x and the variable t), the Kuhn–Tucker conditions of (29). That is

$$\begin{array}{l} -\dot{u}_i + p^T \nabla f_i(x) - \dot{\delta} + \dot{\rho}_i = 0 \qquad i = 1, \ldots, m \\ \sum\limits_{i=1}^{m} \dot{u}_i = 0 \\ \dot{u}_i \geq 0 \quad , \dot{\rho}_i \geq 0 \quad \text{if} \quad i \in NC(x) \\ \dot{u}_i = 0 \quad \text{if} \quad i \in NM(x) \\ \dot{\rho}_i = 0 \quad \text{if} \quad i \in N(x) \\ \dot{u}_i \dot{\rho}_i = 0 \quad \text{if} \quad i \in NC(x) \end{array} \tag{32}$$

But we can see that these are the Kuhn–Tucker conditions of the following quadratic programming problem.

$$\begin{array}{rl} & Sup \left(\frac{1}{2} \dot{u}^T I \dot{u} + \dot{u}^T p^T \nabla f(x) \right) \\ \text{subject to} & \sum\limits_{i=1}^{m} \dot{u}_i = 0 \\ & \dot{u}_i = 0 \quad \text{if} \quad i \in NM(x) \\ & \dot{u}_i \geq 0 \quad \text{if} \quad i \in NC(x) \end{array} \tag{33}$$

in variable $\dot{u} \in \mathbb{R}^m$, taking $\dot{\delta} \in \mathbb{R}$ and $\dot{\rho} \in \mathbb{R}^m$ as the associated Lagrange multipliers (x and p fixed).

To obtain $\dot{u}$, from system (33), we just consider the following cases:

a. $NC(x)$ is an empty set. We have in (33) an equality constrained problem, whose solutions are

$$\dot{u}_i = 0 \quad \text{if} \quad i \in NM$$
$$\dot{u}_i = p^T \nabla f_i(x) - \dot{\delta} \quad \text{if} \quad i \in N(x)$$

where $\dot{\delta}$ is such that

$$\dot{\delta} = \frac{1}{|N|} \sum_{j \in N} p^T \nabla f_j(x) \qquad (N = N(x))$$

We finally get

$$\left[\begin{array}{ll} \dot{u}_i = p^T \nabla f_i(x) - \frac{1}{|N|} \sum_{j \in N} p^T \nabla f_j(x) & \text{if } i \in N(x) \\ \dot{u}_i = & 0\text{if } i \in NM(x) \end{array}\right. \tag{34}$$

b. $NC(x)$ is a non empty set.

The solution of (33) can be explicitly written as

$$\dot{u}_i = 0 \qquad \text{if } i \in NM(x)$$

$$\dot{u}_i = \max(0, p^T \nabla f_i(x) - \dot{\delta}) \qquad \text{if } i \in NC(x)$$

$$\dot{u}_i = p^T \nabla f_i(x) - \dot{\delta} \qquad \text{if } i \in N(x)$$

where $\dot{\delta}$ is such that

$$\sum_{i \in N} p^T \nabla f_i(x) + \sum_{i \in NC} \max(0, p^T \nabla f_i(x) - \dot{\delta}) = |N| \dot{\delta}$$

In order to get a simpler expression for $\dot{\delta}$, we have to define another set

$$\overline{NC}(x,p) = \{i \in NC(x) \quad | p^T \nabla f_i(x) - \dot{\delta} > 0\}$$

Unless it is necessary for the context, the parameters x and p of the sets N, NC and $\overline{NC}$ will be omitted.

If we call $M(x,p) = N(x) \cup \overline{NC}(x,p)$
then $\dot{\delta} = \frac{1}{|M|} \sum_{j \in M} p^T \nabla f_j(x)$ and the general expression for $\dot{u}$ is

$$\left[\begin{array}{ll} \dot{u}_i = 0 & \text{if } i \in NM(x) \\ \dot{u}_i = \max(0, p^T \nabla f_i(x) - \frac{1}{|M|} \sum_{j \in M} p^T \nabla f_j(x)) & \text{if } i \in NC(x) \\ \dot{u}_i = p^T \nabla f_i(x) - \frac{1}{|M|} \sum_{j \in M} p^T \nabla f_j(x) & \text{if } i \in N(x) \end{array}\right. \tag{35}$$

The set $M(x,p)$ can be characterized directly, from the following proposition.

Proposition 1: For x and p fixed

$M(x,p) = \{i | \exists \epsilon > 0 \text{ such that } f_i(x+tp) + v_i - \delta(x+tp) > 0, t \in S(0,\epsilon)\}$

Proof.

It is easy to see from the definition of $N(x)$ and $NC(x)$, that for all $i \in N(x) \cup \overline{NC}(x)$, an $\epsilon > 0$ exists such that $f_i(x+tp)+v_i-\delta(x+tp) > 0 \quad \forall \quad t \in S(0,\epsilon)$

Let i be such that $f_i(x+tp)+v_i-\delta(x+tp) > 0$ for all $t \in S(0,\epsilon)$ and some $\epsilon > 0$.

-if $u_i > 0$ then $i \in N(x)$ and the proposition is proved

-if $u_i = 0$, the definition of $u_i(x,v)$ implies that $f_i(x)+v_i-\delta(x) \leq 0$, but according to our assumption we must have $f_i(x)+v_i-\delta(x) = 0$ so that $\rho_i = 0$ then $i \in \overline{NC}(x)$ and the proposition is completed.

2.2. Directional Derivative of $u(x,v)$

Taking into account that the solution $u(x,v)$ of problem (11) , is the solution of $P(0)$, we have $Du(x,v;p) = \dot{u}(0)$ and according to (34) and (35) it can be explicitly written as follow

$$Du_i(x,v;p) = \begin{cases} 0 & \text{if } i \in NM(x) \quad (36) \\ p^T\nabla f_i(x) - \frac{1}{|M|}\sum\limits_{j\in M} p^T\nabla f_j(x) & \text{if } i \in N(x) \quad (37) \\ \max(0, p^T\nabla f_i(x) - \frac{1}{|M|}\sum\limits_{j\in M} p^T\nabla f_j(x)) & \text{if } i \in NC(x) \quad (38) \end{cases}$$

We can observe that the point marked (1) in Figs. 1 and 2 corresponds to (36) for $u_2(x,v)$ and (37) for $u_1(x,v)$ and $u_3(x,v)$. Also the point marked (2) corresponds to (36) for $u_1(x,v)$ and (38) for $u_2(x,v)$ and $u_3(x,v)$.

It is important to note also that if $NC(x)$ is empty, *i.e* the strict complementarity condition is satisfied in (11), then the $u_i(x,v)$ are continuously differentiable for all i and

$$Du_i(x,v) = \begin{cases} 0 & \text{if } u_i = 0 \\ \nabla f_i(x) - \frac{1}{|N|}\sum\limits_{j\in N} \nabla f_j(x) & \text{if } u_i > 0 \end{cases} \qquad (39)$$

3. Second directional derivative of the Regularized Function.

Once we have proved that $u(x,v)$ has directional derivatives, we can get an expression for the second directional derivative of the regularized function.

The general expression will have the following form:

$$\begin{aligned} D^2\varphi_v(x;p) &= D\Big(\nabla^T\varphi_v p\Big)(x;p) \\ &= \sum_{i\in N(x)} u_i(x,v)p^T\nabla^2_{xx}f_i(x)p + \\ &+ \sum_{i\in M(x,p)} Du_i(x,v;p)\nabla^T f_i(x)p \end{aligned} \qquad (40)$$

where $\nabla\varphi_v$ is defined in (7) and (10) and $Du_i(x,v;p)$ is defined in (36), (37) and (38).

Introducing the following matrix notation

$\mathbf{1}_M$ as a $|M|$–vector of ones and J_M as a $n \times |M|$ Jacobian matrix, i.e $J_M = [\nabla f_i]_{i \in M(x,p)}$, we can express (40) in the form

$$\begin{aligned} D^2\varphi_v(x;p) &= \sum_{i \in N} u_i(x,v)p^T\nabla^2_{xx}f_i(x)p + \\ &+ \sum_{i \in N} \left[p^T\nabla f_i(x) - \tfrac{1}{|M|}p^T J_M \mathbf{1}_M\right]\nabla^T fi(x)p \\ &+ \sum_{i \in NC} \max\left(0, p^T\nabla f_i(x) - \tfrac{1}{|M|}p^T J_M \mathbf{1}_M\right)\nabla^T f_i(x)p \end{aligned} \tag{41}$$

If strict complementarity holds in (11), $NC(x)$ is empty, ($M(x,p) = N(x)$ for all p) and φ_v has continuous second derivatives. The Hessian of φ_v is the following matrix:

$$D^2\varphi_v(x) = \sum_{i \in N} \left\{ u_i(x,v)\nabla^2_{xx}f_i(x) + \left(\nabla f_i(x) - \frac{1}{|N|}\ J_N\mathbf{1}_N\right)\nabla^T f_i(x) \right\} \tag{42}$$

to simplify notation, if we take $\mathbf{1}_{N\times N}$ as the $|N| \times |N|$–matrix of ones, we get that when $M(x,p) = N(x)$, for all $p \in \mathbb{R}^n$

$$D^2\varphi_v(x) = \sum_{i \in N} \quad u_i(x,v)\nabla^2_{xx}f_i(x) + J_N\left(I_N - \frac{\mathbf{1}_{N\times N}}{|N|}\right)J_N^T \tag{43}$$

It is important to note that $J_N(I_N - \mathbf{1}_{\frac{N\times N}{|N|}})J_N^T$ is a $n \times n$ positive definite matrix and:

$$rank\left[J_N\left(I_N - \frac{\mathbf{1}_{N\times N}}{|N|}\right)J_N^T\right] = \min\left[rank(J_N), |N| - 1)\right]$$

This allow us to prove the following proposition

Proposition 2: In the convex case (f_i convex functions) the Hessian of $\varphi_v(x)$, when it exist ($M(x,p) = N(x)\ \forall\ p$), is a positive semi definite matrix and also $D^2\varphi_v(x;p) \geq 0$ for all $p \in \mathbb{R}^n$.

3.1. Optimality Conditions for the Regularized Function

In Gígola and Gómez [1] and [3] we minimize a sequence of regularized functions φ_v^k, for $v^k \in U$ to obtain a sequence $\{x^k\} \subset \mathbb{R}^n$.

We solve at each iteration the following unconstrained problem:

$$\min(\varphi_v(x)|x \in \mathbb{R}^n) \qquad v \in U \tag{44}$$

The optimality conditions for (44) are expressed in the following propositions.

Proposition 3: (Necessary condition)

If $x^* = x^*(v)$ is a local minimum of (44) then $\nabla_x\varphi_v(x^*) = 0$

We can easily see now that the regularized function φ_v is a semismooth function according to the results given in theorem 2 in Mifflin [9]. Second order optimality conditions can now be given using the directional derivative of φ_v.

Proposition 4: If $x^* = x^*(v)$ is a local minimum of problem (44) then $\nabla_x\varphi_v(x^*) = 0$ and

$$D^2\varphi_v(x^*;p) \geq 0 \qquad \text{for all } p \in \mathbb{R}^n \tag{45}$$

The conditions are also sufficient if we have strict inequality in (45) for all $p \neq 0$.

See Chaney, [10] and [11], for further details in second order optimality conditions for semismooth functions. Refering these results to the explicit form (41) of the directional derivative of φ_v, we have

Proposition 5: Optimality conditions for (44) are satisfied in x^* if

$$\sum_{i=1}^{m} u_i(x,v)\nabla f_i(x) = 0$$

and the matrices $\nabla^2_{xx} f_i(x)$ for $i \in N(x)$ are positive semidefinite.

Proof: From (41) the terms having only first derivative of f_i, are always non–negative.

4. Conclusions

We have shown that the regularized function has second order directional derivatives. This allows us to give second order optimality conditions for the Regularized Function.

In order to prove that the Regularization Methods converge to a local minimum of the original problem we need these second order conditions.

In a paper in preparation in ref [12] we will show that the regularized function must be modified by a penalization parameter $r \neq 1$ in order to have a minimum of φ_v^* in a local minimum x^* of the original problem (1), when we take the dual parameter v^* that is, an optimal Lagrange multiplier of (1) associated with x^*.

References

1. C. Gígola and S. Gómez, " A Regularization Method for Solving the Finite Convex Min–Max Problem". SIAM J. Numerical Anal. Vol. 27, no. 6, (1990) pp. 1621–1634.
2. C. Gígola and S. Gómez, "Relation between the regularization and the multipliers methods for the min–max problem". Advances in Numerical Partial Differential Equations and Optimization. S. Gómez, J.P. Hennart, R. Tapia (eds) SIAM (1991) pp. 299–319.
3. C. Gígola and S. Gómez, "Two second order regularization methods to solve the finite min–max problem". Advances in Numerical Partial Differential Equations and Optimization, S. Gómez, J.P. Hennart, R. Tapia (eds.) SIAM (1991) pp. 320–331.
4. V.F. Demyanov and V.N. Malozemov, "On the theory of non–linear min–max problem". Russian Math. Surveys, 26 (1971), pp. 57–115.
5. A. V. Fiacco and G.P. Mc.Cormick, "Non linear Programming: sequential unconstrained minimization techniques". John Wiley, New York (1968).
6. J.H. Bigelow and N.E. Shapiro, "Implicit function theorems for mathematical programming and for systems of inequalities". Math. Programming 6 (1974) pp. 141–156.
7. R.W.H. Sargent, "On the parametric variation of constraint sets and solutions of minimization problems". Written version of the talk presented at the 10th International Symposium on Math. Programming (1979).
8. K. Jittomtrum, "Solution point differentiability without strict complementary in nonlinear programming". Math. Programming 21 (1984) pp. 127–138.
9. R. Mifflin, "Semismooth and semiconvex functions in constrained optimization" SIAM J. Control and Optimization. Vol. 15, No. 6 (1977) pp. 959–972.

10. R.W. Chaney "Second order necessary conditions in constrained semismooth optimization" SIAM J. Control and Opt. Vol. 25, No. 4 (1987) pp. 1072–1081.
11. R.W. Chaney, "Second order necessary conditions in semismooth optimization" Math. Programming 40 (1988) pp.
12. C. Gígola and S. Gómez, "Optimality Conditions for the Regularization Methods to solve the min–max problem". Paper in preparation

MIX
Papier aus verantwortungsvollen Quellen
Paper from responsible sources
FSC® C105338

If you have any concerns about our products,
you can contact us on
ProductSafety@springernature.com

In case Publisher is established outside the EU,
the EU authorized representative is:
Springer Nature Customer Service Center GmbH
Europaplatz 3, 69115 Heidelberg, Germany

Printed by Libri Plureos GmbH
in Hamburg, Germany